FORCE AND NATURE.

FORCE AND NATURE.

ATTRACTION AND REPULSION:

THE RADICAL PRINCIPLES
OF ENERGY, DISCUSSED IN THEIR RELATIONS TO PHYSICAL
AND MORPHOLOGICAL DEVELOPMENTS.

BY

CHARLES FREDERICK WINSLOW, M.D.

Philadelphia:
J. B. LIPPINCOTT & CO.
1869.

LONDON:
R. CLAY, SONS, AND TAYLOR, PRINTERS,
BREAD STREET HILL.

TO

The British Association for the Advancement of Science,

AND TO

The Scientific Students

Of the United States of America,

THE FOLLOWING PAGES

ARE RESPECTFULLY DEDICATED

BY THE AUTHOR.

LONDON, *August* 1868.

CONTENTS.

CHAPTER VII.

CHAPTER VIII.

CHAPTER IX.

CHAPTER X.

CHAPTER XI.

FORCE AND NATURE.

INTRODUCTORY.

1.

THE history of speculative thought and of solid scientific acquirements is well known. The publication of Newton's "Principia" first woke the slumbers of the scholastic world, aroused the mathematicians of various countries to new and great themes of contemplation, and immediately opened fields for numerical researches in all cosmical directions. These themes and these fields have continually extended in virtue of successive discoveries; and the laws of gravitation are found to be as correct in principle as they are universal in their application. Certain facts and phenomena of great physical importance appear, however, to have been strangely lost sight of, from age to age, by the most illustrious men—rather, it seems to me, on account of the magnitude and splendour of the developments to which a knowledge of the laws of gravitation has led, than to any natural tendency of cultivated thought consciously to neglect or avoid fields of fruitful inquiry. Nevertheless, the anomalies not seldom encountered in nature, or detected in consequence of the application of the mathematics to physical facts and observations (the latter frequently demanding treatment through empirical formulæ, and even resisting every form of analysis) have often excited a query in the minds of many astronomers and physicists whether some subtle principle antagonistic to attraction does not also exist as an all-pervading element in nature, and so operate as in some way to disturb the action of what is generally considered by the scientific world an *unique* force. The aim of the following work is to set forth this subject in its broadest aspects, and in such a manner as to invite thereto the attention of the learned, and particularly that of the British Association for the Advancement of Science—since to that Institution I am indebted for many facts of transcendent importance to philosophy, and for those systematic observations which, combined with my own personal travels and researches, have enabled me to undertake and elaborate the discussions, the results of which I now entrust to their consideration and the final judgment of the scientific men of this and other countries.

I have often thought that Newton would have extended his own philosophy, and filled the void in cosmical physics which is at present felt by many scientific men without knowing what is wanting, had the requisite data existed in his time in addition to those presented to his peculiar genius by the previous labours of Galileo, Tycho Brahe, and Kepler. At least, I honour the genius of that remarkable man so highly as to believe this; almost persuaded that his own penetration, stimulated by the more inquisitive talent of Leibnitz for material pursuits, would have solved many problems not yet unfolded by his followers. Newton, however, (as we shall see further on,) had no disposition for mere speculations, and only indulged his mathematical faculties in determining the *laws* of material motion, avoiding the search into the nature and final constitution, so to speak, of forces and causes.

The progress of knowledge upon these abstruse subjects has been slow, delayed by various circumstances, but by none more frequently than by a natural and submissive respect for the *position* of authorities whose opinions or influence would otherwise be of minor value. Experimental physics is, however, exciting scholars of every class at the present time to the profoundest inquiry into the causation of phenomena, and into the final nature and action of the chemical and celestial forces.

We know much, it is true. But much—how infinitely much! —remains to be discovered. Veins of truth embedded in darkness or error, branch in every direction; nevertheless, the ablest engineers will ever be impotent to reach the precious lodes without suitable instruments. The mathematician may *assume* data, and lead the world into error, as well as into truth. The physicist, however, must have matter and the forces themselves, so to speak, in his very hands. The fact that the force of gravitation may be proved by the calculus to act with certain fixed laws, does not by any means disprove the existence, nor even imply the non-existence, of another force. It is the existence and phenomenal capacity and power of *this other force* which I propose to discuss in the following pages, tracing it into its broadest expansions, applications, and connexions with other forces and all forms of matter.

For an undertaking which may appear bold to many scientific men, I, of course, can invoke no special indulgence. Every close thinker has more and more perceived that with the advancement of experimental discovery and solid knowledge, physical theories have grown perplexing on account of the insufficient development of fundamental principles. In many fields of research investigators instinctively feel oppressively hemmed in, chained, and almost stifled by some inexplicable deficiency of means in final inductive analysis.

While conscious that this synopsis may appear unnecessary to some very learned men whom I highly respect, but who are too strongly wedded to an exclusive idea, or to their own opinions, to admit evidence from distant and unheard-of quarters, I, nevertheless, shall not attempt to offer any apology for its publication. Indeed, any explanation I could make, if one were necessary, for the free and honest expressions of fact and inquiry, would not be satisfactory even to myself, and certainly would have no weight with opponents. Excited by a love of Nature to untiring observation of its manifold wonders, having compassed many seas and lands, explored the heavens, listened to the most accomplished professors, and possessed myself of the views and speculations of both ancient and modern philosophers, I have ever felt unsatisfied with the physical foundations upon which some of the branches of science have been so long compelled to rest. Comprehending fully the magnitude of my labour, I have nevertheless been undismayed, because persuaded of the solidity of my positions by the number and strength of accumulating facts. If, however, I have in any way failed to accomplish my undertaking, I will seize this occasion to solicit those who come after me to take courage, and to make more solid the ground and sure the steps where I may only have reached uncertain results, or wholly failed. The field which I have here ventured to enter and lay open, is fully ripe and "white already to harvest."

If here and there I appear to falter and become dizzy, there will arise scholars, I am confident, who, endowed with fitter powers, will reap like giants in time to come.

2.

There is a certain phase of Science to appear, now and then, in the following work—a phase always inadmissible in experimental and mathematical inquiries—to which I invite particular attention; and upon which a few remarks may be necessary, in order to establish a clear understanding in the outset between the reader and author. It is this. No system of natural or positive philosophy can reach its legitimate boundaries, comprehend nature entirely, and unfold by successive sequences into its grandest possible developments, and yet be wholly material and physical. A true and enduring system must embrace both physics and metaphysics. However material our postulates may be, they must insensibly extend the argument into the immaterial; inasmuch as *force* is immaterial. A true system must, furthermore, embrace geometry and the algebras—not their mere physical and symbolical terms, but their high, deep, and purely intellectual principles, which, appertaining to psychology and exposing an absolute universality of mind and purpose, lift us freely and positively into studies of the Infinite. I have, therefore, not hesitated to trace nature to first principles through severe inductive processes, in opposition to the materialistic notions of certain writers. If I have not paused altogether at the point where experimental and mathematical researches are compelled to pause in individual specialities of inquiry, it is because the design of this work will not permit such an omission. For, in a generalization exclusively embracing the discussion of *forces*, in a development of the system of nature as it becomes traceable towards its origin, and in a search after causations based wholly upon forces, as *abstract yet demonstrable principles*, Science becomes transformed into a unit; and the motive spring of the Cosmos cannot be overlooked at any point, nor for a single moment, as an omnipresent energetic principle of a unitary character. No positive system of philosophy, no grand generalization of grave facts and of final irreversible inductions in experimental and natural science, and in universal thought and numbers, can be, or pretend to be, perfectly unfolded and yet exclude ethics and metaphysics. When, therefore, any reference—open or implied—is made in the following work to the existence in nature, or action in matter and force, of a subtler geometric presence which, in default of other terms, may be called pantheistic or monotheistic, the terms are to be understood in a strictly technical sense as the only *scientific* expression for the omnipresence of creative energy, the final excito-motor and intelligent principle upon which all physical energy hinges and upon which every condition of matter and form rests.

3.

While not specially asking indulgence from physicists and naturalists, I must nevertheless solicit the *patience* of every scholar who may attempt to follow my discussions and positions from step to step. It will be soon perceived that my task is in no respect elementary. The opening of the subject may not only be dry and unattractive, but also positively repelling to those whose convictions upon accepted *theories* of various sorts have been long formed and are now unchangeable. There is nothing more certain, however, than that human opinions change from age to age; and that we are at present upon paths of solid progress. I request only that candid and profound consideration for the entire field of research as it may be gradually unfolded, which so vast a generalization actually and properly demands. In positive science nothing can be assumed. I, therefore, shall consciously permit no assumption or hypothesis to enter into any inquiry or argument; for the end would surely be to weaken every sequence, and render no point or conclusion reliable.

4.

The subject of this work, though new when treated *in extenso*, has, it is proper to say here, a certain historical aspect; and justice to the living, and especially to the dead, demands a recognition of their respective labours and merits.

It is in this connexion, and in this sense, that I would refer firstly in a special manner to the memory of Roger Joseph Boscovich, an Italian scholar who published in 1745—1763 a work on Natural Philosophy, in which he presented a *theory* of repulsion, and attempted its mathematical development in a manner which seems to have made no impression upon the learning of his own or of subsequent times. As far as I have been able to comprehend his speculations (for he always calls the point his *theory*—"theoria mea"—in opposition to, or rather as a *middle* range of speculation "*between the theories of Newton and Leibnitz*") they seem to have been acute and prophetic, but only, after all, what he declared them to be, pure *hypothesis.* His aim appears to have been to reduce all the forces which he supposed to exist in nature, under a single theoretic law; "theoria philosophiæ "naturalis redacta ad unicam legem virium in natura exis- "tentium." The doctrine of *molecular repulsion* is as old as the ancient Greek philosophy, proven then by observation, and experiment even, to be a positive entity as clearly as experiment proves it to be so to-day. But as late as the time of Boscovich, although knowledge had so increased that the imagination had more solid food for digestion, still the age of exact inquiry had not opened and philosophy had not established the inductive method by which alone its grandest developments can be successfully reached. Scientific men in those days were content, and compelled also, to advance and labour upon *à priori* conditions—a method always uncertain. But the mind of Boscovich, although hedged in by his sacerdotal connexions, was evidently unsatisfied; and, himself a Jesuit, he obtained permission of the Church to publish his theory ("superiorum permissu, ac privilegio"), careful of course not to molest dogmas, and fettered by the despotism under which he had been educated.

Inasmuch as no attempt will be made in the following work to support any theory, facts alone forming the foundation and development of its entire structure, we shall have no occasion to refer hereafter to the ideas of Boscovich.

I must state, however, in justice to myself, and in reference to the doctrine of *cosmical repulsion* which I have aimed to unfold and establish in the sequel, that the existence of Boscovich's theoretical views were unknown to me until my treatise upon "Repulsion as a Cosmic Force" was printed in 1853. So far as I have since been able to comprehend them, it may be as frankly stated that they have seemed to me deficient in practical and universal application to both cosmical and chemical philosophy. I must furthermore state that the "Theoria" of Boscovich has never been in the smallest degree even suggestive to me of fact or law, either during my observations and researches, or during the discussions embraced in the following pages.

In order to show that even Newton, several years prior to Hooke's observations (points of practical importance to be subsequently alluded to), had himself conceived of the possible existence and methodical working of a force unlike (and the opposite of) gravitation, although his views were strictly conjectural, it will be only necessary to quote a few lines from his letter to Boyle, dated Cambridge, Feb. 28, 1678–9. He says: "I have so long deferred to send you my thoughts "about the physical qualities we spake of, that did I not "esteem myself obliged by promise, I think I should be "ashamed to send them at all. The truth is, my notions "about things of this kind are so indigested, that I am not "well satisfied myself in them; and what I am not satisfied "in I can scarce esteem fit to be communicated to others, "especially in Natural Philosophy, where there is no end of "fancying. . . .

"It being only an explication of qualities which you desire "of me, I shall set down my apprehensions in the form of "suppositions, as follows. And first, I suppose that there is "diffused through all places an ethereal substance capable "of contraction and dilatation, strongly elastic, and, in a "word, much like air in all respects, but far more subtil.

"2. . . . .

"3. . . . .

"4. When two bodies moving towards one another come "nearer together, I suppose the ether between them to grow "rarer than before, and the spaces of its graduated rarity to "extend further from the superficies of the bodies towards "one another; and this by reason that the ether cannot "move and play up and down so freely in the strait passage "between the bodies as it could before they came so near "together. . . .

"5. Now from the fourth supposition it follows, that when "two bodies approaching one another, come so near together "as to make the ether between them begin to rarefy, they "will begin to have a reluctance from being brought nearer "together, and an endeavour to recede from one another: "which reluctance and endeavour will increase as they come "nearer together, because thereby they cause the interjacent "ether to rarefy more and more. But at length when they "come so near together that the excess of pressure of the "external ether which surrounds the bodies, above that of "the rarefied ether, which is between them, is so great as to "overcome the reluctance which the bodies have from being "brought together, then will that excess of pressure drive "them with violence together and make them adhere strongly "to one another, as was said in the second supposition: . . . "and the endeavour which the ether between them has to "return to its former natural state of condensation, will "cause the bodies to have an endeavour of receding from one "another. . . .

"When any metal is put into common water, the water "cannot enter into its pores to act on it and dissolve it. Not "that water consists of too gross parts for this purpose; but "because it is unsociable to metal. For there is a certain "secret principle in nature, by which liquors are sociable to "some things and unsociable to others. . . .

"From this supposed gradual subtilty of the parts of the "ether, some things above might be further illustrated and "made more intelligible; but by what has been said you will "easily discern whether in these conjectures there is any "degree of probability, which is all I aim at. For my own "part, I have so little fancy to things of this nature, that "had not your encouragement moved me to it, I should "never, I think, have thus far set pen to paper about them." (MACCLESFIELD'S *Collection of Letters of Scientific Men. Oxford Edition,* 1841, vol. ii. pp. 407—419.)

Thus it will be seen that there lay (floating, but undetermined) in the imagination of Newton himself, the idea of a principle the opposite of gravitation: and that, so far as he theorized, he conjectured these two principles—one causing bodies to recede from, the other causing them to approach, one another—to reside in an ether which he fancied to fill space, and to penetrate also the innermost constitution of all bodies; and that matter was itself, or at least masses were *per se* inert.

If the simple conjectures, however, of great philosophers were to be historically traced, we should also find Lucretius presenting the idea of a principle the opposite of gravitation in his remarkable poem, "De Rerum Naturâ." Therein the entire system of nature is, to be sure, fancifully unfolded, but with far greater acuteness and subtlety than is discoverable in the most learned disquisitions of any author or thinker subsequent to his time, and prior to our own age of experimental and observational study. Lucretius, indeed, boldly sets aside and declares absurd all opinions upon gravitation—a point in which his contemporaries evidently believed—and places in the strongest foreground of his vast picture of nature the idea of an absolute void in which atoms (or, as Munro translates the word, "the *first-beginnings*" or "seeds" of things) and repulsion are the fundamental agents of creation.

"And herein, Memmius, be far from believing this, that all "things as they say press to the centre of the sum, and that "for this reason the nature of the world stands fast without "any strokes from the outside, and the uppermost and lowest "parts cannot part asunder in any direction, because all "things have been always pressing toward the centre (if you "can believe anything can rest upon itself), or that the heavy "bodies which are beneath the earth all press upwards and "are at rest on the earth, turned topsyturvy, just like the "images of things which we see before us in the water. . . . "Things cannot therefore, in such a way, be held in union, "o'ermastered by love of a centre."

"Space I have already proved to be infinite. . . ." (*Book* i. MUNRO'S *Translation.*)

". . . we will now go on to explain what remains to be "told of motions.

"Now methinks is the place, herein to prove this point "also that no bodily thing can by its own power, be borne "upwards and travel upwards; that the bodies of flames may "not in this matter lead you into error. For they are be- "gotten with an upward tendency, and in the same direction "receive increase, and goodly crops and trees grow upwards, "though their weights, so far as in them is, all tend down- "wards. And when fires leap to the roofs of houses and "with swift flame lick up rafters and beams, we are not to "suppose that they do so spontaneously without a force "pushing them up. Even thus blood discharged from our "body spirts out and springs up on high and scatters gore "about. See you not too with what force the liquid of "water spits out logs and beams? the more we have pushed "and forced them deep down and have pressed them in, "many of us together, with all our might and much painful "effort, with the greater avidity it vomits them up and casts "them forth, so that they rise and start out more than half "their length. . . .

". . . Wherefore in seeds too you must admit the same, "admit that besides blows and weights there is another cause "of motions, from which this power of free action has been "begotten in us, since we see that nothing can come from "nothing. For weight forbids that all things be done by "blows through as it were an outward force; but that the "mind itself does not feel an internal necessity in all its "actions and is not as it were overmastered and compelled to "bear and put up with this, is caused by a minute swerving "of first-beginnings at no fixed part of space and no fixed time.

"Nor was the store of matter ever more closely massed nor "held apart by larger spaces between; for nothing is either "added to its bulk or lost to it. Wherefore the bodies of "the first-beginnings in time gone by moved in the same "way in which now they move, and will ever hereafter be "borne along in like manner, and the things which have "been wont to be begotten will be begotten after the same "law and will be and will grow and will wax in strength so "far as is given to each by the decrees of nature. And no "force can change the sum of things; for there is nothing "outside, either into which any kind of matter can escape "out of the universe, or out of which a new supply can "arise and burst into the universe and change all the nature "of things and alter their motions." (*Ibid. Book* ii.)

Others among both the ancients and moderns have uttered kindred notions in one form or another. But the mere expression of suppositions or fancyings, as Newton properly calls such speculations, however ingeniously or poetically similar ideas may have been set forth, has done little to advance physical science, or lift the origin and phenomena of the celestial motions or terrestrial changes from the realms of doubt or of absolute darkness.

Of course, there must come an end, sooner or later, to all speculations upon physical questions, albeit the period may be yet far distant. The days of indulgence in theories, and of being satisfied with fancyings, are happily passing away. True philosophy, the exact sciences, demand unquestionable facts and principles upon which to found severe discussions, and will be content with nothing short of these.

Fortunately, a remarkable class of celestial objects has always excited the uncommon attention of even *casual* observers. In an especial manner the singular and often startling appearances presented in the tails of the great comets which from time to time have swept within visible ranges of our planet, have during all ages, and even in nations only partially enlightened, incited the most lively and subtle inquiries into the nature and action of the causes which produce them.

The phenomena presented by them have given rise of late to much discussion and inquiry as to the presence within, or rather as to the action upon, them of some principle of force besides that of gravitation.

Dr. Robert Hooke, as long ago as 1682, made such remarkably acute observations upon several extraordinary comets which appeared during his career, as to discover the fact, and express the idea as plainly as any of his successors has done, that some agency seemed to proceed out of the sun, which drove the particles of comets' tails in a contrary direction to that which gravitation would cause them to take. He supposed this agency to be *light*. It is only just to the memory of Dr. Hooke to say that no observations subsequent to his own, advanced knowledge or even conjecture upon this subject, or suggested approximative explanations of cometary phenomena, until Bessel, availing himself of the experimental researches of the present century and the terms adopted in modern physics, applied an indefinite idea of *polarity* to the visible influences exercised over the formation and phenomena of the tails of comets by the solar energies. Hooke even applied the name of "Levitation" to this solar action, upon which he theorizes as follows:—"For though (as I have "before mentioned) the exceeding swift Dartings and Flash-"ings of Light do rather seem to be made by a successive "Accension, like a Train of Gunpowder (and, as I have sup-"posed, if not proved, Lightning); yet to maintain such a "Succession of Steams, there must needs be an exceeding "quick Supply from the Head, even to the utmost extremity "of the Blaze: and it looks not unlike to a kind of actual "Levitation, or a driving outwards of certain halituous Parts "from the Sun, as if the Rays of Light of the Sun were "carried with a local Motion rapidly swift, and in their "Passage by the Star of the Comet, did carry with them such "kind of halituous or fiery Steams. The often observing "with my bare Eye these sudden Dartings of Light, which "seemed to pass almost in a Second from the Head even "to the End of the Blaze, did often make me consider, "whether the Rays of Light might not thence be supposed to "be moved away from the Sun with this exceedingly violent "and rapid Motion; especially since there are some Obser-"vations of Mr. *Romer's* about the Eclipses of *Jupiter's* "*Satellites,* which seem to make probable such a Theory." (*Posthumous Works,* 1705, p. 170). The indefiniteness of Hooke's ideas, however, upon the nature and action of force is clearly shown in his teachings that "Matter" and "Motion" were the real and the only *fundamental* principles of the cosmos. "These two I take to be two single "Powers, which co-operate in effecting the most of the "sensible and insensible Effects of the World. . . . These "are, as it were, the Male and Female of Nature, from the co-"operations of which the most of Natural Productions are "effected. The first is, as it were, the Female or Mother "Principle, and is therefore rightly called by *Aristotle* and "other Philosophers, *Materia,* Material Substance or *Mater;* "this being in itself, abstractly considered, without Life or "Motion, without form, and void, and dark; a Power in "itself wholly inactive, until it be, as it were, impregnated "by the second Principle, which may represent the *Pater,* "and may be called *Paternus Spiritus,* or hylarchick Spirit, "as some call it, without whose Conjunction nothing, or no "alteration, can be produced: For neither can Matter without "Motion, nor Motion without Matter, produce any Effect. . . ." Speaking of those "who have gone the other way and have "been dabbling in this or that particular subject," he says: "The four Elements, the 3 Chymical Principles, Magnetism, "Sympathy, Fermentation, Alkaly and Acid, and divers other "Chimeras, too many to repeat; which having been em-"braced, nothing else will be heard, or go down with them; "whereas, alas, Nature perhaps knows no one of these for a "Principle in any sense, much less in that which they "understand it. . . . But to proceed, this supposition of the "two fundamental and primary Powers, to wit, that of "*Matter,* and that of *Motion,* which I have here delivered, "seems to me very consonant to the sense I understand of "the History of the *Genesis* of the World, delivered by "*Moses,*" &c. &c.

Thus it will be seen that Hooke himself, who states in the course of his Dissertation upon Comets, that he was the first observer of the outstreaming motion of light in the tails of comets in a direction opposite to that of gravitation, a motion which he called Levitation, whose essence was the same as that principle which directs the flame of a candle upwards from the earth, and which he embraced in the name and action of "Light" as one "of the two great Laws of Motion," gravity being the second;—thus it will be seen that Hooke himself, the earliest telescopic observer (so far as I know) of the physical fact of cosmical repulsion, had no exact idea of its character or numerical value as an abstract and radical principle, although he was fully impressed with the conviction, in virtue of his personal observations upon the motions of comets' tails, that there was another great "Law of Motion" beside gravity. "The first is that of *Light*, the second is that of Gravity." His motive impulse or principle of light was "Levitation." Hooke's system of forces and Natural Philosophy embraced an illimitable "*Expansum*," filled with a diffuse material "*Medium*" called "*Æther*," which ether, *per se*, notwithstanding its extreme tenuity, he declared to be heavier, "Bulk for Bulk," than the heaviest Gold. Though this doctrine might "at first hearing seem a little paradoxical, if not absurd," he would nevertheless "be "able to make it somewhat more plausible, if not positively "and undeniably demonstrate it so to be," by the fact that this ether penetrated throughout the interstices of the densest gold; "so that the Vacuity, as it is commonly taught, or erroneously supposed, is a more dense Body than Gold as Gold." In consequence of this Gravity of Ether or Vacuity in the *Expansum*, light darted through it away from the sun as the blaze of a candle darts away from the earth; and the tails of comets were formed by this levitation, carrying with it the "Volatile Nitrous Spirits" and "halituous Steams" of the comet's head into "the vast *Expansum* of the World" which filled "the whole Interstices between the greater globular solid Bodies thereof," in the same way as the levitation of light carried with it the "Steams issuing out of the Wieke" and Unctuous Substance of a candle during combustion. The two phenomena were identical, differing only in magnitude and situation.

Such is a glimpse of the ideas and philosophy prevailing among the learned in the time of Newton, a period when a host of great lights in Germany, France, and England, burst upon the world with a suddenness only equalled by the wondrous comets which awoke them to sublime contemplations of nature and the mathematical study of her laws. If quaintly and crudely uttered, they are not destitute of suggestiveness, and will at least remind scientific men of the laborious works and prying discussions which, though long since shelved or forgotten, prepared the way for the present achievements of physical observation and experimental discovery.

Nor will I overlook the similar conjectures—shapeless ideas, however, scarcely meriting the title of hypotheses—of Kepler, Euler, and others, that the light emitted from the sun might be a repulsive force *per se;* or those of Olbers and others, that the sun ejects *electricity* which exercises repulsive influences directly upon distant bodies; for it is only necessary to submit these theories to the crucible of existing knowledge in order to reduce them to dross, and demonstrate the necessity for the precedent activity of some principle profounder still than either light, electricity, magnetism, or heat, *per se,* as a universal base of mechanical energy.

The speculations of Bessel, resulting from the observations of that illustrious astronomer upon the appearance of Halley's comet in 1835, and especially upon the development and phenomena of its tail, are of a higher order than the preceding, and demand definite attention, because they are founded upon the recognition of the fact of a cosmico-repulsive force as a simple but positive phenomenon—a phenomenon connected with experimental physics by certain specific indications—and an effect produced in some unexplained way, by a solar action the opposite of solar gravitation. His Memoir upon this subject may be found in the *Astronom. Nachrichten,* vol. xiii. No. 300—302, pp. 185—232. The gist of Bessel's opinions and speculations upon this subject may be fairly presented in the following extract from that Memoir: "Meine Vorstellung von der *Möglichkeit* einer Verbindung "aller, an den Kometen beobachteten Erscheinungen ist in-"dessen die folgende. Jede Wirkung eines Körpers auf "einen anderen kann in zwei Theile zerlegt werden, deren "einer für alle Theile des letzteren gleich ist während, der "andere aus den Unterschieden der Wirkungen auf verschie-"dene Theile entsteht. Wenn die Wirkung in sehr grossen "Entfernungen der Körper voneinander, sehr klein ist, so ist "der erste Theil derselben derjenige, welcher, bei einem "Uebergange von diesen Entfernungen zu kleineren, zuerst "merklich wird; der andere kann erst später eine merkliche "Grösse erlangen. Im Falle eines Kometen, welcher in sehr "grosser Entfernung zu der Sonne herabkömmt, zeigt sich "also zuerst die allen seinen Theilen gemeinschaftliche Wir-"kung: ich nehme an, dass sie in einer Verflüchtigung von "Theilchen bestehe, welche der Sonne feindlich polarisirt "werden. Der andere, später merklich werdende Theil der "Wirkung allein, kann eine Polarisirung des Kometen selbst, "so wie eine vorzugsweise Ausströmung nach der Sonne uz, "zur Folge haben. Zeigen die Beobachtungen wirklich diese "Erscheinungen, wie bei dem Kometen von 1744 und dem "*Halley*'schen der Fall war, so kann nicht geläugnet werden, "dass die Auströmung, indem sie aus einem der Sonne zuge-"wandten, also ihr freundlich polarisirten Theile der Ober-"fläche hervorgeht, auch dieselbe Polarisirung besitzt, welche "die ausströmenden Theilchen der Sonne zu nähern sucht. "Dass die ausgeströmten Theilchen dennoch von der Sonne "zurückgestossen werden, wie die Beobachtungen zeigen, kann "vielleicht dadurch erklärt werden, dass die Ausströmung in "einem Raume stattfindet, welcher schon mit ihr feindlich "polarisirter Materie gefüllt ist und fortdauernd damit gefüllt "wird, wodurch die entgegengesetzten Polaritäten sich "ausgleichen und die ausströmenden Theilchen desto mehr "von ihrer ursprünglichen Eigenschaft velieren und desto "mehr die entgegengesetzte annehmen, je weiter sie sich von "dem Kerne des Kometen entfernen.

"Durch diese Ansicht werden alle Erscheinungen, welche "ich an dem Kometen wahrgenommen habe, untereinander "in *Verbindung* gesetzt."

"My conception, then," says Bessel, "of the *possibility* of a "union of all the phenomena observable in comets, is the "following. Every action of one body upon another may be "decomposed into two parts, one of which has the same value "for every particle of the body, while the other consists in "the differences of the actions upon its different portions. "If the bodies are at great distances from each other and "the action small, the *one* part is that which, upon a transi-"tion of these distances to lesser ones, becomes apparent "first; the *other* only assumes an appreciable value *later*. "In the case of a comet approaching the sun from an

" immense distance, a reciprocal action evinces itself at first " in all its parts. I suppose this to result from a volatiliza- " tion of its particles which are oppositely polarized to the sun. " The other action, which only becomes apparent later, may " have for its effect a polarization of the comet itself, as well " as an outstreaming toward the sun. Should the observa- " tions indicate these appearances—which, with the comet of " 1774, and that of Halley, was actually the case—it cannot " be denied that the outstreaming, inasmuch as it proceeds " out of a portion of the surface turned toward the sun (and " therefore its friendly-polarized—"freundlich polarisirten"— " portion) possesses that polarization which seeks to identify " itself with the outstreamed portions of the sun. That the " outstreamed particles are nevertheless repelled on the part " of the sun, may, perhaps, be explained upon the supposition " that the outstreaming takes place in a space filled with " oppositely polarized matter with which the discharged par- " ticles of the comet become continually saturated, in conse- " quence of which the contrary polarities neutralize each " other, and the outstreamed particles lose so much the more " of their original property, and take on so much the more of " the opposite one, in proportion as the distance between " them and the nucleus of the comet increases.

" Upon this view of the subject, all the appearances which " I have been able to notice in comets are rationally united " with one another."

This translation (with which I am favoured by my friend William H. Wahl, of Philadelphia, U.S.A.) expresses at least with approximate clearness, the opinions of Bessel upon the phenomena in question. By this explanation he endeavoured to account, also, for the return of the tails into the nuclei. His telescopic observations of Halley's comet impressed him with the idea that some other function of force probably existed in the sun than that of gravitation, inasmuch as the phenomena noticed by him indicated the presence and action of a contrary element of power. This he denominated *polarity.* But what problem does that simple word solve, when applied to the mechanics of the heavens? The unsatisfactory terms in which even the great Bessel stated his views of its action show plainly that he had no definite idea of the character of stellar and planetary repulsion as an abstract principle of energy, nor of its universal association with gravitation as a central—or rather a decentral—force in all cosmic masses, and as the veritable working power of nature. Professor J. B. Nichol, of Glasgow, in a learned essay upon comets, speaks of Bessel's memoir in the following manner; and from that time (1860) to the present nothing has been added to this department of inquiry: "One remarkable contribution to positive knowledge must, "however, be mentioned—viz. Bessel's memoir on the phe-"nomena attending what seemed the process of *the formation* "of the tail of Halley's comet in 1835. This distinguished "person had rare opportunities to notice these phenomena, "and they led him to the conviction that our central orb "develops a *polarity,* or exercises a *polar force* over these "diffuse meteors. What that polar force is, remains of course "unknown; but, considering it akin to *magnetism,* it will "readily be seen that it renders comprehensible the disturb-"ance of the comet's form, its position towards the sun, and "the fact that the motions of the body are not disturbed "by the development of the tail. The whole of the memoir "in question is worthy of more attention than it has hitherto "received." How much Bessel's hypothetical explanations of the phenomena attending "the formation of the tail of

Halley's comet" has assisted my researches and discussions, and what influence they have exercised upon the developments of the following chapters, the reader will be able to determine for himself, after the perusal of this work. It would not be out of place, nevertheless, to say here, in justice to myself, that I was ignorant of the existence of Bessel's memoir until I had for many years been occupied with other series of phenomena which are discussed at large in the sequel;—not, indeed, until 1858, after the publication of Mr. Bond's celebrated observations of appearances in Donati's comet similar to those observed by Bessel in Halley's, and after Professor Norton's mathematical discussions of certain physical elements involved in the formation and orbit of its tail, were put forth to silence the pretensions of another party to originality of discovery. It was at this time that Bessel's memoir, which seemed to have been long forgotten by astronomers, was first brought to my knowledge.

In 1858-60 M. Faye, Member of the French Academy of Sciences, revived the idea, also, as a consequence of the telescopic observations of Mr. Bond and others, including his own, that a definite repulsive influence was exerted by the sun upon Donati's comet and upon its tail, and upon comets in general, but not upon the denser cosmic bodies. M. Faye elaborates his theory in a communication (No. 1240) to the *Astronomische Nachrichten* dated "Paris, 1860, February 4;" and inasmuch as this is a subject which must very soon engage the earnest attention equally of physicist, astronomer, and mathematician, I shall set forth M. Faye's hypothesis in his own language, and there leave it upon its exact grounds, to be adjudged according to its merits, and in the light of present and of increasing knowledge:—

"La brillante apparition de la comète de Donati que

" j'ai étudiée avec un télescope en verre argenté construit " par M. Foucault, a vivement fixé mon attention sur la " théorie des comètes. . . .

"D'après l'ensemble des faits observés et des discussions " qu'ils ont fait naître, on peut établir, comme conditions " premières du problème, les points suivants :—

"1. Le soleil exerce visiblement une répulsion sur la " matière des comètes.

"2. L'accélération du mouvement d'une comète est con- " nexe avec la formation de sa queue.

"3. Les phénomènes plus particuliers (multiplicité des " queues, secteurs lumineux et leur balancement, enveloppes " concentriques du noyau, etc.) doivent être expliqués, non " dans leurs moindres détails, ce qui serait assurément trop " exiger d'une théorie quelconque, mais dans leurs traits les " plus généraux, sans doter la matière cométaire de propriétés " toutes spéciales.

"4. Une force ne peut être introduite hypothétiquement " dans le système du monde, qu'à la condition de ne pas " troubler sensiblement l'harmonie actuelle basée sur les lois " de la gravitation et sur l'observation.

"5. Il convient de n'accepter, dans le système du monde, que " des forces connues ou des forces susceptibles d'être vérifiées " expérimentalement jusque dans le mode d'action supposé.

"La théorie à laquelle je me suis arrêté ne met en jeu que " les propriétés les plus générales des corps, la gravité, la " chaleur solaire, une force répulsive née de cette chaleur et " exercée au loin par la surface incandescente du soleil. . . .

"Réduisons d'abord la question au rôle de la gravité seule : " ce sera le cas des comètes qui circulent autour d'une soleil " éteint. . . . Mais les choses se passent autrement autour de " notre soleil incandescent. . . .

"Introduisons maintenant la chaleur du soleil. Sur son "influence, la formation des queues sera évidemment activée "par la dilatation progressive de la nébulosité, ou plutôt des "couches qui forment le noyau, et peut-être par un change- "ment d'état physique dans une partie de sa matière. . . .

"Introduisons enfin, avec la chaleur, la répulsion que la "surface du soleil doit exercer au loin, d'après l'hypo- "thèse. . . .

"Pour aller plus loin, il faut définir la force répulsive. De "ce qu'elle est exercée par la surface du soleil, je conclus "qu'elle est indépendante de sa masse, tout comme les "radiations solaires. . . .

"On peut montrer enfin que la quatrième condition est par- "faitement remplie, car, par la nature même de cette force "répulsive, les planètes et leurs satellites échappent à son "action à cause de leur densité considérable. . . . J'appelle "sur ce sujet les méditations des physiciens. On a souvent "cherché si la chaleur qui produit ou exalte la répulsion "moléculaire ne produirait pas aussi une répulsion à dis- "tance: ne serait-il pas intéressant de voir un phénomène "astronomique nous révéler la cause de l'insuccès de ces ten- "tatives si rationnelles au fond, et nous apprendre à quelles "conditions on peut réussir à mettre cette répulsion en "évidence? *Ne serait-il pas plus intéressant encore de re- "trouver dans le ciel la dualité des forces opposées qui régissent "la matière autour de nous?*"

In order to render ample justice to M. Faye, I will express my conviction, after a careful study of his views, that he, like Boscovich, Bessel, and Bond, had become clearly satisfied that some principle of energy contrary in its action to that of gravitation is requisite to develop certain cosmical phenomena, and to fulfil the physical necessities observable in the aspects of cometary bodies and in the formation and direction of their tails. While Euler conjectured that some occult principle of this character depended upon solar light, Olbers upon solar electricity, Bessel upon the opposite polarity of a resisting medium to that of the cometary particles (and I might include Nichol's suggestion of magnetism, as another and final speculation), M. Faye explicitly declares his opinion of the origin of this force to be, that it is born or springs out of heat—solar heat—the incandescent surface of the sun: "une force répulsive," he says, "née de cette chaleur (*la* "*chaleur solaire*) et exercée au loin *par la surface incandescente* "*du soleil.*" Let the reader bear in mind his explicit terms, for the purpose of deciding how far, and in what respect even the latest hypothesis or conjectures may have assisted me, or influenced the progress or final results of my inquiries. Let it be borne in mind at this point where M. Faye declares himself to stand ("la théorie à laquelle *je me suis arrêté*"), that if his cosmic repulsive force is born of solar heat ("née de cette chaleur solaire"), it must follow that heat itself is a more radical principle than the one to which it gives birth. A particular reference to this part of the eminent French astronomer's theory is the point I wish to fix upon the reader's mind, in order that he may recall it when his interest may have become enlisted in the following discussions. The simple statement that my first demonstrations of "Repulsion as a planetary, solar, and universal Force," were published in March, 1853, will be sufficient to establish priority of research and publication upon the subject at large. It has been well stated by a French *savant* that "when "questions of priority interest only persons, cotemporary "science, which often presents history imperfectly, does not "preoccupy itself with them; but it is not so with scientific "facts, or with their interpretation." (*Comptes Rendus*, vol. lxiii. p. 994.) Further remarks in regard to the historical branch of this subject are scarcely necessary, inasmuch as even those who have admitted the observations and opinions of Bessel and Bond to be of value mathematically have not elaborated them, and have extended their application no further than to the formation, direction, and *conjectured* molecular motion of comets' tails.

This statement presents a fair *exposé* of what has been observed, suggested, done, or recorded in scientific works or memoirs of a worthy and reliable character upon the subject of cosmical repulsion up to the moment of my writing, so far as I am aware.

5.

A word upon authorities. Considering the endless store of experimental and incontrovertible facts, and of authentic observations in physics, astronomy, and other departments of research, which have been accumulated from the days of Thales to our own, and especially during the last three centuries—considering these stores of individual facts in the light of a stupendous museum, open freely to all who seek knowledge, or would attempt to classify them, or learn their origin, connexion, differences, interdependence, and general history—similar, indeed, to a universal exhibition of natural objects and of ancient and modern art, free to common study—I have not hesitated to handle and use these facts where they represent generic principles, or would solidify and perfect that work of generalization which has associated itself legitimately with every step of my undertaking, and which is, in truth, the object and sum and substance of the work itself.

As it would be beyond my power, and indeed quite unnecessary, to cite in detail the names of all authors, treatises, and articles to whom and to which I am indebted for facts and knowledge, as industrious compilers are habituated properly to do, I must refer at large to encyclopædic records, and to academic and periodical reports of scientific discovery and discussion for what may not be instantly acknowledged to be original with myself. Claiming no contributions to knowledge which will not be readily accorded where such contributions may deserve recognition, I shall leave the entire field of inquiry, fearfully vast as it is, to unfold itself gradually, without diverting the student's mind therefrom by lengthy comments, or by annotations and references of a minute character.

It gives me pleasure, however, to allude particularly to an able memoir of Professor Rankine, upon the "Outlines of the Science of Energetics," read to the Philosophical Society of Glasgow, May 2, 1855, and to express here a due consideration of its value. While Mr. Rankine has taken an important step in the right direction, it may be stated nevertheless, without unfairness, that his views have been hitherto theoretical, and, so far as I am aware, undeveloped even by himself. At the same time that my labours are, and have been, wholly independent of his studies, the latter have a very conspicuous bearing upon the principles unfolded in the following chapters. The term "Energetics," applied by Mr. Rankine to his Essay, is both happy and suggestive; and I have adopted it in the course of my discussions as a pregnant and significant symbol of the principles and philosophy which I shall present in full, and endeavour to establish throughout the sequel. But even this word—indicative of the most abstract physical conditions of nature, is probably only a temporary "term" in intellectual progress. Like "phlogiston," "caloric," and many others which have shaped, and, like imperial magicians, have presided over the ideas of their day, it also may be supplanted by others as science mounts upward step by step to its lofty destination and fullest developments.

After making the preceding general remarks upon authorities and authors, it is nevertheless permissible for me to say that if special references have been made in the course of our discussions to those great lights who, from time to time, have shed lustre upon science and thought by transcendent discovery, or by observations and experiments directly leading to discovery; such references have been not only essential and proper, but also most grateful duties. The names of illustrious men and assiduous scholars are both types of principles and golden links in the chain of intellectual and moral progress. They ennoble history, adorn nations, incite the young to the emulation of daring thoughts and successful deeds; and are invariably conspicuous in whatever attempts we make to classify facts or trace and unfold general laws. All such scholars are true high priests of nature, and while their labours shed blessings upon conquering science to descend through endless generations, they have at the same time an inalienable right to exact tithes of acknowledgment and gratitude even after having vanished for ever from mortal sight.

6.

There are two illustrious philosophers whose names will here and there adorn the following pages, the recollection of whom demands from my pen more than a simple reference, where the opportunity permits me, as it does here, to acknowledge in grateful terms their friendly interest in the results of my researches. I allude to the late Michael Faraday, and to Baron Justus Von Liebig.

To the latter I am indebted not only for much information derived from his works during a somewhat active professional life, but also for kind attentions and encouraging words since I have been honoured with his personal friendship, and been engaged upon the following discussions.

To the former I have felt deeply grateful since the year 1858, at which time a generous and encouraging letter from him incited me to persevere in those observations and pursuits which have resulted in the present volume. He believed in something more than the simple (yet admitted and accurate) numerical law of gravitation; and he never despaired that the time would arrive when some new light would penetrate even to the bottom, and illuminate every crevice and fault, of the deep and dark mines in which he was such an indefatigable and successful explorer. And in thus referring to the name of that unpretending but truly great philosopher, I would utter these tenderest words of grateful remembrance as I would lay a perennial garland upon his statue or fragrant flowers upon the fresh sod which covers his honoured remains.

7.

The preliminary points to which it has been necessary for various reasons to refer, having been stated, I will now invite the profound and candid attention of both the learned and general reader to the subject of this work—alike to facts, treatment, deductions, and generalizations—which I have the honour to lay before them, trusting that the results of many years' labour may not be wholly devoid of importance to my contemporaries, and hoping that the laws which I have endeavoured to unfold may prove of permanent value in every department of physical and morphological analysis. That this work is incomplete will be only too apparent to the appreciative student of nature; but, incomplete as it may be, that it will receive the careful study which the magnitude of its inquiries demands and which the importance of these inquiries deserve in this age of thought, experimental knowledge, and solid progress, I cannot for a moment permit myself to doubt.

CHAPTER I.

SPACE.

1.

"ABSOLUTE SPACE," that is to say, *immensity of space* conceived of as an absolute void—a state destitute of all matter and all force—would undeniably be *nothing*; or, a condition free of all resistance to the action or reaction of matter and force, whenever they should begin to be, and to move in mutual relations. Such an idea of space might be precisely expressed in the words *illimitable vacuum.*

2.

The origin of space, if it had any, is beyond thought or conjecture. As we know nothing, we can only contemplate it as a vast and bewildering fundamental fact, or ideal primordial depth, in which all phenomena of creation began and have been successively developed, since the beginning of time.

3.

Some suppose space to consist of an ether, a peculiar kind of matter of indescribable tenuity, in which the grosser forms of matter are suspended (an idea originated by the ancients, and revived by Descartes in order to explain his theory of the propagation of light); and they further describe this ether to be necessary as a medium of transmission for the forces which play between these grosser forms, and as a cause of resistance to the motion of at least one of them. Among recent thinkers, Mossotti and Faraday, perhaps the most profound and sagacious of modern experimental physicists, have believed this ether to be so attenuated as to be imponderable, and still to be material; while others of less eminence have insisted that it is of a ponderable character, and that there is essentially no empty space within the boundaries of creation. Dana has supposed such an ether unnecessary as a medium for the intercosmical transmission of certain phenomena, and has replaced it with what he denominates "pulsating molecular force." If "pulsating molecular force" be communicated by one body to another when *infinite distances apart in an infinite vacuum*—that is, without intermediate molecular conditions—the fact has not thus far, to my knowledge, been intelligibly demonstrated. The conception is pregnant, however, with embryonic truth; and, as a *hypothesis,* is an important step in advance of prevailing ideas, as will hereafter more clearly appear. But it has not been presented in a manner plainly to elucidate, so far as I know, the mode in which "molecular force" is transmitted through space; nor can it be expected, indeed, that the ablest physicists will succeed in establishing ultimate truth, when important elements of knowledge are ignored or discarded from physical inquiry.

All these different doctrines of modern theorists—including the conjectures of the ancient Greek philosophers, which continued to prevail through the Middle Ages—exhibit the penetrative restlessness of human thought to extend itself into the widest realms, and find out the deepest secrets of the universe. And no age since the revival of learning has been stimulated by such encouragements and advantages for profound investigation, with such sure prospects of satisfactory rewards, as the present. But, so far, no solid knowledge has been reached in this direction.

The speculations of the earlier thinkers relative to Space may, on the whole, when the character of our facts is fairly weighed, be regarded as nearer the truth than those of modern philosophers—excepting those of Dana, who appears to believe clearly, without proving it, that space is vacuum, and perhaps Helmholtz, who declares that if the ether is material its resistance cannot be *nil.* The former had, indeed, no definite ideas of space, but supposed it to be a vacuum in which cosmical bodies had been projected by some divine mechanism, and in which they continued to move by some *vis inertiæ.* The theory, nevertheless, has long prevailed in molecular physics—at least, ever since it has been reduced to science—that an ether, or attenuated form of matter, exists between atoms which prevents their contact; that attraction draws molecules into a certain degree of juxtaposition until repulsion is awakened, which prevents actual contact; and that intermolecular spaces are penetrated throughout the largest and densest masses by this hypothetical ethereal fluid.

4.

The wonderful discovery made by the astronomer, John Francis Encke, that the comet of shortest period, or of $3\frac{3}{10}$ years, now bearing his name, had been retarded in its return to perihelion by some cause independent of ordinary planetary perturbation, led him to adopt the opinion that the interplanetary spaces must be *likewise* filled with the same attenuated substance. Thus a bold conjecture of Descartes seemed to be demonstrated as a fact, by Encke.

It appeared to him that some *ponderable* means of resistance must have been opposed to the motion of this comet, or it would have completed its revolution at the moment calculated from its well-observed elements. A declaration of such importance, by so distinguished an observer, was startling to scientific men, and received their ready assent in consequence of the high mathematical authority of its author. This presumption of Descartes, sustained by the mathematical conclusion of Encke, has been seized by every speculator upon cosmical phenomena; and it is now adopted for the solution of all problems concerning light, heat, colour, magnetism, electricity, the aurora borealis, the zodiacal light, and even earthquakes. And the idea that space is filled with attenuated but ponderable matter against which the earth, comets, and all other bodies strike in their movements, is now so general as to become a basis for the wildest theories; and it is apparently established, as a scientific fact, as much beyond discussion as gravitation itself.

There was probably never a grosser error introduced into physical science than the ethereal theory; and its influence in retarding solid progress, and in excluding the deductions of experimental physics from their possible application to the causes of light as a general phenomenon, and as a result of repulsion (repulsion being viewed as an agent in the local generation of cosmic light, electricity, &c.), has been greater than at first appears. While the discovery of Encke respecting the anomalous motion of his comet was, doubtless, an important and suggestive fact, the conclusion announced from it was premature and supererogatory. Inquiry after some explanation of so remarkable a phenomenon was, indeed, natural; but the imperfect development of cosmical physics rendered its discovery, at that moment, impossible. More recent observations of comets have established beyond question the truth, that a repulsive principle plays between them and the sun; and it is quite possible that a more thorough discussion of physical forces and of their demonstrable influences in celestial mechanics, will elucidate similar anomalies without laborious or fallacious efforts of the imagination. Solid knowledge is only acquired by slow and positive steps; and exact science can never be advanced by insisting upon theories at the expense of truth.

The fact discovered by Encke simply *suggested* to his sagacious mind, that the resistance existing in the interplanetary spaces must be *material*, because gravitation was considered by astronomers the sole force in the universe acting upon moving masses, and because, if gravitation was counteracted by any other principle, it must be by some attenuated matter, diffused throughout space, acting mechanically upon the advancing comet, and thus retarding its motion. No repulsive force acting between the sun and its revolving system of planets, the antagonist of attraction, was then admitted by astronomers, or imagined to exist—notwithstanding Bessel, ten years previously, had observed phenomena in Halley's comet, which impressed him with the belief that it was influenced by some repulsive principle *proceeding from the sun.* That such a force does exist, is now admitted by some astronomers, as an electrical or thermic element, and future observation may establish REPULSION to be *an independent principle,* as universal in its influence as *gravitation* itself; and when its functions in celestial dynamics shall be thoroughly understood, the prevailing opinion that space consists of a resisting medium would be discarded as a fallacy and absurdity. We have given reasons for supposing that the basis of the speculation is of a very questionable nature. Nothing is absolutely known upon this subject. It is so profound that science is destitute of even a ray of light upon it, except what may be derived from analogy and inductive reasoning. All opinions upon the conditions of infinite space are the merest hypotheses; and in the midst of conjectures, that would be the most probably correct which presumed space to be *a perfect vacuum.* This would accord in analogy with all that is known of the character of ultimate principles, and permit the freest motion to cosmic masses; for every principle or general condition of nature presents *dual* qualities, or opposite methods of expressing its functions or character,—as attraction and repulsion, paramagnetism and diamagnetism, positive and negative electricity, heat and cold, &c. So, also, in the most fundamental of all the conditions of nature, the material should appear in opposition to the immaterial; matter and vacuum should be simultaneously and necessarily associated. The latter is necessary for the existence of the former.

As far as one argument can combat and supplant another, and thus far influence opinions upon a subject so unsupported on either side by direct, demonstrable, and conclusive proofs, it may be said with fairness that M. Faye, in an able paper upon "*The hypothesis of a resisting medium,*" published in the *Comptes Rendus* (p. 68, vol. i. 1860), has invalidated Encke's ideas, if not wholly overthrown them. This assumption of the illustrious German astronomer having been proved invalid, there is not an existing fact to my knowledge upon which any rigorous physicist can establish a conclusion that the interstellar spaces are filled with molecules, however minute their division; or, in other words, that space in any sense of the word is a *plenum.*

If not a *plenum,* it must be a *vacuum;* and if so now, was so in the beginning of things, and has continued so during the course of natural events.

The association of matter and vacuum, of entity and non-entity, implies unobstructed motion, and the necessities for such motion; and these conditions are the only possible opposites of one another.

5.

Space abstractly considered may, therefore, be legitimately and inductively determined to be an *infinite vacuum;* that is to say, it would be so if deprived of those embodiments of molecules or atoms constituting planets, suns, and comets, which are scattered throughout its illimitable depths.

Space, consequently and finally, is nothing. Or, if it be an illimitable ocean of abstract force, in the bosom of which matter was originally dissolved and disseminated, was subsequently specialized and differentiated, and is now suspended in definite compound masses—all in consequence of the action thereupon of an infinite intellectual energetic principle—it is probable that that force is immaterial, and metaphysical in contradistinction to physical, and that it is an unobstructive medium of communication between one body and another, as influences are interchanged by one with the other.

In either aspect of the subject, space, if destitute of all material creations, would be, in every sense of the word, a *vacuum*. Without matter there would be no magnetism, no electricity, no heat, no light, no actinism, no forms of any sort. All would be darkness and void. Nature would not exist. There would be absolutely nothing. *Hence, space would be nothing*—or an entity without a name—or, so to speak, only a *capacity* for holding what might be placed in it without the power to interfere upon motion or existence.

6.

My object in endeavouring thus to set forth in the beginning of this treatise an exact idea of the nature of space, as an abstract, and at the same time a positive, comprehensible scientific fact, and if possible to demonstrate its non-entity in contradistinction to its entity, as an attenuated ethereal condition—a point upon which the opinions of the learned have been in all ages, and are at this moment, at variance;—my object in this respect is to show how the discoveries of the immortal Julius Robert Mayer, almost simultaneously announced by Dr. Joule, and established by his experiments and calculations, and how the knowledge recently acquired in molecular and terrestrial physics, can be extended into the widest ranges of distance, time, quantity and motion; —into ranges wholly cosmical, embracing the sum of matter and force demonstrably existing and constituting the machinery, the very working powers and agencies of nature.

Matter and mechanical force being both fixed quantities, limited within fixed and possible fields of action and reaction, having definite purposes linked to their united and inseparable functions both in their minutest and in their greatest aspects, Nature exposes herself as a vast laboratory and workshop, dissipates the possibility of chaos, and puts accident out of the question.

Indeed, since the present state of knowledge permits us to announce with certainty that neither matter nor force is increased or diminished in its absolute quantity, and that all of each which exists is entirely conservable, and must in some way be correlative, and be so adjusted to or combined with the other, as when subdivided into masses or volumes to hold such reciprocal relations with each other as not to vary in original, fixed, and absolute quantities, powers and functions; since all this is positively known to science, it becomes self-evident that infinite space is only a void condition of nature where masses of molecules saturated with opposing forces may be wholly free to move, and where molecules and masses of different chemical constitutions and charged with these antithetic principles must mutually influence each other dynamically, either directly or indirectly, however widely separated these masses may be from one another.

Such a state of things being a fundamental physical fact, force must be considered potential in its essence as a whole quantity, which, when subdivided among masses and united with matter, becomes actual and entirely phenomenal working power, indispensable to material nature, and incapable of being abstracted or spared from such chemical masses as the earth, sun, and stars, and radiated from them as surplus energy in any form into space and be lost in an infinite wasteful void.

The more this subject is studied as a speciality, the more clearly and conclusively it will demonstrate its own truths. And in view of the mechanical principles recently enunciated, and which will be applied to cosmic phenomena in their broadest applications in the last chapters of this work, all speculations upon the ethereal conditions of the interstellar spaces may be safely set aside, space be regarded as a vacuum in every sense of the word, and some other mechanical element than *ethereal* vibrations sought after by physicists in order to arrive at exact truth respecting the methods by which light is transmitted or interchanged between cosmic bodies.

CHAPTER II.

MATTER.

1.

MATTER is the opposite of vacuum. It is *something.* While space consists of ideal points, matter consists of positive and material atoms. While space is illimitable, matter must have some boundary, because it exists within space. Our first *knowledge* of matter is derived from aggregations of molecules such as the planets, moons, comets, and suns, which are distributed in space, and widely separated from each other. We learn this before knowing the existence of the molecules which compose them.

2.

It is necessary to consider, in the first place, and in this connexion, these known forms of matter in relation to the medium in which they are suspended, simply for the purpose of distinguishing one from the other, and of impressing upon the mind the magnitude of both, and the grandeur of all philosophical inquiry.

Persistent research by the aid of telescopes, chemical and spectral experiments, and mathematics, has greatly increased our knowledge of the distributions to which we have referred, and of the phenomena developed by them. A striking fundamental fact established by two eminent observers, William Struve in the northern hemisphere, and Sir John F. W. Herschel in the southern, is that these cosmical bodies are condensed into a comparatively thin stratum or zone, rendered evident to us by the Milky Way, or Galaxy, which extends in circumferential directions to indefinable depths into space, while in its lateral or polar directions the number of stars progressively diminishes, in about an equal ratio, until they become most sparse in those regions of space called the galactic poles. This wonderful fact, as vast in its significancy of the movements of matter and the action of universal forces in pre-terrestrial time as it is suggestive of future discovery when physical laws shall be better understood, must not be longer overlooked.

Since the days of Magellan and the earliest Portuguese navigators, mariners have spoken, with deepest interest, of three conspicuous spots in the antarctic heavens, two of which are silvery white, while the third is *intensely black.* In my own early explorations, when sailing for several months, at two different periods, through high southern latitudes, around the Cape of Good Hope, Van Diemen's Land, and Cape Horn, encompassing the entire globe, I often observed them with telescopes, and was impressed with equal curiosity. It is now well known that the white spots are nebulæ, while the *black one* is an area of great extent, *where no star exists,* where the eye may peer into remotest space unobstructed by any material obstacle, and where there is apparently an utter void and endless indefinable night.

An accurate numerical study of our Universe has indeed resulted in demonstrating an accumulation of stellar masses from the regions of the galactic poles to the galactic equator; and in showing that the matter is *limited* in space in at least two directions, while its unknown extension in the silvery zone called the Milky Way demonstrates with equal clearness the direction in which the universe has its largest dimensions. An inference may therefore be drawn from these premises with scientific rigour, that one of two conditions primordially existed: that, in pre-universe time, matter existed either in the state of molecules endowed as definite chemical elements, crude, chaotic, and concentrated in a single mass, which, bursting in consequence of the birth of *repulsive force,* scattered its fragments through space; or, that matter existed in the state of molecules universally diffused, as imagined by Sir William Herschel and Laplace, which condensed in consequence of the birth of *attractive force.* In either case, molecular forces, with their manifold physical capacities, must have lain dormant and embryonic in the womb of passive and undeveloped nature, until Omnipotence touched the central molecule with its overflowing and everlasting creative energies.

The first universal positive act of nature, as far as physical events can be traced into the obscurities of the unknown and infinite past, appears to have been of a purely mechanical character;—an act by which matter, either as molecules or masses, arranged itself around a given point in space, as molecules fall toward, or diverge from, each other in crystallizing, when held in solution: for that symmetry of distribution so palpable to the eye demonstrates the action of laws, in the arrangement of the heavens, as definite as those which preside over the formation of a prism. Fortunately for solid inquiry, our system is like a watch-tower in the midst of the Milky Way, and we can look out upon celestial nature in all directions, as if we were an intelligent atom in the centre of a double elongated prism. In the shorter axes and diameters, we might be able to count the other atoms arranged between ourselves and vacant space, while in the longer axes and diameters their endless numbers would render all computation impossible. This fact may be illustrated by the condensation of vapour around our stand-point, upon any part of the earth. The eye can penetrate the mist above, because the stratum is thin, and grows successively rarer from the horizon to the zenith; but as the vapour spreads along the horizon, a dense and impenetrable fog appears in consequence of its greater compactness and indefinite lateral extensions.

This state of stellar distribution, which the studious observations of two illustrious astronomers have precisely determined, and which is undoubtedly one of the most important astronomical discoveries of recent times, establishes the fact that masses were moved to effect the physical arrangement of the heavens, by the same forces, for the same harmonious purposes, and through the same fundamental and organic laws which, as physical inquiry permits us to know, preside over the condensation of molecules into crystals, the formation of the earth and other planets, and the creation of all things within the range of human investigation. As force acts upon molecules now, so it appears to have acted upon cosmical *masses* in the earlier periods of creation, and produced effects *ad infinitum* throughout boundless space, which only recent acquisitions of the human intellect have allowed us to know.

3.

It becomes manifest, then, that if matter has been condensed from remoter regions of space, in order to form our galactic system, or has been projected from an unique central mass, ending in the same result; in either case, the stars which have been distributed outside of this immense zone, and which diminish in number when counted towards its poles, are all parts of the same system. And this truth becomes the more palpable, since all, so far as we know, move in the same direction, and since all are bound together by such chains of force and mutual influences that the keenest astronomers have detected no disturbances of harmony in celestial nature.

But the discovery of these facts and laws, coinciding so completely with those presiding over the condensation and mechanical projection of molecules into crystals and planets, leads to the recognition of another great cosmo-historic fact, viz. that molecules must have been endowed with embryonic, that is to say, with positive nucleated or germinal powers to constitute and establish systems of planetary bodies; or that systems at least must have been specialized with embryonic or central forces, prior to their projection into the present arrangements of the galactic universe. There is no such entity as accident, and no such thing as accidental creation in nature.

4.

By our inquiries, therefore, we are led backward to a period when molecules existed in conditions and relations to each other entirely different from their present stellar and planetary combinations. If not at first specialized with specific capacities and functions as chemical elements, condensed, and combined so as to form crystals, and ready to be projected as perfect worlds and systems, from an unique mass, they must have been in such diffuse conditions and chaotic relations as to be wholly beyond our comprehension. They could not, however, have been in existence for a single instant, as they are now conditioned, possessing affinities for force, and moved by mutual relations and reactions, without the violent development of multifarious phenomena, dynamical, chemical, and morphological. The dawn of creation, primal motion among molecules, the primordial action of molecular forces, would be instantly and inevitably followed by the manifestation of heat, light, magnetism, and electricity, the universal subordinates and handmaidens of attraction and repulsion; and thus the entire realm of chaos would be quickened into activity and life, and the fashioning of all things would have its beginning.

CHAPTER III.

INERTIA, FORCE, AND MIND.

1.

HOWEVER instructive it may be to investigate the speculations of the ancients, it is no less so to study the ideas entertained by our immediate ancestors upon subjects which constitute the bases of existing philosophical theories, and which have descended into and shaped the mechanical doctrines of the present day.

The atomic theory, as far as traceable into antiquity, was first proposed by Moschus some time before the Trojan war, and was revived and set forth in systematic form by Epicurus. Aristotle and his followers adopted it, and they enlarged the system by affirming "that in every particle of "matter there is inherent a sort of mind, the *φύσις* and "*ὥσπερ ψυχή* (of Aristotle), which they called an ELEMENTAL "MIND, which is the cause of *all* its motions and changes." Leibnitz believed in this doctrine, and extended it systematically, supposing every particle of matter not only to be active, but also "to have individuality, and a sort of *perception* of "its situation in the universe, and of its relation to every part "of this universe." This atom, thus endowed, he called a *Monad.* He affirmed that "particles of matter are continually "active, and continually changing their situation in virtue of "this principle of innate indefinable perception."

Newton entertained no such idea. He held, indeed, an opposite one. He believed matter to be inert, and that force lay outside of atoms, acted upon them *ab extra,* and existed in an ether which surrounds matter. When this force was active and communicated its action to bodies, it was *vis viva,* or living force. When passive, it was *vis mortua,* or dead force. Matter, itself inert, possessed nevertheless at the same time a positive power of being so, the *vis insita;* and also of resisting action upon itself of any principle existing within itself, and this was his *vis inertia.*

Newton's astonishing discovery of the laws of gravitation gave his ideas ascendency, and impressed them forcibly upon natural philosophy; and the system of mechanical doctrine at present prevailing is the necessary and normal development from, and the legitimate product of, the speculative assumption so important to the production of the "Principia."

The assumption that matter is inert (whatever might have been the mental reservations of Newton upon this point, he simply *reviving* the doctrine, for the speculation itself is an ancient one) was necessary for the application of mathematical formulæ to bodies in motion; and natural philosophy receiving thereby a new, and indeed its first solid impulse, Newton's followers gradually adopted the idea, without reserve, as an absolute fact, capable of unlimited extension.

Thus an abstract and problematical idea soon concreted into a sort of general law, and was set forth as a precise condition both in molecular and celestial dynamics; and for a long time it exclusively controlled every department of physical study.

With a slight modification in terms, the same definition of matter set forth in natural philosophy over a century since, is adopted still. In a work upon the "Principles of Mechanics, explaining and demonstrating the General Laws of Motion," printed in London in 1758, this principle or idea is enunciated as follows: "VIS INERTIÆ is that innate *Force of* "*Matter* by which it *resists* any change, and *endeavours to* "*preserve* its present state of motion."

Of late years, however, the views of many philosophers have been undergoing gradual changes in regard to this idea of Newton, and much reserve is felt upon the doctrine of the *vis inertiæ* of matter.

The definition of the principle, nevertheless, is still unchanged, and the term "vis inertiæ," as still employed in philosophy, assumes or indicates matter at rest to be without ability within itself to move, and matter in motion to be without ability within itself to arrest its motion.

Such a principle or entity we may now safely affirm (and most philosophers will sustain the view) has no existence in nature. It is purely an abstract assumption, entirely imaginative and theoretical, without essence, or existence in fact —although without such an hypothesis dynamical inquiries would be almost impossible. The term must be regarded as a relic of a period of philosophical speculation when the subtle spirit of inquiry *assumed* in discussions errors for facts, while scholars were positively ignorant of the true nature and relations of matter and force.

2.

No such dead thing, or principle as *inert "vis,"* exists in matter. No molecule or atom of any element exists without force within itself; or so interwoven with its minutest imaginable essence or condition as not only to enable, but also compel it to exert action, and be acted upon by the internal forces of other particles, or masses of particles.

In the present state of physical knowledge, matter cannot be demonstrated nor imagined to exist independent of immanent, active force. A molecule has power within itself and of itself to move, and to move other molecules near enough to it to receive the influence of its own living force; else matter of itself has absolutely no existence, and force alone exists. Chemical researches into the nature of the elements and the laws of atomicities, prove not only the existence of molecules, but also the positive and inalienable presence therein of active energy. Matter, then, does exist, and is endowed with living and indestructible forces. Force could not exist and be sensibly active without matter; nor could matter exist and be in any way active, without force. Neither could exist and still be inactive, or dead.

3.

The study of crystallogeny and morphology leads, furthermore, to the truth that a secret creating agency lies within matter, or within the force with which every molecule is endowed; and that the movements of all are controlled by laws which have been predetermined, and which startle the mind by their subtlety, universality, order, and fixity. This final principle is, at least, co-extensive with matter and force. And since we know, by microscopic and chemical inquiry, into what profoundly atomic realms these primal entities extend on one hand, and, by telescopic, spectroscopic, and mathematical research, how infinitely they spread through space on the other, we cannot even conjecture any thing or principle in nature to be devoid of positive innate energy or a capacity for geometrical development.

Thus the speculations of Plato, Aristotle, and Leibnitz appear to be sustained by modern experimental and inductive inquiry; and I have reason to think that the materialistic ideas which are now engrafting themselves upon the present theoretical status of rational and celestial mechanics must finally yield to those of a purely metaphysical character. Leibnitz foresaw and predicted that the mechanical doctrines of Newton would finally lead social thought into materialism. But Newton never pretended that his discoveries and assumptions were other than simple steps in scientific progress.

This final principle, the true *vis viva* and *vis formatrix*, being demonstratively coextensive with matter and force, and still more subtle, is, in every sense, universal and infinite, filling worlds with influences from which no molecule can escape, and amid which no impulse of force, or life, or thought can be lost. This principle being universal and transcendent in subtlety and potency, is so far traceable as to appear to the human mind supreme, embracing and controlling all matter and all force, and manifesting itself everywhere through them as a geometrical power and predominating agency. Apparent everywhere, and being geometrical in its sensible and material expressions, it is admitted by mathematicians and the profoundest thinkers to exhibit in this manner the highest type or most abstract conception of reason or pure mind and thought. Thus physics and metaphysics are inseparably linked with each other. With the latter, however, we shall have nothing to do in the present work, except to recognise the existence of the principle as a final, intellectual, all-pervading essence.

4.

There is, therefore, as all will agree, no such thing as *inertia* in the universe, all metaphysical speculations to the contrary notwithstanding.

I would submit, however (a point which must be clear to every one), that if mistaken principles lie at the foundation of any investigation, the analysis, throughout its developments, must be defective, and errors will finally result, monstrous and momentous in proportion as the problems under solution are vast and complicated.

CHAPTER IV.

MOLECULES.

1.

IN the preceding chapters I fairly presented the field into which all physical objects, forces, and phenomena have been introduced for eternal action and reaction.

Any system of physics which ignores the capacities for reaction in molecules—or the forces of, or central life, so to speak, in molecules, in any of their combinations or massive conditions, whether they be in motion or at rest—might be excused for inexactness were this the dawn of discovery, and an age of mere speculative inquiry; but when successive new revelations have overthrown ancient ideas, and science has assumed solidity and precision, it becomes unpardonable error in any physicist to cling to false hypotheses, even if no reasonable substitute be presented instead. We know that no such thing as *vis inertiæ* can exist in molecules; but that every molecule is endowed with positive living forces. In a succeeding chapter I shall present the development of molecular forces into massive and cosmical powers, when I trust no uncertain conclusions will follow the discussion of this important branch of the subject.

2.

The existence of matter in the form of ultimate molecules is known and admitted by all. But the shape of these ultimate molecules, if they have uniform and special shapes, is so attenuate that no ultimate atom can be described from actual observation of identity and form. Chemistry, crystallography, and morphology, however, all demonstrate in manifold ways and with wonderful clearness the positive existence of individual molecules, and prove them to exist as the elementary bases of larger forms and as elementary products in the compositions and decompositions of every known substance. At this point of inquiry matter becomes absolutely as profound in our attempts to fathom its conditions, as if it were a purely ideal and metaphysical subject of study. But it is positively material and solid, and we shall soon perceive it to expand into the grandest fields of visible form and substantial development.

Whatever the shapes of atoms or the extent of their divisibility, they never lose their individuality nor their capacity to embrace and identify themselves with force, and to follow the movements of its currents and affinities as if they were not alone vehicles loaded with force, but as if they were identical with force itself. One system of molecular physics reduces matter to force alone: another attempts to demonstrate the birth of atoms and the development of material forms from pre-existing mechanical force: another asserts that matter had no origin, but has existed from eternity:—and so on. However numerous or ingenious the systems of metaphysicians, theologians, or physicists, it is not my design to discuss nor refer to their relative value or correctness. To avoid error upon my own part, I shall endeavour to walk closely with nature alone, looking backward only where the light of positive science illuminates the path, and holding steadily upon physical facts and inductive reasoning, from the beginning to the end of my discourse.

We know that molecules exist. This is our first solid starting-point and stepping-stone.

3.

Numerous disquisitions have, also, been written upon the *shapes* of ultimate atoms by many eminent philosophers from the days of the earliest Greek thinkers to our own time—all more or less ingenious, depending upon some special theory of creation or experimental research into the final nature of things, and presenting a fair picture of the state of knowledge which existed when they appeared. It is equally unnecessary to analyse the correctness of these different opinions, or decide whether molecules possess hooks, or are cubes, spheres, ellipsoids, or of various other shapes whereby to be dovetailed and packed together—or whether they are surrounded with one or several atmospheres of ether or electric fluid, or any other unknown or fanciful condition, wherewith to fill up the spaces between their irregular attachments and peripheries. All these fancies have had their hour, and some are still the central point in this or that molecular hypothesis. It matters little at this stage of our study what the shapes of ultimate atoms may be. It is only important to know, in the first place, that they exist as separate individual points, and as positive physical entities.

This is admitted by all physicists and naturalists at the present time; and the fact is so far beyond discussion that further remarks upon the subject would be supererogatory.

CHAPTER V.

MOLECULAR FORCE.

1.

THE existence of molecules as a first step having been established, the next inquiry is as to their mutual relations.

The same sciences which have watched the combinations and segregations of molecules so pertinaciously for so many centuries, have ascertained, and demonstrated incontrovertibly, that molecules, however highly divisible, and whatever their shapes, can never be kept far distant from each other, and will, when in each other's neighbourhood, draw towards one another. This mutual disposition to unite or adhere to each other is called *attraction*. This principle of attraction is found to exist in all molecules. The fact that attraction is embodied in every separate and individual atom of matter, is as positively proven, and as much beyond discussion, as the fact that molecules themselves have existence. The evidences are so numerous and so well known, that it would be superfluous to introduce them at length here, since there is nothing elementary in the design of this inquiry. It can be safely asserted, however, from all that has preceded, that no molecule is, or can be, dead, inert, or destitute of activity, since it is never destitute of attractive force, can never be deprived of the same, and persists in exerting such force upon every other molecule, in virtue of the secret influence lying within its own bosom, and proceeding out of its own natural capacity and endowment.

But while the force of attraction exists, and every atom mutually attracts every other atom, it is proven with equal certainty that molecules never touch each other, and that they are kept asunder by another force, the opposite of attraction. This is called *repulsion*. A thousand experiments also prove this well-known fact to all who study the first principles of natural philosophy. There is not a substance in nature, so far as we know, that is not reducible in bulk to a greater or less extent by compression or cold, and that will not enlarge its bulk and occupy more space when heated, and return again to a less bulk when its attractive and repulsive forces are permitted to assume a normal equilibrium. Countless researches and experiments of chemists and physicists have established beyond question the universal fact that repulsion is an absolute force, dwelling in all molecules, which exerts as constant and mutual an influence from one upon the other as does that of attraction. It dwells and lives within the molecule, and proceeds out of it, in like manner as attraction, to act upon its neighbour, except that it acts and plays in an opposite direction to the latter force.

So much is positively known and acknowledged by every student of nature, from the chemist to the astronomer.

2.

We will pause here, then, for a moment to note our solid steps, starting-points of knowledge and induction, beacon lights and monuments of truth to which we must constantly recur, for bearings and guidance when navigating the infinite ocean of matter and force upon which we are soon to be launched. These points of positive truth seem at first almost metaphysical, when studied in their profoundest conditions; but in their developments they become so stupendous and wonderful as almost to bewilder the astronomer and physicist, and lead them to doubt whether molecular entities can expand into planetary forces, and thus become subjects of cosmical laws.

These solid steps are the recognition of three definite elements, each of which is as mysterious as thought itself, viz.:

Molecules,
Molecular attraction, and
Molecular repulsion;

not theoretical abstractions nor assumptions; not mere words meaning nothing; but the three fundamental terms of an infinite problem, the solution of which can only be achieved by the laborious and comprehensive researches of many generations of experimental observers.

CHAPTER VI.

UNION AND INSEPARABILITY OF MATTER AND FORCE.

1.

HITHERTO, for the purpose of clearly comprehending the scope and fundamental conditions of nature, we have taken a general glance at space, matter, and force. All are mysteries. Whatever or whenever their origin, the infinite intelligence of a Supreme Being is the active principle above and within them all, and upon this they depend for their continuance and development into endless forms.

2.

Our discussion of molecules and of attraction and repulsion has presented them as distinct, separate, and positive entities, the existence of which, as the working material and dynamical powers of nature, lies at the bottom of all things. It is desirable to receive these facts so deeply and clearly into the mind, that they may never be lost sight of in our future study of their diversified relations, combinations, and reactions. All science is built upon them, and it must be necessarily more or less imperfect as the value of either of these elements is overlooked or disregarded in any department of physical research.

While, however, all these three things, molecules, attraction, and repulsion, are separate and definite objects and subjects of themselves, and can be demonstrated to be so, it is equally true that they all exist in combination, and are inseparable one from the other, wherever matter exists within our observation and study. But our senses, or *material* organs, recognise only the *material part* of these three mysterious elements; and through this alone is made known to our intelligence the existence of the other two, which appear not to be material, but as subtle as mind itself.

While it is positively true that all three exist in combination, and that molecules, as we know them, could not exist without being part and parcel, so to speak, of these forces, or loaded with them as an orange is distended with juices which are sweet and sour at the same time, and without all three of which—pulp, and sour and sweet flavours—the fruit would not be an orange; while it is thus certain that molecules cannot exist, as we know them, without active union with attraction and repulsion, and that enormous aggregations of atoms embody proportionally enormous aggregations of force; many physicists question whether the entire amount of attractive and repulsive force in the universe is exhaustively *combined with*, and *concentrated in*, the masses of matter which we behold as planets, and number as stars throughout the realms of space. Force is not matter: yet it is always manifested to us through matter, and it can never be manifested through matter unless it combines with that matter. Molecules, attraction, and repulsion are all three distinct creations. Yet neither can exist as a manifest entity and dynamical agency without an inseparable union with the other two, albeit this union may be in different and fluctuating proportions of each to the other; and although force may flow into, unite with, and part again from molecules, as food unites with our bodies and separates again to give place to other increments, it is nevertheless for the time being a part of each molecule, and endows it with life and activity. Moreover, from all our studies of molecules, we know that they are never annihilated and never destitute of force; and that their nature and mutual relations are such that they never can be in a state of inertia. While Dr. Louis Büchner and his followers would define force in every sense as a simple attribute of matter, I should define matter to be an equally active and universal attribute of force—in the same light, for instance, that God is as much an attribute of omnipresence, omnipotence, and omniscience, as these three inseparable principles are attributes of God.

A molecule lies in a certain sense as a neutral yet sensitive element between the other two elements, which are of themselves positively antagonistic to each other; both, however, having equal affinities for matter (and matter, as atomicities, unequal but positive affinities for them), and manifesting their peculiar characteristics and activities through matter, as a third and necessary condition, whereby to effect the creation of the secondary forces of light, heat, magnetism, &c. and of all mineral, vegetable, and animal forms which constitute nature. Thus is laid the foundation of that state of things whereby matter and force, combining, decomposing, and recombining, have slowly, gradually, and persistently unfolded the stupendous mysteries and endless complexities of physical and organic being.

3.

Notwithstanding all this, it is generally supposed that light and gravitation also exist in the state of positive and conservative currents of force in the interplanetary spaces, constantly playing from each to every other cosmical body; thus exchanging the mysterious effluences and influences of all, and contributing in part to sustain the mechanical movements so demonstrable in the universe. Whether this be so, and the manner in which force acts between bodies widely distant from each other, may be more clearly determined, as the subject develops itself under subsequent treatment.

For the present, however, I would confine attention exclusively to the point, that *attraction* and *repulsion* exist in permanent and active *union* with *molecules*, and are *inseparable* from them.

CHAPTER VII.

NATURE AND ACTION OF FORCE.

Attraction.

1.

WHILE nothing is, or probably will ever be, known of the *final* nature of force *per se*, a knowledge of its character and action is entirely within the boundaries of human research. Force, as already described—that is, primitive, phenomenal, or creative force, the force from which all other forces, all forms, and all combinations of force and matter spring—is, I hold, of *two* kinds. They are represented by the words *attractive* and *repulsive*. As they are dissimilar, and positively antagonistic, in character and action, no discussion can embrace both within a single treatment. Each must be analysed by itself, and in such a manner, that no confusion or doubt may remain relative to the laws and method by which the action of these forces upon, and their combination with, molecules are effected and governed. By rigorously pursuing atoms, whether they aggregate or segregate, we shall as rigorously grapple force, and at last become so familiar with its movements as to begin to detect the secret of its nature, and be able to discuss it as an object almost palpable and wholly distinct from matter itself. So clearly will the knowledge of force, as an entity, a positive though invisible and immaterial something, become impressed upon the understanding, that, as the subject develops, we shall find it less important and instructive to contemplate molecules and masses as mere matter, than the active and motive principles with which they are endowed. Matter, when viewed as a neutral agent—that is, molecules viewed as neutral points—will almost vanish for a moment in an active or positive aspect, so that we shall scarcely wonder that Newton and other mechanical philosophers have at any time treated matter as an inert object, *per se,* and that certain thinkers have sometimes altogether rejected the idea that matter is a substantial entity.

2.

Since the importance of the subject demands a separate and exclusive treatment for both attraction and repulsion, from their incipient molecular conditions to their grandest cosmical developments, it will promote a clearer understanding of their action and mutual relations if the analysis be opened with the consideration of the former. This is the force the influences and action of which have been the most assiduously studied, and the astronomical expansions of which were reduced by Newton, toward the close of the seventeenth century, to definite numerical laws. There is every advantage to be gained, besides, in attempting to explore the realms of uncertainty and of the unknown, from scientific points which have already been definitely established.

In the chapter on "Inertia" it was stated that atoms and masses are never dead or inert, but constantly charged with active central forces; and it will be important to look back for a moment, for the purpose of showing how conclusively this point is proved by the wonderful discovery of gravitation. I use the word "gravitation" *here* to express simply the *numerical law* which presides over the *force* of attraction. The discovery of Newton has determined this force to act in its cosmical expansions in *direct proportion* to the *quantity* of matter—that is, the *number of molecules*—in masses, and *inversely* as the squares of the distances of these masses from one another. This is a pure mathematical discovery of *quantity,* without discovery or demonstration of the nature, characteristics, or mechanism of physical force, or of its secret method of action, whereby to constitute planets or to produce diminutives or multiples of power; nor does it pretend to open clearly the mind to a comprehension of the method—or *modus operandi*—by which matter combines so as to effect cohesion, adhesion, or any other form of attraction by which atoms are aggregated into masses and maintained in a system of moving spheres. But in this immortal discovery of Newton lies the hidden law of potency in molecules, as well as in planets. Attraction, as an active, indwelling, central, and indestructible force, *must exist* in every *atom,* since it exists in the *whole earth,* or in the entire mass of any body; for no demonstration can prove a *planet* to be endowed with the power of drawing other bodies to itself, and of responding to the attraction of the sun, in *direct proportion* to the *quantity* or *number of molecules* embraced in it, and still allow scope for admitting its *separate molecules* to be *destitute* of innate force, or to be in any sense *inert.* Molecules, and planets, and fragments of planets, therefore, wherever existing in the realms of space, are replete with the same force which the genius of Newton reduced to exact and unchangeable law.

It may seem supererogatory to reproduce these points in this connexion; but as they have the widest bearing upon the graphical developments of all the sciences, it is important to cement these so firmly, that no succeeding step shall lack stability in consequence of uncertain or inexact antecedents.

Thus are the foundations laid for the development of a solid and harmonious philosophy: vacant space for unimpeded motion; molecules, the bases of all material forms; attraction and repulsion, opposite fundamental forces always in combination with molecules; and molecules themselves so instituted primordially with an occult geometric energy that both attractive and repulsive force have an equal and perfect affinity for them, and play through them and between them in obedience to laws so definitely exact that masses and distances, whatever their magnitude, may be weighed and measured with as much accuracy as the mass and angles of natural crystals.

Such is the field of exploration—so vast that no single mind can embrace its endless developments, and yet at foundation so limited that the principles implanted in the germs of things appear to be extremely few and simple.

The points already established are, that all masses are made up of molecules; that each and every molecule is possessed of an individual force of attraction through which it attaches itself to another, and that all these individual molecular forces combine in such a manner as to assume ponderous units, or a central attractive force, in proportion to the number or quantity of molecules in any particular body.

3.

The point to be next demonstrated is the secret principle through which molecular attraction so acts as to cause cohesion of atoms, adhesion of masses, the falling of bodies to the earth, the gravitation of planets to the sun, and the rolling of the sun and of all its attendant spheres through the immensity of space—in a word, through which the augmentation of molecular force is effected so as to produce units of attraction, whose single central power becomes proportional to the number of molecules in such masses. This is one of the most interesting subjects of research, and is pregnant with the grandest consequences in every practical department of knowledge, as will be hereafter apparent when it shall be more fully unfolded and completely understood.

It is manifest that the juxtaposition of molecules effects such mutual changes in their individual characters, that each loses identity and individuality in the magnitude of all, and that each becomes of greater importance in proportion to the number of others with which it is associated. When separated they are comparatively nothing: when joined they exhibit energetic functions. We are now only contemplating *simple* atoms and simple attractive force; not the universal and compound energies of nature which were aroused when attraction and repulsion, both playing alike upon and through molecules and suns, and interchanging their respective energies throughout creation, had unfolded the present grandeur and glory of the universe. It is the simplest beginning of things we are attempting to analyse; and as we trace the aggregation of molecules into masses, we discover that the attraction of the latter augments in such a manner as not only to bind the molecules into conditions of solidity, producing various degrees of hardness and specific gravity, but also to manifest an augmenting and still an unique volume of force in proportion to the number of molecules. The individual forces appear to run together, lose their separate individuality, and coalesce exactly as we know that particles of vapour coalesce to form a drop of water, and exactly as we see two drops of water coalesce and become one, or various particles of quicksilver run together and become an unique drop, with a single force acting from its centre to its surface to give it sphericity, in like manner as a smaller unit of force acted from the circumference to the centre of the several original particles to give each of them a similar sphericity.

These illustrations might be indefinitely multiplied by examples drawn from the coalescence of fluids and gases, from the coalescence by induction of the secondary forces, magnetism, electricity, &c. of similar sorts; and they will serve to convey my conception to other minds, and communicate the idea and truth of the secret capacities expressed, and the modes of operation and coalition which act between the molecular forces—capacities and methods through which, indeed, the latter unite to become planetary powers, and assume gravity in infinite quantities. It is at this point that the mind may temporarily drop the recognition of matter, forget it entirely for a moment, and discuss force alone. Thus, force—the *attractive* force—one of the two mighty and universal powers of nature—a veritable dynamic and motive principle, embodied in planets and suns—is seen to be capable of flowing like a fluid, of coalescing and unifying, of being separated and recombining, like water or quicksilver, in particles, drops, or oceans; and it will be further perceived that this very property of molecular force, as thus demonstrated, saturates worlds to repletion and thereby becomes the secret principle of gravitation itself. Indeed it is *the active living force* in each particle of water, mercury, or other matter, which coalesces; and *not the particles* themselves. Molecules simply cohere through this tendency for unity inherent in attraction. Multitudes of molecular forces change into a single unit of force in a cohering mass of atoms, and each atom becomes bound to the rest by the accumulated energy acquired through the union of all. Particles of matter, under this treatment, become mere particles of force, or, at most, merely neutral and plastic elements suspended in the closest juxtaposition in a volume or sphere of force. By way of illustration, they may be represented as saline molecules, at first diffused and imperceptible by solution in a transparent medium, which subsequently becomes solid and visible through crystallization. As force condenses into definite volumes, atoms follow, or rather move with its movements, and are constrained into narrower limits in obedience to inherent law. Thus is cohesion effected. Force possesses—that is to say, molecular *attractive* forces possess—the inherent property of running together, mixing, centralizing, and magnifying their energies in certain numerical ratios. Atoms, always alive in their affinities for attractive force, or for combinations with it, follow its currents into the nearest juxtaposition. It appears to be this property of attraction, as a specific and fundamental principle, to combine exclusively with its cognate, and to flow from the extremest boundaries of space to seek its like, in order if possible to coalesce and be united into one single concentrated volume of invisible energy, which constitutes gravitation; and planets and suns, saturated with this essence, are moved by its tendencies to unite mutually with one another — which tendencies, encountering the antagonistic influences of repulsive force already developed also into cosmical quantities, extending its energies through equal areas, and exerting unequal and fluctuating degrees of power, like gravitation, produce the paths and orbits of the celestial masses which are composed of heterogeneous elements.

All this cosmic action we shall perceive, further on, to result from the chemical constitution of these bodies.

4.

I do not mean to anticipate in this connexion the development of points reserved for subsequent consideration; but simply to throw in occasional striking facts, illustrations, or contrasting conditions in order to bring out more strongly the great principles which I am attempting to delineate, and to establish my present position upon irrefutable foundations. Returning from this broad consideration of force and fact to the narrow point of molecular attraction, I repeat that it is by this tendency and capacity of force to unite intimately, and remain in unity, that the thousand common phenomena of cohesion, adhesion, and visible attractive movements of all sorts are effected. Various illustrations will prove this. Sheets of ice whose surfaces are in coaptation become absolutely one body; all points of contact are soon obliterated; and every stratum closes into a single mass. This union of matter and force is as perfect as the cohesion of condensed vapour; and the centre of gravity of the entire mass changes with the addition or abstraction of every sheet, or any atom. Plates of glass, placed together, in like manner present a similar phenomenon. The attractive force embodied in each immediately coalesces, and the plates adhere; in process of time, they become almost welded into one, and cannot be separated without fracture. Capillary attraction is but another act of this same property of molecular force to coalesce with that of its neighbouring molecules, and to carry its molecule with it, where free to do so; and it lies also at the basis of every act of endosmosis and dialysis, and of a multitude of other phenomena which have puzzled philosophers and naturalists. Floating substances, placed within given distances, in a basin of water, will attract, and adhere to, each other; and be attracted by, and adhere to, the sides of the basin. Ships on a calm sea, within limited distances of one another, will also reciprocally approach and collide, in consequence of the molecular forces in each exerting their effort to coalesce and form one volume. These various acts are not results of electric or magnetic affinities, strictly speaking; but are dependent, rather, upon a principle still more profound, which enters into the secret of the polarity of force; polarity being considered as the subtle *law* or expression of the function or powers of segregation and *segmentation*, and of *assimilation*, which exists in the two abstract forces of attraction and repulsion, and which makes itself sensible through molecules and atomicities, simultaneously with the development of thermic, electric, and magnetic forces—*secondary phenomena*, which spring up (as we shall learn in a future chapter) when the radical forces play antithetically through molecules, and when all nature has become animated with compound and heterogeneous influences, the oscillating action and reaction of which end in developing phenomena of both the subtlest and grandest characters.

5.

This explanation of the nature and action of attractive force embraces also the secret of the cementing and consolidation of rocky strata and conglomerates, and of the preservation of fossil organisms, and of many other geological phenomena observable throughout the crust of our planet. The longer the attractive force has poured into and through molecules in cosmical currents and with accumulated volumes, the closer their juxtaposition, and the denser and heavier do masses become;—in other words, the greater their affinity and assimilation, and the stronger and more intimate their union with themselves, and with the planetary volume with which they are connected. The attraction in everything which falls to the earth or which in the solar system is in any way connected with it, is endeavouring to combine with the like principle in the earth, and *vice versâ*, attraction being reciprocal, and in direct proportion to the quantity or number of molecules in each object. It must be borne in mind that it is not matter which, under present aspects of discussion, combines with matter. The recognition of this was temporarily dropped, and we now behold force alone combining with force; and molecules reduced to mere plastic agents whose affinities, or powers of absorption, association, assimilation, tension, and displacement, we shall subsequently discover to be alive to every modification of force, and whose inherent nature is to concentrate and retain force within themselves, and magnify its manifestations in proportion to their number and juxtaposition.

Thus the mighty volume of attractive force in the earth holds such affinity for and control over the same force in everything upon and near its surface, even over that which fashions and fills the moon (an object nearly one quarter of a million of miles distant from its centre), that the molecular forces of all these things are being drawn into it as if it were an absolute vortex. And it is not improbable that the speculative attempts of Descartes and his disciples to explain the secrets of creation by vortices of force might have been based upon this idea, which occurs to me only as a figure of speech to illustrate the tendency of separate volumes of attractive force mutually to unite into a single body and to combine in all proportions and quantities. In whatever light it may be delineated or discussed, such is the certain action of attractive force. Every point of its invisible and immaterial entity holds affinity for every other point, and wherever existing it is striving to combine and create larger volumes, as physics plainly teaches when bodies fall to the earth, and as astronomy confirms with remarkable emphasis where the planetary forces exert such mighty movements to effect union with a kindred force embodied in the sun. This union of all planetary volumes of attraction in our solar system with that of the sun would be as certainly effected as the union of falling bodies with the earth, were there not an antagonistic force in nature (to be discussed hereafter) to constrain cosmical masses into definite paths of motion, and maintain them at various distances from the central body.

6.

It is manifest that the combination of the diminutives of attractive force constitutes massive attraction; and it will soon become equally clear that the disposition to combine existing in all masses constitutes gravitation, and that these words are synonymous, whether applied to attraction in molecules or suns. If the act of gravitation or of mutual attraction were not counteracted by sufficient cause, any body near its surface would fall to the earth and coalesce with it. Even if counteracted in space, as the earth, for instance, is checked in its fall to the sun by some opposing cause, it will nevertheless perpetually strive to effect union, and thus the planet loaded with force will for ever move in a definite path around the attracting centre. Gravitation is the same everywhere. Its character and action, defined by a simple word, is the *weight* of a body in comparison with that of another body. In exact terms, "weight" or "gravity" is the aggregate, represented numerically, of the molecular forces in any body, which are striving to unite and coalesce with similar forces in the earth, and which are prevented from doing so by some mechanical obstacle, the value of which may be determined by certain graduated mechanical inventions. In physical astronomy it indicates the quantity of the molecular forces in the earth or any other planet, which are striving in vain to combine with like forces in the sun. The attraction of the earth's centre acting upon any thing, causes the molecular *forces* in that thing to move toward it, thereby drawing in order to absorb them into its central sum, and unite the thing with the solid mass; and this effort of the smaller force to unite with the greater, involves the actual movement of matter and force, which mutually embrace each other. If this movement be prevented by such conditions of resistance as are capable of accurate mathematical measurements, the thing is said to be "weighed." Specific gravity is only another name for the same force when more accurately calculated by definite comparisons with a given amount of force. Weight and gravitation are equivalent terms, in so far as both represent dynamical action, or an effort of the molecular attractive forces, inhabiting different masses, to combine and make one volume of matter and force. It is well known that all forms of matter—that is to say, both atoms and masses, whatever may be their physical, chemical, or organic conditions—are subject to this fundamental principle, and obedient to its influence and movements. It becomes manifest, then, that it is *force* which moves matter, and not the matter which moves itself, or which acts upon other atoms and masses. Matter possesses inherent susceptibilities and affinities for the attractive principle, but is of itself a mere vehicle—a neutral agent—an obedient, subservient slave. The fundamental nature of atoms is to possess and hold force as a bladder holds water; a sack, meal; a cask, beer; a ship, cargo; a stomach, food; a cell, or a man, life. Their final function is to obey an all-pervading power more profound, subtle, secret, and commanding than attraction and repulsion themselves. Through this they spring into geometrical and organic substance and form; and, thus creating a complex system of nature by their dynamical collisions, combinations, and disintegrations, they impress sense and consciousness by their multifarious capabilities and devělopments, and connect finite mind, the crowning elaboration of molecules and force, with the eternal principle which originated, exists in, and controls all things, and which is, indeed, absolute mind itself, filling space and eternity, and in comparison with the vastness of which all else is as nothing.

7.

Such is the nature of matter; such the action of force. This subtle, immaterial, mysterious, active entity moves and flies, so to speak, with inconceivable and accumulative rapidity; and atoms, aggregated into masses of all forms and magnitudes, are swept along, charged with it and enfolded in its grasp. Atoms and force are inseparable. Neither moves separately. One does not follow the other, except in our own conceptions of their correlative action. This differential conception is only apparent, and springs from the succession of our ideas. Both, in fact, move simultaneously, although force is the inherent motive agent in a molecule, in like manner as volition is the real man—neither being of itself inert, but volition the power, and man its obedient representative and dynamical agent.

Under the light of the preceding study, the reasons become apparent why matter moves in such directions as "gravitation" transports it. It is, in a word, because volumes of attractive force, as living entities, sensitive to and possessing affinities for other volumes of similar force more or less distant elsewhere, strive mutually to combine in order to become a unit, that many phenomena exhibited by matter in motion have awakened the attention of the human mind, and led to great discoveries in astronomy and dynamics.

In the succeeding chapter we shall discuss the character and developments of *repulsion,* and attempt to discover the action of a force opposite in all its tendencies and effects to that of attraction.

Still later, we shall perceive how both these forces act upon matter, and thereby produce those compound motions of cosmical bodies whose distances are so nicely measured, and whose quantities are so exactly balanced with each other, that the heavens become an unique machine, whose clockwork is so regular and harmonious that the ingenuity of man can there detect no error (except it be his own), and whose perfectness of design and majesty of execution proclaim the universality of a Being so veiled in the stupendous glory of His own acts, that the profoundest philosopher bows, like Moses on Sinai, before the passing presence, and lifts his eyes to behold only the shadowy spectre of some Living and Almighty Power.

CHAPTER VIII.

NATURE AND ACTION OF FORCE.

Repulsion.

1.

IN the preceding chapter I have endeavoured to trace the unfolding of attractive force from its molecular functions to its cosmical magnitudes; to establish the identity of character, action, and origin of both these conditions of force; and to elucidate by discussion the basis in nature of the numerical law discovered by Newton, which has for ever determined the two facts, that all bodies attract each other with a force directly proportionate to their respective masses—that is to say, to the number of molecules embraced in each mass—and inversely proportionate to the squares of the distances between these bodies.

I have endeavoured to explain why everything falls perpendicularly toward the earth's centre throughout its surface, and why everything in our solar system, planet and atom equally, is constantly striving to fall into the sun, upon the ground that it is the nature of the attractive force of molecules, independently of the molecules themselves, to become absorbed in and transformed into the attractive forces of all other molecules which they approach, and with which they come in contact, in order thereby to form one volume of force and one body of matter. If atoms enough were free to move in infinite space, so that when brought together they might constitute a mass like our earth, I merely assert that the inherent nature of the attractive force in each molecule is such that it would combine with that of every other, and become continually more potent as a single and centralized power in proportion as the mass increased by increments of additional molecules; and that as the molecular force acts from the centre of every molecule, so the accumulated forces become a central function in every mass, and in the growth of every mass, until the earth itself be formed with a centre of gravity determined by the union and centralization of every molecular force embraced in the original atoms as they existed before they were swept into the mass. I would be understood to mean, speaking generally, that only the force in each original atom now combined to form this planet constitutes its planetary or cosmical force of attraction; and that this, as a unit, is represented by the potency of its central gravity, and that all these original molecular forces respond now, as a single central force, to the solar attraction; and finally, that this is the true description of the elements of the numerical law of gravitation, propounded and demonstrated by Newton in his "Principia." The action of the molecular forces which converge and react through infinite radii in a round drop of water or of quicksilver, and which associate and condense or centralize themselves, so to speak, in order to present a unit of force and of weight, and thereby to constitute also a spherical volume of matter whose centre of gravity becomes pregnant with power by every increment of atoms—this action is the exact representative and counterpart of that which takes place in a planet or sun when atoms and force are permitted to move freely in space and coalesce in infinite quantities. Cohesion as a physical phenomenon is necessarily involved in this result, but the chemical and vital forces so modify its action that we are not permitted to observe its simple effects aside from complex conditions. These complex conditions—a subject to be hereafter discussed—produce the endless irregularities of form, differences in density, and other general properties of bodies observable in nature, and so act upon different things, that while certain chemical and complex compounds of unyielding density, like rocks for instance, lie at angles in relation to the centre, others, like water, a mobile element, present no impediment to the action of that law which prevails throughout nature to arrange molecules into spherical masses, whether they be spherical particles of vapour or globes of atoms as large as the sun or Sirius. All objects are grasped alike by this adhesive force, for while the water clings to the globe in a broad sheet diffused equally over the terrestrial ball, the rock is bound just as strongly throughout its angular bulk in proportion to the number of atoms it may contain. It is this disposition of attractive force in all bodies, whatever their form or composition, to run together, coalesce, centralize, and become accumulative into a single volume to any extent or degree ; to hold irregular masses together ; and to sweep atoms along with it so as to create drops floating in the air, or globes rolling through the heavens, which I have endeavoured to trace and illustrate, in the preceding chapter. These facts lie at the basis of Newton's universal law.

Immediately opposed to this force appears another of equal magnitude and importance, which has heretofore been ignored by physicists, but which as certainly exists, and plays as important a *rôle* in nature as gravitation itself.

This is *repulsion.*

2.

Heretofore, gravitation has been considered, and it is still taught to be, the *only* FORCE which controls the movements of the heavens and determines the stability of universal nature. Newton, in order to arrive at his great numerical law, *assumed* that a planet was of itself an inert body, subject to attraction, and set in motion by some infinite unknown impulse ; and that, if undisturbed by any other body, or by any outside force, the planet thus in motion would move for ever in a right line in consequence of its original momentum. Two or any number of bodies, thus projected into space, were supposed to have been mutually diverted towards each other with degrees of attraction differing directly in proportion to their masses, and inversely as the squares of their distances asunder: and it was assumed, further, that they would finally come together, unless their respective initiative impulses had been so different as to impress them with specific but various degrees of motion. These bodies, moving in right lines with varying momenta, and influencing each other with various degrees of attraction, thus mutually modified their respective movements, and in this manner their original right lines of motion were changed into curved ones; and the planetary orbits with their different degrees of eccentricity became established in virtue of the united action of these two conditions —the projective impulse and the gravitative force. Orbits once established by the potencies of these two functions—the first assumed, the last positive and known—there appeared to Newton no reason why they would not remain the same for ever.

Thus the present theory of celestial mechanics was initiated; and the numerical laws of gravitation were brought to light by the mechanical inventions and mathematical inspirations of the immortal Newton. This wonderful achievement, based half on fact, half on assumption, having been accomplished, the human intellect, as if exhausted by the grandeur of so vast a scientific step, paused for another century. Such lofty lights of thought have dawned from time to time upon science with a periodicity so remarkable, that devout students of nature have often become impressed with the belief that Divine Wisdom matures the conditions, and kindles the flame which is to illuminate them from the fires of its own altar. With Copernicus, born 1473, the heavens began for the first time to unveil the infinite glory of the One True God, and to reveal the secrets of eternity and space to common humanity. The following century Tycho

Brahé (1546), Galileo (1564), and Kepler (1571) opened still more widely and conspicuously the wonders of this vast arcanum. Another centennial ushered in Newton (1642) and Leibnitz (1646), who may be said to have vanquished the powers of the sun and stars, insomuch that by the application of numbers to the magnitudes, distances, and orbits of bodies already discovered, a method of knowledge and investigation not possessed by former astronomers, they adjusted the balances of the universe, established the numerical functions of planetary attraction, and weighed the sun, moon, and planets, as a merchant weighs feathers or iron with his steelyards. And now, a century later, appeared Laplace, another profound explorer of the universe, who, struck with the uniformity of many astronomical phenomena, conceived and propounded the idea that the solar system in preplanetary periods had existed in nebulous conditions as a vast volume of molecular dust and vapour, and that from this form of matter, set into revolving motion from west to east by some unknown cause, had at last been evolved, by successive coolings and condensations, comets, planets, moons, asteroids, aërolites, and, ultimately, the sun. The almost uniform westerly-eastern motion pursued by all these bodies upon their respective axes and around each other, was asserted to be conclusive proof of this doctrine. The conception of so vast and simple a scheme of creation was indeed grand, and worthy to be ranked with that most ingenious postulate of Newton from which had been elaborated the universal law of gravitation. So, astronomical knowledge has been developed by successive intellectual movements of a centennial character, each step dependent upon the preceding, and each leading to clearer insight into the profoundest secrets of nature.

The projectile impulse assumed by Newton, and so necessary to play one of the important parts in his theory of celestial mechanics, was thus supplanted by a new theory of centrifugal motion—a theory purely mechanical also, like the Newtonian conception, but differing in so far that a mechanical origin could be accounted for in the latter, while the earlier assumed the heavenly bodies to have been already created and completed when projected into rectilinear paths.

It will be seen that in both these hypotheses their illustrious authors founded their views of celestial motion upon conjectural bases, *assuming in both cases some foreign impulse to have impinged upon great masses,* although under different circumstances, and that they did not found their deductions upon the fundamental forces and functions of the molecules which constituted the collective and individual bodies of the solar system. These hypotheses, although fortified by the tacit support of all astronomers and mathematicians, and taught as academic truisms, remain nevertheless to this very day only simple theories, as in the day of their invocation from the abysms of chaos. They have, indeed, accomplished the invaluable work for which they were invented. For, the quantity of one positive term being known to the calculus with others more or less established, celestial problems of great complexity may be solved with astonishing accuracy.

3.

Now it so happens, that two opposing forces, acting more or less unequally upon bodies in motion, will compel them to deviate from rectilinear or move in curvilinear paths. Attraction as a principle, as a something influencing matter, was known from observations of magnetic and electric phenomena before the time of Newton; but the numerical value of this attraction—that is, its power, or force of gravity, or weight, in proportion to masses of matter and their distances from each other—was not known until Leibnitz and Newton applied their new methods of notation to this subject, both having discovered the higher powers of the calculus independently of each other and at the same time, as if by inspiration. The sizes and orbits of certain celestial bodies being known, and the distances, for instance, of the earth, sun, and moon being also given, it may seem simple enough to mathematicians in these days to determine the numerical force of gravitation. But it was not so easy before Leibnitz, Newton, Horrox (a gifted youth, who died at the age of twenty-three), and a constellation of other remarkable men appeared, as if by a miracle, at about the same time, to unlock those new and profound secrets of the Infinite mind. Nothing comparatively was known in those days, of experimental physics, of chemistry, or of the character of atoms which constitutes the basis of the latter science. Alchemy had not advanced much beyond the general speculative opinions of Thales, Plato, Aristotle, and other ancient Greek philosophers; and even the nominal forces of atoms, that is, attraction and repulsion, were not taken into account in any of the rude scientific processes of those days. Everything done, generally speaking, was wild experiment based upon wilder theories of the properties, abilities, and capabilities of gross matter. How, then, can it be expected that notions upon so speculative a subject as ultimate atoms should have entered into mathematical calculations involving the movements of appreciable masses and immense globes of composite chemical elements whirling through space in consequence apparently of some mechanical impulse, and flying around each other in virtue of an innate *vis inertiæ* with a thousand times the rapidity of the strongest wind? Nevertheless, fundamental ideas were not wholly disregarded by Newton; but he failed to satisfy himself that any other principle than attraction of mass for mass, each of which was set in motion by some conjectured exterior impulse equal to the attractive force, was necessary to establish planetary motion, and maintain the stability of the solar system. This was the goal he sought, and nothing more, in this direction. It was sufficient for him and his day. He imagined gravitation to be the *only positive, active, universal force* existing and directly operating upon the heavenly bodies; and that this gravitative attraction acting (in virtue of some principle of pressure in the ether—a principle altogether *ab extra*) upon the *vis inertiæ* of planets already under a momentum assumed to have been previously communicated, is the entire and only force necessary to explain every phenomenon of planetary form and orbit, and to maintain the general equilibrium of cosmical nature. This doctrine, that gravitation is an unique and the only universal active force, has not only prevailed from Newton's day to ours, but is so strengthened by the approval of successive generations of illustrious astronomers, mathematicians, physicists, and teachers, that to shake and expand it seems almost as impossible as to move the world. No other function than this *assumed* momentum may indeed be absolutely necessary to the calculus as applied to physical astronomy; and since solutions of celestial problems so constantly comport with theory, it naturally is held sufficient to satisfy the demands of this science. But many facts in natural philosophy, geology, and other branches of study, are difficult of explanation, and apparently inexplicable without a development of this hypothesis. Indeed, so wide a chasm now exists between molecular and cosmical physics, and is likely to continue to exist, so long as this doctrine is held in its present limited form and in such exclusive favour by astronomers, that any attempt to expand theoretical principles so far as to embrace every class of phenomena in a simple system of explanation, must be most welcome to those who feel the present necessities of science.

The point assumed by Newton was of incalculable importance. It cannot be over-estimated. It was the happiest of conceptions, pregnant with the grandest of consequences. He boldly assumed a datum—which happily proves itself to be numerically an equal, although opposite, force and function to those which appear in cosmical attraction—and from this datum he evolved the laws of gravitation, the most ingenious and wonderful discovery of all time. The principle of the Newtonian theory is the simple assumption of a body in motion with a given momentum. It ignores the nature and origin of the force which imparts the impulse, and the method by which the motion is initiated. In this theory of celestial mechanics, its founder and his followers have no occasion to regard the one or to discuss the other. It was only the gravitation of various bodies toward each other according to their magnitudes and distances, which Newton discussed mathematically. His treatment of the subject was not graphical and general. All this he left for subsequent developments of fact and philosophy. Most fortunately for science, he not only established the present laws of gravitation, with their numerical quantities and with unalterable accuracy, but he also laid foundations in their very construction for developments of discussion and discovery which I hold his followers have been slow to comprehend. Indeed, without himself comprehending the immense value of his discovery, or the force of his own terms, he has settled principles from which must spring a perfect theory of celestial motion, and, in the progress of science, a complete and harmonious connexion of universal nature. His law settles the essential point that attraction as a cosmic force exists in direct proportion to the amount of matter, that is, to the *sum of atoms*, constituting a mass. Mathematicians confirm, and astronomers and physicists will all agree upon, this point. There is no escape from its truth. It is a point fixed in nature and knowledge ; and is a nucleus from which, and to which, all physical research must radiate and converge. The fundamental truth in Newton's law, viz. that attraction is in direct ratio to the amount of matter—that is to say, the number of atoms in combination—settles at least *five* radical facts in physical science. In the first place, it determines astronomically and synthetically the absolute existence of individual *atoms*, a fact which chemistry sustains by all analyses and logical deductions: it, secondly, establishes the existence of *attractive force* in atoms as *individual molecular entities*, which, combining, create gravitation in every planetary and stellar mass : it, thirdly, puts the fact beyond contradiction that force is *not* a principle exterior to matter, and acting *ab extra* upon planetary and stellar bodies : it, fourthly, demonstrates clearly that the nature and action of attractive force are alike in the infinite and in the infinitesimal and identifiable everywhere as one and the same thing: it shows, in the fifth and last place, that attraction is, *per se*, an ultimate force capable of all degrees of subdivision and combination, and also of centralization wherever its phenomena are studied, from single molecules to masses of molecules as large as the Earth, or Jupiter, or the Sun. Here, then, are five important facts settled upon immoveable foundations connecting molecular with cosmical physics.

It is upon these foundations I stand as I proceed to unfold the nature and action of *Repulsion* as a general and cosmic force, cognate in origin, equal in value, quantity, and extent with that of gravitation, and going hand in hand inseparably with it into atoms and into masses, and throughout infinite expansions into the dynamics of celestial motion, and playing an equal part also with attraction as a primary agent in the production of all physical, organic, and vital phenomena.

5.

We have already seen it established in preceding chapters that molecules are endowed with two positive and inseparable forces, *attraction* and *repulsion*. In the last chapter we traced molecular attraction into cosmical developments; and in all that precedes we have discussed the subject in such a manner as to show exhaustively that it is the accumulation of individual molecular forces of attraction into great volumes which constitutes the veritable force of gravitation, and that an inseparable consequence of this coalescence of the molecular forces is the juxtaposition of the molecules into bodies like planets and suns, with accumulative central powers which have been reduced by the mathematical genius of Newton into exact numerical functions. It remains now to follow *molecular repulsion* into its *sensible* and *cosmical developments*, and to treat this branch of our subject in such a manner that the existence of this force as a cosmical fact of co-ordinate value with that of gravitation can be comprehended by every mind, and be rendered incontrovertible by physicists or mathematicians. So far, indeed, as the application of the calculus to this subject is concerned as an *experimentum crucis* of the existence of repulsion as a cosmical force, that has already been accomplished by Newton, incidentally, and without special cognition of the effect of his own act. By establishing the numerical value of gravitation upon an assumed *vis inertiæ* and momentum, he proved a co-ordinate of force to be embraced in the initiative impulse (whatever its origin or character) impressed upon the bodies under mathematical treatment: for the problem of the ellipticity of planetary orbits with attracting bodies in one of their foci reduced to circles with the same attractive bodies in their respective centres, will show an exact equilibrium of the opposing forces. If there be questions upon this point, other Newtons will no doubt arise to settle them. My object is simply to point out here the existence of repulsion as a general and cosmical power hitherto overlooked and ignored, and to prove its infinite importance, as the source of most wonderful phenomena, the origin of which has not, up to this time, been clearly comprehended.

6.

There is no question of the existence of *molecular* repulsion. It is known to exist as conclusively as molecular attraction. There is no need of proof nor argument upon this point. All philosophers, of every name and grade, agree upon it. But as soon as we pass beyond molecules, the consideration of this force is dropped by scientific men, while attraction is carried through all acts and conditions of molecules to their grandest cosmical combinations and phenomena, until it is determined by the calculus to possess definite and progressive quantities and functions.

Now, although matter is not visible throughout its *entire* extension, it will be obvious without much argument that it is limited in quantity; and since all bodies are composed of molecules, it follows that molecules also must be limited in space. As matter and force are united and inseparable when practically considered, it follows that the individual attractions of molecules, considered as abstract and active quantities and powers, are also limited; and, as a final corollary, that gravitation, as a quantitative force, is limited in amount and action. The deductions of Newton confirm this conclusion with the severest mathematical certainty; and since space is boundless, while cosmical bodies are confined within positive although unknown boundaries, it can no longer be questioned that Newton's laws define and confine force as well as matter within limited boundaries also, and that mathematical discussions must sustain deductions derived from graphical treatments of the subject. So far as matter and *attractive force* are concerned, this is conclusive. However vast space may be, matter must be circumscribed in it like water within a lake, or wine within a cask—or more graphically speaking, a planet within the sky;—and since force is limited in its fields of action to matter, directly as the masses and inversely as the squares of the distances, the strength of gravitation as an active quantity must dwindle, vanish, and end in the indefinable depths of unoccupied space. Hence matter is limited within the fields assigned to it by Infinite Intelligence, and the *attraction* of its molecules uniting into cosmical functions and planetary spheres of gravitative force is equally circumscribed within the same field. All beyond this field of matter and force must, upon this reasoning, be empty space—nothing more—nothing less—neither matter nor force—absolute space—darkness—vacuum—nothing. The character of this field has already been described in our first chapter, and is only referred to here in order to bind preceding facts with present discussion and immediate sequences.

Now it is apparent that no *attractive* force is lost, or escapes from a single molecule in nature so as to be lost in space, or can escape and be lost there; but on the contrary, that all—the entire quantity of attractive force of molecules—must unite with the molecules and be inseparable from them, and become an eternal working power in the universe.

It is equally apparent that no atom of matter can drop out of the universe, or be lost, or annihilated. Neither can a particle of attractive force vanish or be annihilated, any more than an atom of matter.

Since, then, we know that molecular attraction and molecular repulsion exist as individual, positive, and coördinate quantities of force at the same time and equally in all molecules; and since we know that molecules and molecular attraction and molecular repulsion are inseparable, eternal, and indestructible; can we admit that *repulsion* is annihilated simply because molecules combine into masses and assume the shape of crystals and worlds, or of trees and animals, or because the respective molecular attractions of all these molecules coalesce into palpable gravitation, and become transformed into magnified powers and vast local spheres of attraction so potent as to influence each other when far asunder? Are we on this account to conclude that repulsion has vanished from nature? or should we not rather suspect that molecular repulsion, like molecular attraction, always accompanying molecules, changes its infinitesimal functions into infinite ones, and that these enter into and become a part of the mechanism and diversified phenomena of all departments of creation? That this question, with all its pregnant sequences, demands candid and exhaustive study, there can be no doubt. The laws of gravitation remained profound secrets until recent times, notwithstanding that ages of observation and study, from Thales and Hipparchus to Copernicus and Newton, had been devoted to the phenomena resulting from the occult operation of this wonderful principle; and even to this day its complete character has not been disclosed. That Repulsion, as a dynamical, cosmical, and morphological force, has been so long overlooked by physicists is indeed strange, and would be inexplicable were not the natural progress of human thought and scientific development well known. Philosophy, indeed, demands another element whereby to effect its largest developments; and the accumulations of facts in botany, anatomy, physiology, physics, chemistry, geology, and astronomy are now so great, that, like the gathering of mighty waters within time-worn and imperfect dams, they are ready to burst the bounds of the theories within which they have been laboriously and patiently garnered by many observers, from Copernicus and Tycho Brahé, to Donati, Bond, Faraday, and Liebig;—most precious treasures of knowledge, for which future ages cannot be too grateful to these times. It is in this inexhaustible storehouse of facts that I discover the evidences whereupon physicists and naturalists may unite in elaborating a system of analysis, in which repulsion shall be seen to combine with attraction as a co-ordinate force, and thereby generate mechanical energy, effect creations of every kind, continue celestial motion, and maintain the stability of universal nature.

If repulsion is not, and cannot be, annihilated when molecules associate, combine, and assume greater or less masses of matter, it must discharge some important and palpable function in the creation and motion of things, and become visible somewhere. If atoms aggregate into masses, and their attractive forces also coalesce and magnify themselves into potencies proportional to their numbers, so also must their repulsive forces assume increasing sums of power, and become active agents of some sort in the developments of nature.

That repulsive force can become not only a palpable entity, but an active and powerful principle *in bulk* and *when concentrated,* in proportion to its accumulation and concentration, we know from its action when gases and other substances are compressed—when it is released from water as this fluid passes into steam, and from carbon when gunpowder, guncotton, and nitro-glycerine explode.

I present these points simply to delineate at a glance the

nature of repulsion as a *pure force* in opposition to that of gravitation. I would present repulsion, if possible, for a moment divested of its material habiliments, as I presented attraction in the preceding chapter, so that the mind may comprehend this force in its *essence*, independent of matter, and contra-distinguished from attraction when this has passed from molecular force into gravitation.

7.

I may here be permitted to say that one of the chief impediments to the progress of modern science, is the error of cultivating natural philosophy from exclusively material points of view; whereas, if all acts of nature were studied more profoundly, it would be discovered that *force alone* is the motive power of all things, and that the study even of the shapes of atoms, upon which so much genius has been spent and is still employed, must be comparatively fruitless, since force is the basic constituent of atoms, and not shape or motion until after force has inoculated them with its special and compound principles. Taking this view of the subject, a profound, definite, and natural system is at once opened for the exercise of ideas encompassing all facts. We shall then treat of the combination of special forces, and of their coalescence and transformation into other forces possessing properties wholly different from their simpler originals, and no longer speak of atoms combining in this or that proportion, or number, or volume, or be compelled to divide atoms into smaller atoms or "atomettes" to meet the embarrassments of every existing theory. This view will permit scientific men to discard further speculations upon the shapes and sizes of ultimate atoms of matter—a point which no experiment or study has thus far approached—and will render it unimportant to determine whether ultimate particles possess special, variable, or similar shapes or dimensions;

since, inasmuch as force is the parent of all form and of all difference of properties in matter, it is this which so moves molecules that, basic ones commencing work, similarity and affinity of force will arrange atoms, and build up geometrical and organic forms upon this base, thus producing crystals and other things nearly or wholly alike in measurements, specific gravities, " atomic " equivalents, &c. while their chemical, internal, and specific properties are found by analysis and experiment to be wholly different from one another.

8.

In what precedes, I have endeavoured to show that pure force lies at the basis of nature, and that it is toward this fact the mind must be drawn, in order to understand and work out the true principles of every department of science. It is essentially so in respect to repulsion and attraction, genetic agents in the production of all mechanical action. In connexion with these the chemical and vital forces act in such a manner as to produce all forms and all substances out of molecules, so infinitesimally minute, as to be considered not only without definite shapes and void of all specific differences of and within themselves, but almost destitute even of material being. Under the light of this treatment molecules become only a sort of plastic and neutral element in nature, devoid of specific forms or properties, and capable of infinite divisibility; while force is the pure, unsubstantial, and supreme element which draws them into masses, and separates and holds them asunder, and changes their conditions according to laws existing in force alone. I am anxious to show that forces alone—many, specific, definite, distinct, ultimate, and potent, capable of every degree of union and divisibility under special conditions and through inherent laws—are the agents which determine the movements of molecules.

Molecules are simply sensitive points, and the working material of force. Planets are only larger sensitive points made up of these same sensitive atoms, and are the neutral crude material upon which the mechanical powers of attraction and repulsion in their cosmical aggregations are expended. Gravitation and repulsion are the supreme forces of the universe; and by their mechanical leverage the motions of the heavens are eternally maintained on one hand, while on the other, the oscillations of their functions bring all those secondary forces into being which work out through chemical and vital channels the numberless diversities of created things.

The subject of force thus developed in this connexion, will lead the mind to comprehend more clearly what I mean by *pure force,* when that term is applied to repulsion. Dynamical force, discussed in its cosmical expansions, appears gross and almost material and tangible in comparison with those subtle conditions of force which the analytical chemist and mineralogist are wont to consider in the cultivation of their branches of science. Nevertheless, all are equally subtle, differing only in quantity and function. Blending into harmony with the chemical and vital forces, which according to most delicate numerical laws select and reject or appropriate molecules at will, repulsion and gravitation grasp all alike, both singly and collectively, in order to accomplish special ends of a purely mechanical character through oscillating intensities of their respective functions. Each of these cosmico-dynamical elements can be demonstrated to possess absolute massive conditions of its own kind from which the opposite force appears to be almost entirely eliminated, although, when studied in connexion with planets themselves, or the entire system of nature, they appear to act inseparably, simultaneously, co-ordinately, and harmoniously, and to be so accumulative as to maintain conditions of perpetual tension, which tension, in consequence of the location of the sun in the foci of elliptical orbits, is constantly fluctuating like the movements of a pendulum, and exchanging quantities in such a manner as to maintain perpetual although unequal motion.

9.

It is evident in the present state of knowledge, that the two forces which maintain the motions of the heavens are equal in absolute amount, although fluctuating from instant to instant in quantity, as is proved by the *ellipses* of planetary orbits with controlling bodies in the *foci* of all these orbits; and it has already been clearly shown that molecular attraction has swelled into massive aggregates to constitute one of these forces, that of gravitation. We are familiar with this because philosophy is constantly discussing it, to the exclusion of all other forms of massive force. When repulsion is clearly comprehended as pure force capable of swelling into bulky conditions—co-ordinates with those of attraction—no theory of cosmic motion will easily reject it from its natural place in the celestial economy.

Certain illustrations are necessary to make this point clear, and they will serve not only to show the subtle relations of these forces to atoms and masses of matter, but also to present them in the light of abstract principles running through molecular gradations into voluminous expansions, in such strong contradistinctions from each other and from matter as to define the function of each and to prove that all elastic force, reactionary force, projectile force, tension, expansion, eruptive force, &c.—the entire nomenclature of modern mechanical philosophy—is nothing more or less than simple molecular repulsion, augmenting by successive increments and modifications of condition, until its concentrated potencies assume co-ordinates of gravitation in planetary motion, and become transformed into cosmical repulsion. Kepler, after discovering the ellipses of planetary orbits and the fixation of the sun in one of their foci, was evidently at one time impressed with the idea that planets were repelled by the sun as well as attracted; but this idea, like many other true conceptions, even that of gravitation, was lost in the midst of his perplexities and doubts, and the sublime novelties which almost bewildered him.

10.

The illustrations by which repulsion may be most clearly set before the understanding may be presented in two forms: indirectly, so as to develop it by way of comparison; and directly, so as to make it appreciable by positive demonstration.

Gold and platinum are among the simplest substances. Their physical characteristics are great density, specific gravity, malleability, and ductility, with small elasticity, and these characteristics distinguish them from everything else in nature. I select these substances therefore from all others, for the purpose of illustrating the abstract existence of attraction, so that repulsion may be the better comprehended by comparison when afterwards referred to. Their malleability and ductility clearly indicate such tenacity of cohesion between molecules, as to exhibit the principle of attraction almost, if not altogether, separated from repulsion in this condition of matter. This is a display of pure *molecular* attraction, while their density shows the same form of force, abounding and concentrated in masses, drawing their particles into closer juxtaposition. Their exceeding gravity is an expression of the same force, responding to similar force in the earth itself; in other words, pure gravitation—mutual attraction acting according to the Newtonian law of direct proportion to sums of atoms, as heretofore described. Their want of elasticity proves absence of molecular repulsion; their minimum fusibility (at least that of platinum), the great difficulty even to impress upon their particles the repulsive principle known to exist in heat, either as chemical or mechanical force.

Now, if in this delineation of the physical characters of gold and platinum we contemplate force and molecules in their infinitesimal conditions, we shall discover the least possible amount of active repulsion entering into their relations, and comprehend the fact that attraction may be a simple, abstract isolated principle even in massive quantities; and if the conditions described exist in a grain or a pound, or in tons' weight, as we know they do, they might exist in masses as large as the earth or Sirius.

Such is the nature of attractive force. The more free from intermixture with repulsive force, the more positive and visible is its action upon molecules and masses. When acting alone it produces intense cohesion, the greatest density, and pure gravitation. But as the two combine or mix naturally together, so to speak, existing as mechanical, chemical, and vital forces, the physical characteristics of bodies are diversified, and numerous quantitative differences of these combining forces produce endless forms of matter. Hence the impossibility of attaining by rapid steps the end of physical knowledge. It is only after mountains of minute facts have been accumulated, that some accidental thread will lead to the unfolding of true systems of classification through which all nature will become transparent to the human mind.

As the study of platinum and gold indicates the nature and action of the principle of attraction in both its molecular and aggregate functions, the study of gases will in like manner make us acquainted with the nature and action of repulsion.

Although gases differ among themselves in rarity and expansibility, they are all more or less compressible, and some to such a degree as to liquify and even to solidify. Of the twenty-eight gaseous bodies known, twenty-five are reducible to liquids and nine to solids. When compression is removed they spring back to their original conditions with great rapidity, and, if experiments be not carefully conducted, with destructive energy. The mere fact of the compressibility of gases from great volumes into liquids and solids, without the slightest change in their absolute gravities and atomic properties, is proof of positive distances between their respective molecules, and of an active inherent force extending its influence from one to the other, and from each to all, which maintains them at given distances asunder. Compressed and existing as palpable liquids and solids under restraint, their sudden and energetic expansion into original, invisible, impalpable conditions when restraint ceases is conclusive evidence that molecular repulsion acting from a central molecule to outer ones in all directions is, like attraction, capable of coalescing and uniting its individual powers into massive sums, and becoming transformed into magnified units of repulsion whose aggregates, as mechanical force, may be numerically computed as easily as that of gravitation. The existence of repulsion isolated so completely from gravitation that it may be recognised as an independent abstract principle, may be studied in aërostatic experiments; and some of the physical facts even, observable in these experiments, will be seen in the sequel to have their counterparts among cosmical phenomena.

Hydrogen, the most attenuate and expansible of known ponderable elements, will, when isolated in a balloon, not only rise in consequence of absolute repulsion between itself and the earth, but this repulsion is also observed to increase simultaneously between its particles, in numerical ratios to its removal from the influence of gravitation as a centralized principle, so that the balloon will not only expand with increasing tension, but even burst at last into many fragments. I am aware that it will be objected to this illustration that the balloon rises in consequence of being *lighter* than the atmospheric gases. But herein lies the occult principle of energy whose positive nature I am endeavouring to elucidate.

This is an illustration of the nature and action of pure repulsion, a principle possessing powers and properties in direct opposition to those of gravitation. As a secret fundamental principle, operating in the largest as well as smallest aspects, expanding into terrestrial and cosmical functions not as a mere reactionary and negative agency, but as a positive mechanical and creative power in the universe, repulsion presents a field for research and discovery not only as ample as that of gravitation, but so rich and exhaustless also that future generations will wonder that it has remained so long unexplored. This force varies its functions and expressions so much, is really so subtle, elastic, and latent in its character, ready to spring into action under the slightest cosmical, solar, or molecular disturbances, that it does, indeed, become the true mechanical, or, so to speak, the organizing element of nature; for neither heat, light, electricity nor magnetism, as will appear hereafter, can be generated independently of its precedent energy exerted in antagonism to attraction, nor can geometrical form, or vegetable or animal existence, be developed without its direct and active agency.

A remarkable instance of the play of repulsive force between the molecules of solid matter is presented in the substance called "elastic sponge." Ordinary sponge, cut into small pieces and submitted to certain chemical processes, acquires such remarkably elastic and expansive properties, that a bale of it weighing a single pound, compressed and held into a space of three inches in thickness by six in length and four in breadth, when cut apart, will expand to over three feet in length.

Repulsion, if studied in its relation to water, will be found to display its function in an equally positive, but different and unique manner. When hydrogen and oxygen combine to form water, the respective repulsive forces inherent in these gases either instantly vanish, or they coalesce as their molecules condense and liquify. These new relations of the attractive and repulsive forces remaining quiescent at common temperatures, they render this form of matter almost absolutely incompressible by any degree of mechanical pressure; and at certain temperatures, absolutely incompressible by cold. These singular conditions are clearly demonstrated by observation and experiment; while other important and suggestive facts are also established by them: for, since cold below 39·2 Fahr. (4° C.) produces expansion of water, it becomes evident that heat is of itself not pure repulsive force any more than cold; that cold embodies some foreign principle of positive force, as well as heat; and that repulsion is a principle whose existence *per se* is alike independent of both heat and cold, and is capable of combining with one as well as the other, in such a manner as to lead to the production of the most powerful mechanical results. Ice and steam as mechanical agents are too well known to require further remark. The point to be established by drawing them into this connexion by way of elucidation, is, that the mechanical lever inherent in both—in the base of both, and in the originals of both—is neither heat nor cold, but pure molecular and massive repulsion, an entirely different and the fundamental principle of expansive and projectile force. Like attraction, it accompanies molecules into all their various combinations, becomes apparent in some forms of matter more than in others, and so endows certain elementary bodies that it may be almost isolated, and studied as independently of matter, and as definitely, as gravitation and molecular attraction in gold and platinum.

11.

Now, since we can demonstrate in many ways that there is such a thing as pure repulsive force of a voluminous character, made up of increments of molecular repulsion, as gravitation in metals and planets is made up of increments of molecular attraction,—and since we know that all gases are held in their expansive and invisible states by repulsion accumulated into greater or less volumes of activity, and carrying those molecules with them whose chemical forces or properties permit such exclusive associations with and absorptions of this force as to hold them at special distances from each other,—since we know these things, and know further that the entire atmosphere, and all the waters and vapours enveloping and penetrating the planet are charged to their utmost capacities with this force, who can doubt that it must and does enter into the general constitution of nature in such a manner as to play an important function as an active cosmical quantity?

Recognising the universal presence and activity of repulsion in the envelopes of the globe—their saturation with it, so to speak—we can moreover trace its energy with equal certitude in the closest association with that of attraction, throughout the mineral and stratified crust. Although not treated by physicists as an active mechanical agent in crystallization, repulsion is without doubt as present and effective as attraction in arranging molecules into geometrical forms, and expanding these forms into their grandest and most perfect proportions. It becomes, indeed, in union with attraction, a formative and developing power; determining directions, limiting bounds, and confining atoms within specific distances of given centres. Were attraction the only force at work in and upon molecules when free to move and arrange themselves in juxtaposition, spheres would be the unique forms in nature, and all matter would be agglomerated into a single ball: were repulsion the only force, an opposite state of things would exist; since it is the property of the former principle to converge to common centres, and of the latter to diverge from such points. These properties are inherent in the fundamental nature of these two basic forces; and although we may not discover the subtle and directing agent incorporated still deeper into their constitution, we know these points to be facts, and must rest upon them as such. The forces unite and mingle their currents without sensible interruption, but their action upon molecules is such as to display their sensible presence and functions in dynamical phenomena, and these phenomena appear to be always generated at right angles to combining and decomposing currents, as will be more clearly seen hereafter.

They appear thus, also, to resolve themselves into the *secondary* forces of electricity, magnetism, &c., and to infuse these new elements into every mechanical, crystallographic and organic act of creation; and they fashion matter by measure and weight as the more subtle and latent chemical affinities measure out combining quantities, and in the same manner as the sledge and anvil fashion metals under the intelligent hand of the artisan. Therefore the form of every crystal—that is, the motion of atoms in order to be appropriated and arranged into form—must result from the compound action of these two forces; and since it is known that attraction is the force which moulds the faces and produces the shortest axes of crystals, it follows that repulsion must shape their salient diameters and angles. Such, indeed, is the case, the chemical forces (not here subjects of discussion) imparting various properties to molecules, and rendering them more or less susceptible of association with, and subserviency to, the two great underlying mechanical forces of nature. Researches upon crystallization clearly reveal the dynamical action of both these forces. When evaporation, for instance, of any saline solution slowly and quietly takes place, active universal motion is observed among the particles as soon as they become visible, as if each was endowed with life and intelligence. They act like so many animalculæ or infinitesimal magnets. They repel and fly from each other in one direction, and attract and attach themselves in another, and finally assume regular, solid, geometrical forms. While attraction appears to be the basis of solidity, repulsion appears to assume a higher function, and to inspire in matter a spirit of selection, adjustment, arrangement, order, and beauty, and to be an ascending step from plastic, neutral atoms, and solidifying force, to intelligent life-like endowment. As we observe the action of this force in the arrangement of molecules into crystals under a microscope, so we know it to act from a basic particle throughout the largest crystals and throughout every crystal in the world; and thus acting it presents the incontrovertible fact that repulsion extends from centres, as a continuous and voluminous energetic, and persists in exercising this function, in huge masses, as a special magnified power, composed of molecular increments of the same principle.

12.

Now, since all amorphous conditions of rocks and minerals are abortive or confused crystallizations; and since even sandstones, shales, and muds are only fragments and solutions of old crystals, or are capable of transformation into new ones ; it follows that repulsion, as an absolute and formative force, has existed from the origin of things, and still exists throughout the known solid portions of this planet. That it does exist, and that it acts in the atoms and germs of things, or shadows the outline around which and into which the chemical forces must play, and gauge their delicate weights and measures, or group molecules after their own special laws, we are compelled to believe from the continuous geological changes transpiring throughout the rocky frame of the globe. In limestones and lavas, sands, clays, and conglomerates,—it matters not what form of rock or metal,—wherever solutions can infiltrate, wherever heat or cold can penetrate, even where no perceptible change of temperature takes place, there unceasingly steals this subtle element of force, and pushes one atom here and another there, while attraction as subtly assists in the transfer by direct or counter impulses; co-workers and co-partners in these greater or less architectural designs amid which enter certain subordinate forces like apprentices and journeymen to dovetail the mechanism, carve the arabesques, and polish the handiwork of nature.

Upon the theory that the earth is solid throughout its unexplored depths,—whether it be a mass of trachyte or granite, as some geologists have fancied,—of soft iron, as Professor Lamont, the Bavarian astronomer, imagines,—or of a sort of water-glass, as is believed by Professor Schafhäutl of Munich,—repulsive force must have discharged, and must still perform, the same function so observable and demonstrable throughout the *known* parts of its crystalline structure.

Turning discussion, however, toward the planet's centre in view of the theory of Leibnitz that the globe was originally a ball of igneous fluid, that its crust has resulted from cooling, and that its interior is yet a molten mineral mass, the evidences of voluminous and active repulsion are still more obvious and conclusive. Without favouring any special hypothesis, or introducing any into this connexion to prolong the discussion, it is impossible to overlook great physical events and facts which display themselves in all hemispheres and in all latitudes and longitudes of the earth, and thereby become recognisable cosmical phenomena. Earthquakes are indisputable effects of repulsive force acting from greater or lesser depths, and their not unfrequent simultaneous occurrence at nearly opposite points of the earth's surface presents strong evidence that this repulsion often radiates from central regions of the planet. To enumerate and describe these terrific events which overthrow cities, lift and fissure vast areas of plain and mountain, swell and move entire oceans with measurable, frequently with destructive, impulses, and shake a hemisphere from the Caspian Sea to Lake Superior and from Iceland to Sahara and Martinique, is quite unnecessary, since the monthly records of this class of subterranean phenomena are known to every physicist. Few will attempt to deny that all these great expansive and undulatory movements of the crust result from tension produced by the molecular repulsive forces, by the voluminous action, indeed, of repulsion viewed as a pure force capable of an accumulative function, and whose nature and action, being the same whether expressed in infinite or infinitesimal quantities, are demonstrated by its functional capacity, so to speak, to segregate, to break loose from its connexion with the planetary mass, and to fly and carry immense sums of atoms in its course (for the reason that matter clings to force) in like manner as molecular attraction, becoming gravitation in masses, carries matter with it in *its* passage to unite with the earth or the sun. Repulsion is active mechanical radiation: gravitative attraction is active convergence toward the centre of anything. Matter is moved in opposite directions by these opposite forces. This is self-evident.

Besides this class of phenomena whose effects are observable at surface-levels, there is another where the mechanical effects of repulsion proceeding from the planetary centre rise far above the surface. To say nothing of extinct volcanoes of enormous altitudes, active ones here and there exist all around the globe, and even in high northern and southern latitudes. It is unnecessary to state that their elevation and activity result from expansive expressions of this force, the voluminous and cosmical character of which I am endeavouring to elucidate. The mere recognition of volcanic eruptions—a comprehension of their vastness and sustained energy—must convince practical and truly philosophical minds that another force as mighty as gravitation, antagonistic to it, and even exceeding it in power for the time being, is at work to transport such immense volumes of matter hundreds and thousands of feet above the general level of the planet. The summit of Cotopaxi, near the equator, is 18,858 feet above the ocean, that of Erebus, near the South Pole, is 12,400 feet, while the channels which conduct from their craters to active fountains of repulsive force may be scores of miles in length. Prolonged researches upon this subject, conducted extensively in order that physical configurations and eruptive phenomena might be most accurately observed and studied, have taken me to many volcanoes, both active and extinct, and to some of the highest and most remarkable upon the earth. Beside listening to repeated fearful rumblings beneath the valleys of Quito, Latacunga, and Riobamba, moving in the directions of Sangai and Cotopaxi, and suggestive of violent movements of immense amounts of material of all sorts in profound caverns of the Andes, I have many times observed glowing discharges projected from one to three thousand feet above the sublime and symmetrical summit of the latter mountain. So prodigious is the display of eruptive force as the igneous rocks emerge from the throat of Cotopaxi, that they appear like a perpendicular column of compact fire; and it is only when, at immense heights, they separate and fall in pyrotechnic showers of indescribable grandeur that these huge fragments of an inner world are distinguishable from one another. I have seen the same subterranean repulsive force vent itself many times in a similar manner from the summit of Isalco, and in a feebler manner from Masaya in Central America; while in Hawaii I once witnessed a molten bubble scatter its igneous spray toward space, as it burst from that ever-restless fountain of Kilauea, whose agitated *surface* is probably never lower than three thousand feet above the surface of the Pacific Ocean.

Definite, numerous, and satisfactory as have been my own observations of this class of phenomena, those of other persons are still more notable. The Rev. Titus Coan, resident as a missionary of the American Board upon the island of Hawaii, has for many years made careful notes on the various eruptions that have burst at different heights from the remarkable volcano of Mauna Loa. On the 27th December, 1855, he addressed me a letter from Hilo in reply to interrogatories upon plutonic phenomena observed by him at Hawaii, and some of his facts bearing strikingly upon our subject, I select them from numberless records to strengthen this class of evidence:—" From Kilauea, in June 1840, there " was a great eruption, the lava of which rolled furiously " down to the sea, in a stream from one to three miles wide." Kilauea is not far from 4,000 feet above the sea-level. " In " January 1843 there was an eruption from the summit of " Mauna Loa, the streams of lava being thirty miles long and " from two to three miles wide." The summit of Mauna Loa is nearly three miles high. " In February 1852 there was an " eruption of unrivalled splendour on the north-east slope of

" Mauna Loa. For twenty days a column of incandescent " minerals was ejected vertically to the height of one thousand " feet, from an orifice nearly one thousand feet in diameter. " This rare pyrotechnical display exceeded all description of " pen or pencil. The stream from this crater was thirty " miles long and from one to three miles broad." This eruption sprang from a point estimated to be eight or ten thousand feet above the ocean. "The last and most terrific " is our present eruption, which commenced on the summit " of Mauna Loa on the 11th of August, and which is still " in full blast after a lapse of 140 days. The stream of " fusion, including its windings, is now sixty or seventy " miles long, from two to six miles wide, and varying in " depth from 6 to 200 and 300 feet." This opening took place about 12,000 feet above the sea, and the stream was visited by me in March 1856, when it was still flowing through the forest at a distance of about six miles from the Bay of Hilo. Another great eruption took place on the 23d January, 1859, on the northern slope of this volcano, at an altitude of about 7,000 feet, a description of which appeared from an anonymous pen in the *Pacific Commercial Advertiser*, published February 12, 1859, at Honolulu. "The lava " does not simply run out from the side of the crater, like " water from the side of a bowl, but is thrown up in con- " tinuous columns very much like the Geyser Springs, as " represented in school geographies. At times this spouting " appeared to be feeble, rising but little above the rim of the " crater; but generally, as if eager to escape from the pent-up " bowels of the earth, it rose to a height nearly equal to the " base of the crater. But the columns and masses of lava " thrown out were ever varying in form and height. Some- " times, when very active, a spire or cone of lava would shoot " up like a rocket or in the form of a huge pyramid to a " height of nearly double the base of the crater. If the " mouth of the crater is five hundred feet across, the per- " pendicular column must be eight hundred to one thousand " feet in height! Then, by watching it with a spyglass, the "columns could be seen to diverge and fall in all manner "of shapes like a beautiful fountain." "Large boulders of "red-hot lava stone, weighing hundreds if not thousands of "tons, thrown up with inconceivable power high above the "liquid mass, could be seen occasionally falling outside, or "on the rim of the crater, tumbling down the cones and "rolling over the precipice, remaining brilliant for a few "moments, then becoming black, were lost among the mass "of surrounding lava." The Rev. L. Lyons, in a published description dated at Waimea, Hawaii, February 4, 1859, says also of this eruption:—"On January 23d volcanic smoke "was seen gathering on Mauna Loa. In the evening the "mountain presented a grand yet fearful spectacle. Two "streams of fire were issuing from two different sources, "flowing apparently in two different directions. . . . In the "morning of the second day we could discern where the "eruptions were. One appeared to be very near the top of "the mountain, but its stream and smoke soon after disap- "peared. The other was on the north side further below the "top. . . . The burning crater seemed to be constantly enlarg- "ing, and throwing up its volumes of liquid fire above the "mouth of the crater—I will not venture to say how high "—and the fiery stream rolled onward and onward, still "adding grandeur and terror as it proceeded, till on the "morning of the 31st, about sunrise, it reached the sea. . . . "The volcanic stream was one mile wide or more in some "places, and much less in others. The whole distance of "the flow, from the crater to the sea, is some forty miles. "Last night the volcano was in full blast." The following April "a long point had been formed running out into "the sea at Kiholo with a depth of water at the outer "edge of sixty-three fathoms, and the liquid rock was still "dropping out seaward." I will not attempt to compute the cubic quantity or weight of matter thus ejected into space from the interior of the planet in these well-noted repulsive outbreaks. But similar phenomena have so frequently occurred at Hawaii and in other volcanic regions that it can be scarcely doubted that the earth must be in a state of constant tension.

On the 27th and 28th March of the present year (1868) the summit crater of Mauna Loa, called Makuaweoweo, was seen to belch fire and smoke. This crater is not far from 14,000 feet above the sea. The tension was so great, that "a rending took place on the southern slope of the mountain, "near the summit, and fiery streams issued from four points "and ran off in diverging lines. The largest stream ran "south about fifteen miles, when suddenly all the moun-"tain valves closed, and fire and steam disappeared to-"gether." While these pages are in type, I am again indebted to the Rev. Mr. Coan (as well as to several other friends) for a detailed account of many phenomena which have been observed during the recent remarkable eruptions in Hawaii. These may be said to have been ushered in by an unusual action in the molten lake of Kilauea, which commenced rather suddenly on the 20th January of the present year. This subterranean tension and local activity continued steadily to increase in violence, as related to Dr. Hildebrand by Kaina, a daily eye-witness of what took place there for five months preceding the terrible earthquake of April 2d. I will only call attention, however, to two or three points, in order to fix upon the reader's mind the existence and the voluminous action of a force the opposite of attraction, and which, though an active and the radical element in the production of heat and steam, is not itself either steam or heat, and which, as we shall see by one of the Hawaiian facts, and especially by our discussion in the chapter on Mechanical Force, may be effectively displayed in expansive phenomena in which no heat is perceptible or demonstrable.

"On the 2d inst." (April), writes Mr. Coan, "at 4 P.M., "such a shock occurred as had no record in the history, the "memory, or the traditions of these islands. . . . Great boulders "were tossed and rolled like a foot-ball; land slides were "numerous, and precipices fell thundering into the sea. The 'earth was rent in a thousand places, and fissures from one

" inch to six and ten feet opened. The streets and yards and " fields of Hilo cracked, and seams opened all around us. . . . " It seemed as if the rocky ribs of the mountains and the " marble walls and pillars of the world were breaking.

" At this moment the ground was rent in Kau, between the " ranches of Reed and Richardson and Mr. F. S. Lyman's, and " an eruption of mud, rocks, and earth (*not heated*) took place " about three miles long, and one to three wide, burying " instantaneously a village with thirty-one natives, and also " 500 or 600 head of cattle. The discharge was so sudden that " there was no escape for those in its range. *It was like the* " *bursting of a bomb.* At the same time the sea rose and fell " rapidly all round the island. In Hilo it came up to Front " Strait, six to eight feet above high water mark. In Puna " it rose ten to fifteen feet. . . . In Kau the tidal wave was " twenty to twenty-five feet high, sweeping away all the " villages and churches along the low shores, and destroying " scores of people," &c.

Dr. Hildebrand, who is known to me personally to be both a reliable observer and a scientific scholar, writes to the *Pacific Commercial Advertiser*, published at Honolulu, a lengthy account of his observations made on the 10th and 11th April last, from which I extract the following data: " On " Tuesday afternoon, April 7, at 5 o'clock, a new crater, " several miles lower down than that referred to, and about " two miles back of Captain Brown's residence, burst out. " Four enormous fountains, apparently distinct from each " other, and yet forming a line a mile long, north and south, " were continually spouting up from the opening. These jets " were blood-red, and yet as fluid as water, ever varying in " size, bulk, and height. Sometimes two would join together, " and again the whole four would be united, making one " continuous fountain a mile in length. . . .

" This was the magnificent scene, to see which we had " hurriedly left Honolulu, and had fortunately arrived at the " right moment to witness as it opened before us in all its " majestic grandeur and unrivalled beauty. At the left were "those four great fountains, boiling up with terrific fury, "throwing crimson lava and *enormous stones weighing a* "*hundred tons* to a height varying constantly from five "hundred to six hundred feet. At times these red-hot rocks "filled the air, causing a great noise and roar, and flying in "every direction, but generally toward the south. Some- "times the fountains would all subside for a few minutes, "and then commence, increasing till the stones and liquid "lava reached a thousand feet in height. The grandeur of "this picture, ever varying like a moving panorama painted "in the richest crimson hues, no person can realize unless he "has witnessed it."

When physicists know that that *Force* which pushes up is the *antithetic* of that which pulls down, and that which produces cracks and fissures and expansion (whether small or great) is the *opposite* of that which affects concentration, cohesion, and adhesion, can they longer doubt that *Repulsion* is really and practically a principle capable, like gravitation, of assuming planetary expressions and becoming an absolute cosmo-dynamic power?

A patient hand has collated the following data which bear directly upon the point under illustration, and they might, if necessary, be indefinitely multiplied from my pregnant records. "Cotopaxi, in 1783, threw its fiery rockets 3,000 "feet above its crater, while in 1744 the blazing mass, strug- "gling for an outlet, roared so that its awful voice was heard "a distance of more than six hundred miles. In 1797, the "crater of Tungaragua, one of the great peaks of the Andes, "flung out torrents of mud which dammed rivers, opened new "lakes, and in valleys of a thousand feet wide made deposits "of six hundred feet deep. The stream from Vesuvius "which in 1737 passed through Torre del Greco, contained "33,600,000 cubic feet of solid matter; and in 1794, when "Torre del Greco was destroyed a second time, the mass of "lava amounted to 45,000,000 cubic feet. In 1697, Etna "poured forth a flood which covered eighty-four square miles "of surface, and measured 100,000,000 cubic feet. On this

" occasion the scoriæ formed the Monte Rossi, near Nicolosi, " a cone two miles in circumference and four thousand feet " high. The stream thrown out by Etna in 1810 was in " motion at the rate of a yard per day for nine months after " the eruption; and it is on record that the lava of the same " mountain, after a terrible eruption, was not thoroughly " cooled and consolidated ten years after the event. In the " eruption of Vesuvius, A.D. 79, the scoriæ and ashes vomited " forth far exceeded the entire bulk of the mountain; while " in 1660 Etna discharged more than twenty times its own " mass. Vesuvius has thrown its ashes as far as Constanti- " nople, Syria, and Egypt; it hurled stones eight pounds in " weight to Pompeii, a distance of six miles, while similar " masses were tossed two thousand feet above its summit. " Cotopaxi has projected a block of 109 cubic yards in " volume, a distance of nine miles; and Sambaua, in 1815, " during the most terrible eruption on record, sent its ashes " as far as Java, a distance of three hundred miles' surface." Facts of this character abound in the records and traditions of all volcanic localities. Every *extinct* volcano even is indicative of the same eruptive action from the planet's interior toward space in times past more or less ancient. It is the same all over the sphere from Jan Mayan 19° of latitude from the North Pole and Kliutchewskaja in Kamschatka, 15,763 feet high, to Erebus in Wilkes' continent, 12,400 feet high, and only 12° from the South Pole; and from the middle of broad oceans at Hawaii 14,000 feet high, to Gualatieri in Peru 21,997 feet, and to the interior of Asia, 1,500 miles from any sea. From the low outlets of Santorini, rising in successive heights of mechanical energy through Vesuvius, 3,948 feet; Jorullo in Mexico, 4,205; Etna, 10,874; Idrapura, Sumatra, 12,300; St. Helen's, Oregon, 13,300; Ariquipa in South America, 20,320; and a hundred other intermediate outlets not enumerated, we discover the same irresistible principle at work *radiating from the centre of the globe*, which, exerted in its molecular and multiform capacities, can with equal facility push an atom, shoot a ball, turn a wheel, raise an island, rift a world or roll a system.

It is all one and the same repulsive force acting from the bowels of a fecund and overstrained planet or of a fecund and overstrained universe. Like gravitation, which throbs equally through an atom or a Sirius, and binds the cosmos together as with bolts of steel, so repulsion, on the contrary, can send an atom whirling from a planet's centre, or a comet from the sun so far that it may lose its way among the stars. It is force, and force alone, which executes all. Matter without force might well be called inert or dead. But our philosophy, like that of Plato, Aristotle, and Liebnitz, can admit of no death, no inertia, no rest, nothing but sensitiveness in matter and universal life, physical, intellectual, and spiritual, eternally excited and attuned by ever-changing and ever-oscillating activities of unstable force. Repulsion upon one side, attraction upon the other, and the sensitive, plastic, submissive molecule between, ready to receive and obey either or both quicker than the quickest thought of man, quick as the quickest thought of God. In proportion to the quantity and conflict of engaged forces is the mechanical character of any material act.

It is obvious without further evidence, or discussion, upon the point, that volcanic and earthquake phenomena are acts of pure repulsive force, and that these phenomena are resultants of the infusion into matter of those subtle principles of energy which act and react in the movements of machines, or in the revolutions of planets. If repulsion plays its part secretly, and proceeds out of the depths of the earth, it is no more secret than gravitation, and it has no other origin. If the working power of one is within the reach of the calculus, so is that of the other ; and if either be a creative essence and an inexhaustible fountain of inductive power, the other is not less so. I hope to make this appear more fully hereafter.

13.

While we see repulsion exert itself in the form of mechanical action from the central depths of the planet outwardly, thus apparently embodying itself into substantial and visible relations with cosmic masses, we observe that in crystallization it acts also by silent methods of persistent leverage in pushing atoms into geometrical forms. In this secret universal system of mineral architecture its function is still mechanical. But it takes, moreover, a wider range even in the crust of our earth. That it exists in a state of cosmic tension, an absolute superabundance of mechanical energy, as well as in the special condition of a molecular and morphological agent, is sufficiently proved by that category of phenomena long observed by quarry-men and miners, where stones spring laterally and upwardly out of place when suddenly detached from continuous strata.

That the earth is charged with an active principle of repulsion, from its central point to its outmost atmospheric atom, in the same manner that it is charged with attraction —that it is thus loaded to absolute tension—and that this principle performs diversified functions, like attraction, in union with the chemical and vital forces, modifying their action and the weight and density of matter—will become, I trust, more apparent as the subject is studied from the point of view whence I have endeavoured to unfold it. That attraction and repulsion united, playing with incessant oscillations of intensity upon and through atoms, are the sources of the secondary forces, light, heat, electricity, magnetism, which in their turn assume new functions, developing the great work of creation, and presiding over the fountains of eternal change and continuance, of conversion and conservation of mechanical and organic force, which prevent nature from drifting into rest or decay, and which create all things anew, and extend life and being from everlasting to everlasting— that attraction and repulsion are the alpha and omega of all these things, universal observation and experiment indicate; and I believe that few physicists or naturalists will question this conclusion, when repulsion shall have been as much a subject of thought, discussion, and research as molecular attraction and universal gravitation, and when it shall take the place of "elastic force" and "latent heat" in solid science. As an element of the highest character in scientific investigations, repulsion has been overlooked by philosophers, and, indeed, rejected altogether from consideration in celestial mechanics. The chemist has recognised it for a moment in his molecular philosophy; but when it begins to expand into mechanical dimensions and organic functions, he abandons the pursuit of it as beyond his field of inquiry, or vaguely comprehends it as a doubtful fact embraced in the meaningless nomenclature of latent forces. So difficult is it to break away from ideas to which the most sagacious minds become enchained by authority and tradition.

14.

In like manner as gases, when condensed and thus made liquid or solid, *appear* to concentrate and confine repulsive force by a sort of chemical or mechanical restraint; so it seems that by the processes instituted by nature to amass atoms into spheres and create solid or liquid worlds, this same repulsive force is held as if bottled up, so to speak, or absorbed to repletion or tension, as water is held in a sponge; or (a better illustration) like vapour mixed with the atmospheric gases, ready to expand or condense as external conditions may influence or determine. The more we study puzzling phenomena of a general character, the more we shall *suspect* this state of things, until we prove and establish it. Even the same class of phenomenal movements which in earthquakes and volcanoes is so impressive and remarkable in the solid sphere, is equally conspicuous, but modified by circumstances, in its aqueous and atmospheric envelopes.

Passing over the general subject of tidal movements the theory of which is by no means perfect, and the remarkable undulations of the ocean, which are at times associated with earthquakes, and at others appear to have no connexion with them, I ask particular attention to a phenomenon of constant occurrence on every ocean, never heretofore, to my knowledge, noticed by physicists, and which is strikingly dynamical as well as meteorological in all its bearings. Every mariner will immediately recognise it by his own characteristic name, as soon as the description meets his eye. I was many years ago so forcibly impressed by this appearance, that I have made it a subject of observation and study, through many extended voyages, upon all the oceans of the globe. It is a phenomenon the most perfectly exhibited at the calmest hours, and when there have been no storms to disturb the sea. I believe now that it depends upon a positive radiation of the repulsive force, as a definite cosmic principle, which acts upon the body of the waters to rarefy, in like manner (although in an opposite sense) as attraction acts to condense them, and that in this method of action it becomes also an evaporating agent. In the night this condition of the ocean is often accompanied with a fiery luminosity. The general term "radiation" expresses the true mechanism, or physics, of this telluric function. The appearance of the phenomenon referred to, in its most active conditions and under the most favourable aspects for observation, is as follows:—A vessel will be lazily fanning along, rising and falling upon the gentle undulation of a smooth sea, and scarcely moved by the slightest breath of air, when, in some direction, ahead or around, the water will appear agitated; and soon the observer is in the midst of a disturbed surface. There is still hardly a breath of wind. Indeed, it often becomes much calmer than at first. All aërial movements die away, the sails flap against the masts, and the vessel will toss as if jarred by adverse and tumultuous impulses all around her bows and stern and beneath her keel. The condition of these spots in the ocean is very singular. They are sometimes small, at other times they cover large areas, and have smooth undisturbed patches in the midst of them. Some physicists might at first sight suppose them to be results of the meeting and intermixture of adverse currents: but the most curious part of the phenomenon is yet to be described. The water appears to be alive. It sparkles and moves with such lightness and spirit as to give the impression that it is charged with gas. The waves absolutely dance, and jump up, and separate into wavelets, and point into innumerable cones, and then burst upwards into drops and bubbles and spray as if they were thrown up by the snapping of millions of fingers underneath the water. Not only do these drops not go back into the sea, but they further continue to break up into the finest sparkling droplets, rising higher and higher till they vanish into mist. Whether they sink immediately back again or become absorbed and carried into the upper strata of the atmosphere, succeeding events will clearly demonstrate. Soon the air becomes hazy and the sky overcast. Dense leaden clouds collect as if by enchantment, and rain falls in a deluge. For a few moments this tumultuous uprising, bursting, sparkling surface of the sea is beaten down, and the vessel will lie and roll in a perfect calm, with floods sweeping across the deck and the scuppers not ample enough to vent the rain. Before the sky has completely discharged its burden, the ocean is observed to be executing the same upheaving performance; clouds condense again overhead and around, and the rainfall is repeated. This state of things may continue for hours, or it may suddenly cease, and the excited surface of the ocean return to its ordinary aspects.

The intensity and definiteness of this phenomenon vary with circumstances; but in all its variety of expression, whether at equator or pole, it is as clear to my mind that repulsion, radiating from the profound realms of the planet, carries up this water from the ocean, as that gravitation brings it in bulk down again. Even the very secret agency of this method of distillation, by which fresh-water may be eliminated from the great natural saline basins, strengthens the probability that repulsion is the elementary principle which causes this phenomenon, since all observations and experiments show that repulsive force has positive affinities for the gases composing water above all other matter, and it is reasonable to conclude that this force selects and seizes on these particles to the exclusion of the crystallizable ones upon which the antagonistic force of gravitation exerts for the time being a stronger power.

When it is understood that these currents of repulsive force possess the same subtlety and power as those of attraction, and that they carry all atoms with them in all directions away from the earth's centre as gravitation brings all atoms over which it exerts the greatest power downwards, we shall the more satisfactorily arrive at an explanation of this phenomenon. But in the production of waterspouts the action of this force cannot be mistaken, since it plays as conspicuous a function in their development out of the ocean into the upper regions of the atmosphere as it does in developing a volcano; and when the facts attending this wonderful phenomenon are delineated, the action of repulsion will become so obvious that further illustration and discussion upon the subject might be rendered unnecessary.

The primal conditions of the ocean attending the formation of waterspouts are such as have been previously described, except that they are more tumultuous. I have watched the development of several perfect and imperfect ones, and have studied them in action for a long time, and, therefore, describe them from personal observation. The ascending current of force is so circumscribed, and, at the same time, grows so strong, that the vapours from the little bursting wavelets already described, collect in such volumes as to become visible in the atmosphere as a sort of cloud near the surface of the ocean. But it would appear that this ascending circumscribed volume of force has been some time at work invisibly mounting from below the surface of the sea into the upper regions of the atmosphere; for, from a gradually forming and outspreading mass of condensing vapour, hundreds, perhaps thousands of feet high, a funnel-shaped projection of a leaden colour is seen gradually to descend, presenting a well-defined attenuated tube with an uneven, still more attenuated end. This tube, vanishing into thinnest invisible air, is directed toward another of precisely the same character, which is now seen to be ascending from the inverted cone of vapour resting upon the agitated surface of the sea. Thus the two extremities, one rising from the water and apparently collecting mist from the surface, the other descending from a cloud, and evidently receiving vapour with great rapidity (for the cloud fills and spreads out and grows leaden-coloured, thick, and heavy, and at a distance is perhaps discharging rain), are distinctly seen to approach each other, while the intermediate portion is still invisible. Little by little, sometimes rapidly, the column extending in both directions at last coalesces, becomes visible throughout its length, and assumes an intense heavy leaden aspect, like the cloud above and the cone below. The colour is now uniform, and the waterspout is finished; and there it stands, a slender, portentous pillar of cloud rooted in the ocean and unfolding its mysterious foliage in the sky, as distinctly defined in shape as a palm-tree in the desert or a molten fountain playing from the bowels of Mauna Loa. It may be from a few hundred to several thousand feet in height; and when examined carefully with a good field-glass it is discovered to be a comparatively hollow tube, since it presents a translucent cylinder in the middle, like the hollow structure of a hair as seen under a microscope, or a barometer tube as seen with the naked eye. Meantime there is no storm around, and a ship might drift or sail into this field of ascending force.

15.

Since it is not my design in connexion with this particular phenomenon to discuss the method by which a submarine principle seizes upon atoms of purest water (generally, but not always, rejecting the salts), the reader's attention is exclusively confined to the active movement of simple repulsive force as a cosmico-dynamic element. It has been already shown that force cannot be separated from matter, or matter from force, although it is proved by abundant experience that attraction and repulsion may, and do, mutually exchange quantitative intensities, oscillating in their influences upon matter by processes of apparent induction and exclusion; and that this function or law of fluctuation, of alternate vanishing and outbreaking, ascends from molecules to masses, and cannot exist in masses except through capacities inherent in molecules: indeed this operation of exchange, of induction and exclusion, of substitution and ascendency of one force over the other, is so instantaneous, subtle, and extensive—as, for instance, in the rising and falling functions of the pendulum—that the mind can only embrace the phenomenon by repeated successions of the fact accomplished. This idea may be further illustrated by calling to mind a similar phenomenon, which is exhibited by spongy platinum in its absorption of oxygen to more than eight hundred times its volume; and this absorption is so rapid, so instantaneous, as proved by experiments upon it with hydrogen, that time cannot be measured during which eight hundred volumes of a known *material* substance can be absorbed and appropriated by a single volume of *other* atoms.

Now if such a phenomenon as this, the absolute *translation of matter*, can be instantaneously and repeatedly effected among wholly *material* conditions, how can we question or doubt the instantaneous *absorption* and *exclusion* by matter of its fundamental *forces* of attraction and repulsion (the radical principles of motion), the action of which are far more attenuated, subtle, and searching than those of light, heat, electricity or magnetism?

16.

We may see by the preceding illustrations, that while repulsion is demonstrated to be as definite and as general a fundamental principle as gravitation, and to be the co-ordinate of it, it is also exhibited, like the latter, in union with molecules : and the illustrations are precisely to the point, since repulsion as a terrestrial force carries outwards in every radius of the sphere the self-same molecules which gravitation brings down, when its molecular increments have again asserted supremacy over repulsion. The secret of this fluctuating function of the attractive and repulsive forces lies in the elementary nature of molecules, a point yet too profound for explanation, except chemical laws are by some method expanded into cosmical correlatives, and both expressions of the molecular forces are studied in their grandest developments as observable in planetary motion. But these phenomena as simple facts cannot be rejected, although their importance may be still longer disregarded in physical theories. Like heat and cold, repulsion and attraction radiate from and pass into molecules and masses to effect equilibrium and engender alternate tensions; except that viewed separately these latter impart no sense of heat or cold, and that their transmissions are so infinitely rapid as to be instantaneous and inappreciable. Like paramagnetism and diamagnetism (using these principles only by way of illustration), which penetrate everything with the quickness of thought, encountering obstacles nowhere, repulsion and attraction rush into and pervade everything, whether atom or sun, but, unlike the former, with immeasurable velocity; existing nowhere in equilibrium; absent from nothing, whether matter, or heat, or magnetism; and embracing all atoms and all worlds and all the other great natural forces within their irresistible, universal, and everlasting grasp. While light, heat, electricity, and magnetism are comparatively local forces, and obviously originate by the antithetic play of repulsion and attraction through molecules and masses—that is to say, molecular motion—requiring time for transmission, and undergoing successive conversions into each other and back again to motion, repulsion and gravitation are universal, are the originators of motion, absolutely require no time for transmission, and never change their identity of nature. In fact, all the secondary forces lie enwombed between these two. They are to everything else in nature what the upper and nether millstones are to grain. As they grind, creation unfolds its manifold phenomena throughout its ample domain ; and from crudest matter spring forth warmth and being.

The immortal Michael Faraday, endowed with all the perseverance of the ancient alchemists during their vain search after the philosopher's stone, has failed to find that magnetism is molecular attraction or attraction of gravitation ; but while he has convinced himself that it is not, he is still constrained to believe that it must in some way be allied to, or hold some definite relations with gravitation, or that gravitation must possess some property or principle not yet ascertained or understood. All observations of magnetic phenomena tend to this same obscure conclusion. In no field of inquiry is progress more retarded by the prevailing theory of the existence of only one fundamental universal force, than in that of magnetism. Neither this, nor any other secondary force, can be generated in any experiment by molecular attraction alone. They never appear, and are never active or functional, unless molecular repulsion is in absolute play with attraction in varying degrees of intensity ; and the secondary force generated is in direct ratio to the amount of molecular repulsion spent upon the experiment. The existence, therefore, of magnetism as a *terrestrial force* implies the absolute certainty of the capacity of molecular repulsion to unite its individual forces into volumes, and thus to expand itself into a function or force with planetary magnitudes ; establishes the fact that molecular motion must be generated throughout the globe as the result of attraction and repulsion at work on a planetary scale and as cosmical principles ; and brings us to the final conclusion that REPULSION is one of the bases and originators of the entire range of natural forces and phenomena of a planetary and mechanical character.

Such is partly the nature and action of repulsive force. Although the subject has by no means been exhaustively treated, a multitude of illustrations being necessarily omitted for the purpose of brevity, yet enough has been presented to open long avenues of progress in various directions. It will be perceived how difficult it is to disentangle attraction and repulsion, as pure and independent entities and forces, from matter, and especially from the chemical forces; for chemical laws, and all the secondary forces, modify the conditions of matter for receptivities of these *radical principles,* and complicate the analysis of this subject. But, once presented in a forcible and distinct manner to the earnest scholarship of coming time, I believe Repulsion will be identified as a positive and imperial agent of nature, and will never be relinquished by the physical sciences.

Attraction draws everything to the earth's centre, and is made known to us by the *falling of atoms.* Repulsion projects everything from the earth's centre, and is made known to us by *the rising of atoms.* A grain of sand falls straight to the centre of gravity throughout *every point* of the earth's surface. A spark flies upward throughout every one of the *same* points. The chemical nature of the grain of sand or of the spark is not in question, only the forces which guide ultimate atoms, as elemental agents—agents far more subtle than oxygen or carbon, or heat itself.

17.

But one of the most important of the laws connected with the nature, and governing the action and effects, of repulsion, is, that when observed to exert itself in a normal manner, the phenomenal results of its action are *always at* RIGHT ANGLES *to attraction, whether it be studied as a molecular, terrestrial, or cosmic force:* and it is in the development of this law that we shall discover the secrets of planetary rotation and of all physical motion.

The announcement of a law of this character and in this form will at first appear doubtful or difficult of comprehension to those who view repulsion as a force acting in a strictly opposing sense to attraction in all its material relations. But it must be remembered that matter and masses can move in no direction except through the action of force —and force alone—upon their molecules, both singly and collectively.

Now the terms in which I express this law are simply explanatory of the axiom laid down by Sir Isaac Newton:—"*To every action there always is opposed an equal reaction;* "*or, the mutual actions of two bodies upon each other are* "*always equal, and directed to contrary parts.*"

The differences in the manifestations of repulsive force, developed as a dynamical element, depend altogether upon the chemical constitutions of the bodies through which it acts, as an abstract entity; and this is a point which cannot be fully discussed in this connexion. But it scarcely requires elaboration, since it must be immediately evident. An illustration or two will set forth this law in its practical or material bearings; and hereafter, when it is unfolded in connexion with the correlation of forces, and the conversion of one form of energy into another, I trust it will be clearly apprehended and accepted.

Two elastic balls, as of india-rubber or ivory, when forced together, rebound. The molecules in each react upon one another, and experiments now prove that all reactionary force, when exerted in the masses of bodies, *proceeds from the centre* of those bodies, into whatsoever point the force may appear as an effective and measurable quantity. The first action in each of the elastic balls when they strike, is to flatten from pole to pole in the lines of their motion, and expand in their equators through the effect of the reactionary principle. This reactionary principle, thus manifested, is the *repulsion* of each molecule united into a voluminous expression of mechanical energy, directed from the central molecule to the equatorial circumference. This is the first step in the phenomenon. But in order to bring about the rebound which is the phenomenon of palpable reaction, as the term and act are commonly understood, these very molecules which were previously projected to the circumference by a centrifugal force, immediately spring back again in virtue of their property of cohesion (a chemical condition), and redevelop the same principle or function of force at the impinging poles, the physical action of which becomes again projectile, in *a contrary direction* to the *primary* impulses. In each step the dynamical phenomena have been developed at right angles: in the first, to the straight lines of projectile motion; in the second, to the centripetal action which simultaneously re-establishes the rotundity of the two balls, and the equilibrium of the antagonistic forces.

The first phenomenon is specially manifested and clearly expressed in an exclusive manner, so to speak, when *plastic* balls strike each other. They flatten and coalesce. The attractive force unites them. The repulsive element expands them, in whatsoever direction expansion or *reaction* takes place.

When brittle substances strike and break, their fragments fly at right angles to the lines of the projectile force.

Action and reaction are equal and contrary, and I doubt if the law, as it is clearly displayed throughout nature, could be more definitely stated than it has been by Newton himself.

These phenomenal effects of masses of matter upon each other are so common as not to excite attention, or suggest a question of the *modus agendi* of the energies at work upon their interior radii, or of the changes which occur in them in virtue of the action of the communicated, or of their respective innate forces, of *attraction* and *repulsion* considered as *abstract* and *separate entities.*

To imagine, in order to explain such results, that attraction is decomposed, and that it resolves itself into reaction or repulsion, does not simplify the difficulty of comprehending the combination of forces with matter, or of their action upon matter.

If dynamic energy be a unit and expressed in the single word FORCE, we see, nevertheless, that this FORCE is immediately separable, by every form of experiment and every method of reasoning and induction, into two elements or functions; and that these two functions are the direct and absolute opposites of each other, when considered as abstract principles, and when lying dormant, so to speak, in nature. But when excited into sensible action, and when matter is subjected to their conflicting agencies, all observavation and experiment show that, while action and reaction are equal and opposite, the phenomenal results, wheresoever molecules can move in the most natural manner and freest sense, are equatorial and at right angles. No demonstration is required to prove that the force which produces cracks, fissures, disruptions, and chasms in dense and hard bodies, whatsoever their magnitude, whether a pane of glass, a pebble, a mountain, or a world, is not that which produces condensation and cohesion. The statement that the two forces which produce these contrary effects are the opposites and equals of each other is a truism which it is unnecessary to repeat; and yet all the physical results of the action of these two forces upon matter, when molecules and masses are free to move in space, are displayed at right angles to each other.

In our chapter upon "Mechanical Force," this branch of the subject will be further discussed; and the discoveries of Oersted in dynamic electricity, as developed mathematically by Ampère, will carry us into considerations of the nature and action of these forces, and of the methods through which all forms of force become correlated and conservable. I will not, however, anticipate the treatment of that branch of our discussion; yet I may be permitted to say in closing this chapter, that we shall find the principle here enunciated to pervade even the morphological development of every class of organisms, thus showing a fundamental connexion between the planet and its inhabitants, and between matter, force, instinct, and mind.

CHAPTER IX.

COSMICAL REPULSION.

1.

In the preceding chapter we have endeavoured to establish the important truth that molecular repulsion, like molecular attraction, is capable of swelling into and assuming voluminous conditions; that it does, as a positive pure force with a distinct, definite, and independent function, move matter in masses away from the earth's centre, in the same incomprehensible manner as gravitation draws masses to that centre; and that it holds such relations to gravitation as to exchange functions with the latter in so instantaneous a manner, as to execute, with fluctuating intensities, all the mysterious movements of molecules and masses, thereby perfecting that harmony and stability combined with eternal activity so visible throughout the realms of nature.

If repulsion assumes terrestrial magnitudes and functions, as it evidently does, since eruptive energy has always been, and continues to be, manifested from the earth's interior in every latitude and longitude; since crystallization is universal throughout its mineral crust; since its surface teems with vegetable and animal forms which could not spring from, and mount upwards, nor be detached from and become locomotive upon the globe, if attraction alone reigned; since its aqueous envelope is incompressible; since its atmosphere is but fluid and solid matter expanded into gases; since winds (phenomena depending upon the *gravitation* of these gases to fill spaces *primarily* rarefied or made vacant by REPULSION) are universal, constant, and often fearfully violent;—if repulsion assumes *terrestrial* magnitudes and functions, as the foregoing facts clearly demonstrate, then it becomes, indeed, a truly *planetary* force, as much so as gravitation, and must be so considered. No movement of a molecule takes place in, or upon, or around the globe, into which this force does not enter and perform its part, as well as attraction. If the latter be a universal terrestrial force, so is the former. Not a breath, not a ripple moves in air or sea, not a drop of water flows along a channel, quivers in a waterfall, or pours upon a wheel, the resistance and reaction to which is not at its foundation repulsive force springing into mechanical energy and productive power. The planet is full of it to saturation; and in proportion as gravitation manifests its special agency, and apparently predominates, repulsion is instantly developed to supplant it, and *vice versâ;* so that both spring alternately into tension, or suddenly vanish, so far as our faculties can detect their action. As to activity of function, they appear to act upon a principle similar to a fundamental one laid down in physics, which declares it impossible for two atoms of matter to occupy the same point of space at the same time. Thus it will be seen that while repulsion charges the earth to repletion, it is at the same time a restless force, and becomes almost palpable as the great, living, working power of nature; and that while gravitation tends to stability and unity, repulsion tends to motion and production.

2.

Now, since repulsion, pervading the globe as it does from its centre to its outmost atmospheric boundaries, is as much a planetary force as gravitation, *it necessarily becomes a* COSMICAL *power*, and must be as capable as gravitation of extending its influences and functions to other cosmical bodies, and of receiving similar influences from them.

But are other cosmical bodies—suns, planets, satellites, comets—also charged with repulsion, as we have proved our own globe to be? It will not be difficult, I think, to arrive at the affirmative of this question if we follow Nature step by step from her subtlest traceable principles into her grandest physical developments. It is admitted that every body in the universe is composed of molecules, and that the natural action and reaction of these molecules depend upon the existence and activity therein of the two well-known molecular forces, attraction and repulsion. These, as I have attempted to show, assume and manifest *planetary* powers and functions; and this fact becomes self-evident, inasmuch as these bodies display *different degrees* of *density.* Besides these individual and local conditions in all, both comets and planets disclose a remarkable *universal law of varying density,* holding inverse numerical proportions to their respective distances from the sun—VARIATIONS OF DENSITY, that is to say of INTERNAL COHESIVE ATTRACTION *of a* FLEXIBLE AND FLUCTUATING CHARACTER, *bearing constant inverse relations to the length and sweep of their respective radii vectores.* Why is this, except that the sun, as a controlling body, and a larger volume, so to speak, both of repulsive and attractive force (not necessarily of magnetic, electric, or thermic force, as we shall be hereafter convinced, but of pure gravitation and repulsion, as distinct, independent, and fundamental forces), holds direct relations with similar forces, or volumes of force, in all these revolving spheres; thereby exerting and exchanging *influences* (not necessarily the forces, the very entities, themselves)—influences which, so far as attraction is exerted, have already been proved to be mutual, and to depend upon elements of mass and distance only? If all these spheres, like our own, be also charged with molecular repulsion, are we to suppose it plays no part in the movements or phenomena of the universe as a physical whole and a complex machine?

3.

Now, the simple facts that the planets exhibit different densities and weights, the numerical values of which hold notable relations to the respective heliocentric positions of these bodies, are so striking as to incite not only attention to this field of cosmical physics, but also a careful inquiry into the action of planetary laws and final causes.

When observation is directed to another class of celestial bodies, to comets, whose paths are more varied and eccentric, and whose translucency favours more accurate internal study, the same conditions are noted ; but instead of tracing this fact by steps of number, weight, and density, so to speak, through successive chasms in space from the Sun to Mercury, Venus, and so on to Neptune, the phenomenon can be distinguished with the naked eye, or with telescopic aid, as a positive, continuous, and connected result of the action of an invariable law, while these spheres advance from the outer regions of our system to their perihelia, or recede again into the immensity of space. We are thus awakened to comprehend this mighty physical truth, that an entire series of heavenly bodies can condense and expand to such measurable degrees, like masses of gold and iron, or volumes of gas, in virtue of solar influences ; and are compelled to believe that a *universal force* of *repulsion*, as well as of attraction, reigns throughout the cosmos, the influences of which can operate among planets and stars as instantly and incessantly, as potentially and positively, as they do among atoms and through the pendulum upon our own whirling sphere.

The ordinary phenomena distinguishable in all comets as they approach the sun, pass perihelia, and recede into space, only require simple description to convince every clear-minded physicist that gravitation is not an unique force in celestial mechanics. When comets are first descried, at remotest distances, journeying toward the sun, they are diffuse, nebulous, misty objects, with or without observable central points of greater brightness than the rest of the body. As they approach the sun, they progressively condense, and a central brightness appears; or if at first apparent, this grows more concentrated and intense. It may be that a tail begins to push out. These phenomena steadily augment, the disk becoming smaller, more defined in circumference, and dense at centre; and by the time a comet has reached perihelion, and for some hours or days afterwards, all its elements of matter and force exhibit the most intense agitation, and the action of a repulsive force becomes as visible throughout its translucent volume as that of gravitation. In the great comets, whose perihelic distances have been smallest, the nucleus glows like a furnace, and boils from centre to circumference with marvellous motion. If gravitation alone existed throughout these masses of cometary matter, not the slightest reactionary motion would be visible in them. But in addition to this molecular agitation, which clearly demonstrates the intense play of antagonistic forces, generating as a consequence secondary forces of light, &c. to be discussed hereafter, another condition arises in these bodies, which illustrates the predominance of cosmical repulsion during their retrocession, as incontrovertibly as attraction was predominant while they approached the sun. When comets have passed their perihelia and commence their backward journeys they are seen to expand, to fade in brightness, to be less agitated throughout, to be less clearly defined in circumference, to withdraw gradually, and at last to lose altogether, their radiating appendages, to become more and more diffuse and misty, and ultimately to regain their original physical conditions.

This is a general description of the physical appearances of this class of cosmical bodies, as they have been always observed, and of late years more definitely studied by Messrs. Bond of Cambridge, U.S. and other eminent telescopic astronomers. The phenomena are quantitative, measurable, and computable; and, as cosmical facts, are entirely pertinent to our subject.

A succinct presentation of three important facts among the most striking in their successive transfigurations will elucidate the point to be established.

When first observed to approach the sun from the remotest distances, comets possess their largest dimensions, and are seen to contract steadily in size and increase in apparent density up to their perihelia and somewhat beyond it. In this fact the predominance of attraction is observed to reign throughout their molecular structure, and to produce, as in the physical constitution of planets, a direct increase of density in proportion to the attraction exerted upon their entire mass by the sun. In a word, the attraction of their individual molecules toward their respective centres, as a quantitative and internal cosmical force constituting the body as a sphere, is in direct numerical proportion to the attraction exerted upon their entire masses by the sun. This is a remarkable condition of things, and cannot too much enlist the attention of physicists; *because it shows that powers inherent in the sun exert absolute and constant control over every atom of matter in bodies revolving around it, however near or distant;* and that there is a universal law regulating incessant interchanges of influence in some way (not, I hold, direct translations of each other's individual forces across space) between the sun and the centre of every planet and comet; and that no body in our system possesses an independent constitution of its own. This is certainly the case with attraction: for we have heretofore traced it ascending from molecular into terrestrial functions; and now it is seen expanding into universal planetary conditions, induced and governed by demonstrable influences of some sort springing from the great centre of gravitative force. Whatever reasons (whether from mathematical or imperfect magnetical data) scientific men may have heretofore had for suspecting direct interplanetary communications, they are proved beyond all question by rigorous analysis of this class of physico-cosmical phenomena. By this remark I do not mean it to be understood that there necessarily ensues quantitative and positive interchanges of each other's allotted sums of energy, or fixed means of work and motion; but mutual susceptibilities, or a sensitiveness, so to speak, to the neighbourhood or distance of each other's masses and contained sums of motive power.

At the same time, it cannot be overlooked that the luminous phenomena displayed in the outstreaming motion of comets' tails, always in directions away from the sun and more or less in a line with the radius vector, are indicative of an influence positively dynamical and repulsive which extends far into the interstellar spaces. This fact, however, will be less difficult of explanation when it is remembered that the sun is constantly interchanging *its* influences (whatever they may be) with all the stars in all directions (as well as with the members of its own system), and that the luminous lines of a comet's tail only follow the lines of the stellar connexions at every point of the paths of these remarkably ethereal and sensitive bodies.

The second fact, to which I invite careful attention, is the remarkable agitation which occurs in comets when in perihelia. Attraction alone cannot create this internal cosmical disturbance. This anomaly, or paradox, forced itself upon the judgment of the late lamented Cambridge astronomers, the elder and younger Bond, when, in 1858, they studied the remarkable agitations of Donati's comet. I was permitted by those eminent observers to examine that wonderful object through the great equatorial telescope of Harvard College; and on that occasion the elder Bond announced to me his conviction that, notwithstanding all that was said of gravitation, a repulsive force was evidently at work throughout that body. These observations confirmed conclusions of my own, derived from studies of earlier data, which were published in 1853, and then affirmed to be of universal application.

It is, indeed, evident that there must be an antagonistic force in conflict with attraction to produce the outward surgings so observable in this translucent class of bodies. Since all the conditions of these phenomena are traceable, from what has preceded, to *molecular* relations of matter and force, we can scarcely doubt that molecular repulsion, tranquilly performing its legitimate function in maintaining the sphere at its maximum of expansibility while remotest from the sun, subsequently yields its predominant influence as solar gravitation grasps and draws the comet onward as if to absorb it into the solar mass; and that at last, enduring this sort of grasp no longer, its concentrated energy is aroused into a positive reflex function. The conflict of force begins in earnest when both attraction and repulsion, in extremest tension at perihelion, are excited into stronger efforts to separate from each other; the former to unite with, the latter to alienate its special form of energy and elongate it by polarity, so to speak, from the solar reservoir of cognate forms of force. The peculiar physical constitution of comets renders the action of these fluctuating forces visible and unmistakeable to the human eye. Striking phenomena and new forces are developed by them. Molecules throb, and currents and oceans of lights and shadows swell, and vibrate, and roll backwards and forwards with alternate intensities of repulsion and gravitation; and it is in this condition that a great comet, always a wonder to mankind in all ages, seems to be presented to advancing science as an angel of light for the purpose of illuminating the hidden depths of nature, and exhibiting a clue whereby to follow up the laws and discover the marvellous secrets of the Most High.

The third fact would appear to remove all doubt from the subject: as soon as a comet has entered upon its journey away from the sun, the internal force which immediately takes the place of attraction and assumes a positive predominance, becomes demonstrable repulsion. This it proved by a phenomenon so visible and constant, that physicists and mathematicians the most wedded to the prevailing hypothesis of an unique force acting upon inert bodies by exterior impulses, must be not only led to question its correctness, but also compelled to extend their views and inquiries. The same agitation before described continues for a while from centre to circumference of the receding body; but, instead of becoming smaller in volume, as when advancing to and passing through perihelion, the comet becomes larger; and although dilating and contracting to measurable extents by a species of vibrating molecular impulses, it steadily increases in dimensions, becomes less agitated, loses the brightness of its nucleus, and its general density, and gradually assumes its original size and misty nebulous characteristics in remotest space. In proportion as the body recedes from the sun, its individual particles become loosened from the force which drew them to the centre as it approached perihelion, and they not only decentralize—that is to say, recede from the centre, as is shown by the comet's nucleus becoming more feeble in luminous intensity, or splitting into several nuclei—but they also recede from each other, at first with violence, and at last with a steady expansive action, just as we observe hydrogen to expand in a balloon in proportion as the balloon recedes from the earth. To call these and similar phenomena simple "reaction" explains nothing; for reaction, as a word, means nothing unless it represents an idea of direct and positive physical force. If it mean anything, then, it expresses a force opposite in action to that which drew the particles toward the centre and increased the density of the comet's mass as it approached the sun. If it be a force—a positive principle exerting action opposite to attraction—it is the equivalent of molecular repulsion, admitted by every writer on molecular physics, and which has already been demonstrated to be susceptible in our own planet of swelling into indefinite volumes, of producing general radial phenomena, and of becoming an absolute planetary force, the properties of which are everywhere similar, and always expressed in functions of a cognate and voluminously repulsive character. And it may well be concluded from all which precedes, that the final principle of nature which, throughout our globe, expands matter, creates the waterspout, guides the currents of vital force and molecules in the palm-tree and pine, lifts plutonic fires thousands of feet high, shakes alike centre and circumference, causes flame to ascend perpendicularly from every radius of the sphere, is no other than a positive living force of repulsion,—a principle of energy the opposite of attraction ; and that it not only performs the same function in every planet and comet, but also is the identical principle which, in every star and controlling central body of a system, determines and guides all radial phenomena in their outward courses ; and in our own system, proceeding from the sun, projects the comet's tail in all its radial aspects in the same manner as repulsion, ever insensibly radiating from the earth's centre, projects the ascending atoms or luminous currents from a burning taper. It is one and the same principle, as universal in its reach as gravitation, and ever antagonistic to it, whether in a molecule or a world. It is the occult principle of energy, the final element of antithetic power, *per se,* in polarity, as will be more clearly seen hereafter.

Upon these demonstrations we might rest with this part of our subject. But as if comets were very messengers of light to physical science and modern discovery, they present other phenomena too pertinent to our discussion to be overlooked.

4.

It will be admitted by all candid physicists, that no planet or other great volume of atoms can divide itself into several parts which shall separate many thousand miles asunder, and so continue, approaching and receding from one another more or less during successive revolutions around the sun, if attraction of gravitation be the only reigning force in matter or cosmical systems.

Biela's comet, an object known to be embraced within the constitution of our own system, whose periodic revolution of $6\frac{3}{4}$ years has been long determined, and whose return to perihelion has been repeatedly observed, was perceived in December 1845, after having passed perihelion, and during great internal agitation, to divide itself into two distinct bodies of nearly equal size. These bodies—now lost—continued to travel together during several revolutions as two clearly definable comets, returning to perihelion with great regularity, and fluctuating, both with regard to distance from each other and to the intensities of their internal molecular motions as they drew near to and receded from the sun; and they mutually exhibited such phenomena as to leave no doubt of constant interchanges of influences, affecting the activities or energies of their respective molecular and general forces;—activities visibly, if not measurably, swelling into cosmical magnitudes, and appearing in all respects dependent upon their relations of distance from the sun.

The extraordinary fact of the further disruption or the final dissolution of these two fragments of Biela's comet, only strengthens the inference which I am endeavouring to draw from that of their first division or segmentation.

Fortunately for science and philosophy, such a remarkable fact does not stand alone in the field of astronomical observation. On the 26th of February, 1860, another double comet was discovered at Olinda, Brazil, by M. Emmanuel Liais, formerly attached to the Paris Observatory, and of late Astronomer to the Brazilian Government. In his communication to the French Academy of Sciences, and also in his volume "*L'Espace Celeste et la Nature Tropicale,*" M. Liais carefully described this remarkable phenomenon: "The " comet consisted of two distinct nebulosities, one of which " was many times larger than the other. The intensity of " both was very weak. The distance between the two, when " first observed, was 27 seconds of time in right ascension, " and less than 1′ 8″ in south declination. On the 29th of " February, and the 3d of March, this distance was found to " have diminished to 23 seconds R. A. and 46″ Dec. respec- " tively, and on the 10th it had diminished again to 21 " seconds and 21″. The smaller comet was always circular, " with a nucleus in the centre; the larger one was elliptical, " with a nucleus in one of the foci, or nearly so; but on the

" 11th of March it appeared to have three, but this changed " again on the 12th, when only the usual one remained " visible, but much larger. On the 13th the small comet was " no longer visible, and the large one was nearly circular, " with greater intensity in the centre, gradually diminishing " towards the borders. After the 13th, M. Liais did not " see the comet again, owing to the state of the weather. " Its elements are stated as follow :—Perihelion distance, " 1·197,342 ; inclination, 79° 35′ 54·5″ ; longitude of the " ascending node, 324° 3′ 25·4″; longitude of perihelion, " 173° 45′ 21·1″ ; passing through perihelion, Feb. 16th, at " 13h. 50m. 65s. Motion direct."

Notwithstanding Liais's comet was *not seen* to separate like Biela's, the facts above recorded will leave no doubt that both these bodies were originally a single one, which divided in the same manner as Biela's, in consequence of the predominance of an internal force, the opposite of attraction, and of gravitation. What can this be but repulsion, molecular and cosmical, in its visible and active phenomenal results ?

Besides these two remarkable facts, so important to a true understanding and solid development of celestial physics, there stands upon record another phenomenon still more extraordinary (and unique so far as my researches extend), confirming the existence of repulsion as a force, whose functional activity may not only partially or locally predominate over that of gravitation, so far as to divide a cosmic mass into two distinct spheres, but even so entirely to transcend it as to split a comet into infinite fragments, and scatter them instantly out of sight, as if by evaporation ; or as a balloon full of hydrogen would burst and vanish in the sky.

The first telescopic comet of the year 1818, when in perihelion, and at the time it was expected to exhibit its greatest internal agitation and its most intense luminous phenomena, suddenly broke up, dissolved, and wholly disappeared, to the astonishment of observers. (*Astronomie de l'Amateur.* Par G. HIRZEL. Geneva and Paris, 1820, p. 269.)

The great antiquity of the observation of Democritus, who thought he saw a comet divide, and resolve itself into a great number of small stars, I will only mention as a curious record, receiving authenticity from the preceding well-sustained facts observed within my own lifetime.

The same may be said of the statement transmitted by Ephorus, a Greek historian, who records that the comet of the year 371 B.C. separated into two parts, which pursued different paths. Notwithstanding that Seneca doubted this event, more modern observations render it not impossible.

Kepler thought and insisted, notwithstanding that Pingré and others ridiculed his opinion, that the second comet of 1618 divided into several parts. But Figueroes at Ispahan, Blancanus at Parma, and a Jesuit missionary at Goa, saw two comets at the same time, in the same part of the heavens, both having direct motion toward the north, and the second appearing suddenly (as if it had had its birth from the first) when the first had been continuously visible several weeks, thus confirming the observations of Kepler: and modern discovery renders his opinion very probable.

The astronomical records of the Chinese previous to A.D. 1000 make mention of several double comets which pursued their course in company.

5.

It is for physicists to contemplate these facts with all their pregnant import. No mechanical theory of the universe, however ancient or honoured, or authoritative its origin, can put them out of sight. No mathematical astronomer, whatever his eminence, can gainsay or reject them. There they stand as physical facts, as positive and durable as the sun, or as the rings of Saturn, with all *their* continuous breakings up and fluctuating changes, and they must be explained upon rational principles. Considered in the light of attraction of gravitation alone, they are inexplicable anomalies, and must excite doubts of the exclusive universality of its reign. Doubts upon the subject of gravitation would indeed be errors, when proofs of its universal nature and action are so numerous and certain. When these remarkable facts are viewed, however, as physical possibilities and phenomena appertaining to, and inseparable from, atoms and spheres, lying between and subordinate to the action of *two* great fundamental forces, co-ordinates of each other, whose functions become excited and augmented by solar influences of an inconstant character, and whose antagonistic energies intensify from centre to circumference in every sphere in inverse numerical proportions of distance from the sun—that is to say, the nearer the sun, the greater the internal agitation, and *vice versâ*—they will no longer rank as anomalies, but receive immediate solution.

6.

The general display of fluctuating dynamical forces, so constant and remarkable in comets at perihelion, augmenting occasionally into exceptional phenomena of unusual violence, are conditions which should arouse every philosopher to question if they be merely local phenomena—that is to say, confined to comets only—or if they be not reproduced, in some manner or degree, in every other cosmical body which enters into the constitution of our system, and in direct numerical ratios to the length of its radius vector. These mechanical conditions and changes are observable and measurable in all diaphanous bodies which form part of our system, or temporarily enter it to wind their eccentric ways around the sun. So far as the records of astronomy present well-observed facts, there are no exceptions to the reign of contrary and violent forces acting from the centres of comets to their circumference, and *vice versâ*, in direct numerical ratios to their heliocentric distances. The progressive density and rarity, or condensation and expansion of these bodies, as they approach and recede from perihelion, as judged by successive transitions of internal phenomena, obviously follow the same law which determines the molecular constitution and internal cohesive force of *planets,* and which governs *their* distribution into their respective orbits. These internal conditions, dependent altogether upon fundamental molecular forces swelling into mighty cosmical impulses under solar excitation, evidently control the planetary motions, modify heliocentric distances, and suggest the only solid theory and natural system of celestial dynamics. These steady transitions of cohesive force—in other words, of molecular repulsion and attraction—which have already swollen into cosmical volumes, and which, drawing atoms into globes, now constitute central planetary forces;—these steady transitions of density in spheres thus constituted, directly dependent as they appear to be upon the nearness of bodies to the sun, and so observable through the elongated paths which some comets trace from orbit to orbit of the globes whose respective mean densities and weights have been mathematically established (at least approximately established), show beyond a doubt that the law which determines this function with its ascending and descending magnitudes in one sphere, must determine it in all, whether opaque or translucent.

Now, since the molecular constitution of all is more or less alike, with densities depending upon inverse numerical ratios of heliocentric distance; and since all revolve around the sun in orbits more or less elliptical or eccentric, with the sun in one of the foci of these orbits (and never in the centre), thereby demonstrating positive uniformity in physical economy, indicating common origin, and proving a single fundamental organic act to have ushered in the plan, creation, arrangement, motion, and mutual relations of all; it follows that the same solar forces which govern the cohesion and central fluctuations of density in one, must govern these phenomena in all. Therefore, if repulsion be a demonstrated force, whose cosmical magnitudes and functions in comets are incontrovertible, it becomes equally so in planets, satellites, and asteroids; and as an element in celestial physics it demands as complete a recognition by science as its co-ordinate of gravitation. While astronomers may accept it as a legitimate means of correcting theories and discussing anomalies, mathematicians will find no substantial reason to reject it as a conflicting element in their discussions, since it simply replaces an assumed and doubtful by a positive and living power. By physicists it should be studied with special application to the great problems—yet so obscure—involved in the production of light, heat, electricity, and magnetism in all their terrestrial and cosmical aspects.

7.

As conclusive and incontrovertible as are the facts of the absolute divisions of comets into several parts, and of their subsequent companionship—thus, as I assert, plainly proving a cosmical force of repulsion to exist in, and to interchange its reactionary influences in some obscure but mutual way between, these bodies—the action of this force *within* them is confirmed by the separation of the nuclei of some into several distinct nuclei, the luminous intensities of which are perceptibly different from the rest of the comet, thus presenting (so to speak) several nuclei suspended in and surrounded by the same diaphanous disc. This phenomenon has been observed in a number of comets, and it demonstrates molecular activities which could not take place if attraction of gravitation reigned alone throughout these cosmical masses. Associating this class of facts with the constant phenomenon of internal agitation, evolving itself concentrically or progressively from centre to circumference, and *vice versâ*, as if the comet were a boiling mass whose most active energies sprung from its centre and returned into the same point;

and appreciating all these varieties of physical condition as phenomenal results of the action of two constant forces whose energies depend upon special numerical ratios of distance of the cometary centre from the solar centre; we can no longer reject the conclusion that cosmical repulsion governs the molecular relations of revolving bodies as well as cosmical attraction, or what is called in more common language, attraction of gravitation. Whatever conditions there may be about these phenomena which we do not understand, the incompleteness of our knowledge and the extraordinary character of the events should only stimulate us the more to study them until they be philosophically and fully comprehended.

The phenomena connected with the development of the tails of comets, as these bodies approach perihelion, and the changes which they undergo during their visible retrogression, might be treated at length in their bearing upon this subject, but I will only refer to them as facts which, when investigated in all aspects, will more strongly confirm the conclusions already derived from studies of the nuclei.

8.

The constancy of all these facts—that is, of the *internal* molecular motions observable in diaphanous cosmic bodies during their contiguity to the sun, the evident submission of these bodies to a constant solar force which develops throughout their molecular constitution such palpable repulsive energies, and which, regarded as attraction or attraction of gravitation, could not effect such contrary results, and which appeared to proceed as an exciting influence out of the sun, and to be transfused in some occult way into comets, or at least to stimulate them into this remarkable repulsive or reactionary state;—the constancy of all these facts led me, many years since, to suspect that similar conditions must be reproduced to a greater or less degree, and in some positively demonstrable manner, in all other bodies revolving in orbits of various eccentricities around the sun ; and unwearied investigations have led me to conclude, not only that such is the case, but also that the eccentricities of planetary orbits and of cometary movements are in some way correlated with and dependent upon the varying density of the revolving masses; in other words, that the orbit and path of every body in our system are determined by the internal molecular constitution and the chemical conditions of that body, and by the play of the internal molecular forces of attraction and repulsion, which respond to similar forces in the sun, in inverse ratios of distance, as may be seen so conspicuously and constantly displayed in comets.

To establish a fact of such magnitude, and of so much importance to physical science, as the existence of constant molecular movements of a repulsive character throughout the mass of opaque cosmical bodies, similar to those so palpably existing in translucent ones, and to demonstrate the intensity of these movements to be specially dependent upon conditions of solar distance, seemed at first impossible. But perseverance in the only sure method of physical investigation (which method is *never to relinquish for a single instant the guidance of nature,* confident that even a single molecule, ultimately traced, will lead to its Creator and Lawgiver) has elaborated phenomenal evidences from the study of our own globe which have ended in the determination of numerical results of a very remarkable character, which I believe prove the existence of physical laws heretofore undiscovered, the reign of which must be as universal as cosmical nature itself.

In the preceding chapter there was occasion to allude to those terrestrial phenomena which develop themselves as earthquakes and volcanic eruptions. When profoundly and ultimately analysed, these phenomena become reducible to so many simple expressions of an internal and general repulsive force, which, proceeding from central points of the planet, not only indicate universal tension of its mass, but also produce positive and constant disruptions of its surface. All these results are, in several aspects, perfect counterparts of phenomena so generally observable in comets. The difference in the physical constitution of these two classes of bodies—the gaseous-diaphanous and the solid-opaque—really constitutes the only difference of condition and appearance in reactionary and eruptive results. These phenomenal evolutions in comets clearly demonstrate the activities of their molecular forces to be developed into cosmical amplitudes; and these activities are so universal as visibly to exhibit evolutions of *repulsive force* reaching from the centre to the circumference of the bodies, and even projected beyond their nuclei: and in this statement I will *specially omit the tails of comets,* upon the physical elements of which scientific opinion may not be entirely united, in order to avoid all conflict with the speculations of Bessel and with the conjectures of his predecessors or successors. When *terrestrial* phenomena in their entirety are studied from this point of view, earthquakes and plutonic eruptions appear to be simple exponents or external indices of the same internal cosmic reactionary activity or molecular repulsive energy assuming massive conditions. Columns of molten matter charged with heat and light, projected thousands of feet above the general surface of our sphere and falling back or spreading out upon it, and the same matter cooling in some places into a solid crust, and then replunging, as in the volcano of Kilauea, and many active craters, into the plutonic fires beneath, evidently represent, or, in more exact terms, are *analogues*—indeed, *the very homologues*—of the evolutions so visible in the great comets, which have often startled observers and perplexed thinkers. The uninterrupted manifestations of earthquakes, and of plutonic movements of one sort or another, statistical observations upon which are now so numerous and authentic as to show their frequent simultaneousness in widely separated areas, and to indicate that, if all could be noted, not a day or an hour passes during which some locality is not disturbed by these phenomena, and probably not a moment in which plutonic matter is not impelled against, or into, or through the crust;—all these phenomena, whatever their class, appearance, condition, or degree of manifestation, are only so many impulsive movements of the interior matter and force of the globe upon its exterior. This is almost the identical language used by Humboldt to express his idea of the cause of earthquakes—an old idea, indeed, but enunciated by this great German scholar in modern scientific terms—"*the reaction of the interior upon the exterior of our planet.*" Although his extensive travels and erudition have added nothing to our positive knowledge of final causes in this field of inquiry, this expression precisely defines the mechanism of internal cosmic repulsive force in developing eruptive phenomena which, in opaque globes, are simple reproductions or analogues of the eruptive phenomena so evident, definable, and under favourable circumstances almost measurable in diaphanous ones.

Mr. Robert Mallet, whose special studies of seismic literature, and whose experiments in impulsive physics (if I may give them such a name) justly entitle him to the highest honours which British physicists can confer, defines an earthquake to be "*the transit of a wave of elastic compression in any direction from vertically upwards to horizontally, in any azimuth, through the surface and crust of the earth, from any centre of impulse, or from more than one, and which may be attended with tidal and sound waves dependent upon the impulse and upon circumstances of position as to sea and land.*" While the definition is similar to that of Humboldt and the more ancient authorities, it is enunciated in such exact terms as to delineate all molecular impulses, and be applicable to every action of repulsive force, whatever their origin or amplitude, *in all cosmical bodies.* Accepting it as the veritable dynamic condition under which all agitations occur in every class of bodies, and as the most exact expression of the scientific thought of our day, we are struck with its equal application, as a phenomenal law, to both comets and planets; and we may furthermore observe that this law is demonstrably applicable to the internal mechanical processes by which physical violence is enacted throughout all cosmical spheres, ending in external disturbances of any sort.

Now, if at times within historic periods the surface of our planet has been violently agitated over vast areas from *foci* of dynamical force, *i.e.* repulsive centres located at different points far below the surface, between the surface and centre, or in the centre itself, it is only a repetition of what we observe to take place in comets so distinctly as to allow no error of judgment as to identity of molecular and cosmical conditions in the two classes of bodies. Besides, there is no proposition respecting seismical physics,—that is, no delineation of earthquake phenomena taking place under the physical laws and deductions announced by Mr. Mallet, or any other author,—which does not represent and graphically reproduce identity of internal molecular force developing into cosmical quantities, ending in identity of results, and producing external cosmical violence equally observable in both comets and planets. Indeed, not a meteor explodes in the sky except by internal developments of this same eruptive and repulsive force; nor has a fragment of meteoric stone or iron ever been found on our planet that is not charged with hydrogen, the most expansible of known gases, or that is not *marked* by the action of the disruptive violence with which molecular repulsion forced it from its prior cosmical connexions. Furthermore, all physical agitations seen, with the telescope, in comets, are observed or felt to be duplicated in our own planet throughout its diversified constitutional structure. Even the miscalled zodiacal light, the terrestrial auroras, and the remarkable brightenings and darkenings of cloudless, moonless nights are, as observation is now teaching, but so many phenomena developed by internal disturbance of cosmical forces and exhibited exteriorly, as similar ones are excited in and developed around comets by direct solar influences, and modified by perihelic distances.

But so far as the earth's crust is observed, its endless

varieties of superficial motion witnessed in plutonic phenomena, embracing earthquakes in their entirety, can only be effected by the most tumultuous and complex expressions of interior violence, springing in some instances from the centre of the planet. The origin and nature of these motions, from those of a gyratory character, evincing a whirling impress of the dynamical agent, to those whose vertical or oblique impact cannot be misunderstood, are evident reproductions, duplications, or analogues of phenomena so visibly developing themselves in cometary bodies with intensities dependent, as before described, upon inverse perihelic distances. While the former can be appreciated by our several senses, determined by instantaneous convictions, and confirmed by experiments and mathematical computations, the latter are clearly detected and immediately determined by sight alone. But we recognise their identity as positively as if the comet had a solid crust whereupon we might stand, observe, deduce and measure degrees of repulsive force by eruptive results. We see elements, constituting the planet's interior, not only thrown out upon its surface from equator to pole, but also projected with enormous force for miles beyond its central molten nucleus. We know these dynamical phenomena to have been of perpetual recurrence, and to have exhibited a sort of eternal function in the physical existence of the planet, so to speak, because geological formations of every age have been disrupted by igneous dykes. Seams, fissures, cleavages, dislocations, tiltings of strata—ten thousand similar indications of subterranean motion—are observable everywhere, wherever human research may extend, all proving ceaseless action in the central elements of our globe. All these data determine limited or local disruptions of the planet's surface, and denote incessant reaction of its repulsive interior upon its mineral crust. But the more striking irregularities of hill and valley, mountain ridge and marine gulf, continental outline and oceanic depression, constituting the grand and universal features of the earth, not only confirm the foregoing conclusions, but also afford irrefragable testimony of still more extensive and revolutionary disturbances, which could only result from the profoundest evolutions of both repulsive and gravitative force; and such evolutions of energy, indeed, as could only spring from excitations of a central and cosmic character. So that if we behold comets disrupted, or their entire mass in tumultuous action, their superficial areas moved inwards toward the centre by gravitation, and their central parts forced outwards by repulsion, and portions even separated and projected from the main mass by the latter force, we discover the same phenomena repeated in our own globe, and only differing in character or result in consequence of the dissimilarity in physical constitution of the two classes of bodies. But we discover that the forces effecting these results are identical in principle and in action, and that both in planets and comets molecular attraction and molecular repulsion are expressed in their infinite amplitudes, as central forces, assuming cosmical functions; and we may safely conclude that the existence of this state of things in one series of cosmic bodies absolutely proves its existence in all others.

Indeed, in order to exhaust this point of our discussion, I will here refer to the entire array of aërolites, whatever their physical constitution, which have thus far been found upon the earth. In considering a subject so grave as this, while aware of certain *theorics* that these bodies are comets which have been, are to be, or were to have been, I cannot entertain them, and it is not my purpose to discuss them. But, on the contrary, I must adhere to facts, and I am compelled to say, after careful examinations of large numbers of these bodies, that every one of them without exception presents indubitable proofs of having been violently separated from some larger cosmical mass: and it is self-evident that molecular attraction, attraction of gravitation, cohesive attraction, or attractive force of any description, could not have caused such violent disruptions; but that the opposite of attraction, *repulsion alone*, must have been the agent engaged to produce such degrees of tension as end in explosions of celestial masses of solid mineral or metallic matter. After our inquiries, then, no name can more appropriately be given to such a central force than that of COSMICAL REPULSION.

9.

This extensive series of facts, whose analogy and identity of origin, character, and result are so manifest—all springing from the activity of molecular forces expanded into planetary magnitudes, exhibited in the eruptive and explosive phenomena which take place in bodies of different physical constitutions, and thus grouped in this new and fruitful field of discussion—must awaken philosophers to the grandest contemplations of cosmic nature, and to more profound and severe investigations of general laws.

The increasing density of a comet's nucleus as it approaches perihelion, its greatest density and the strongest excitement of its molecular and central forces at perihelion, and its diminishing density as it recedes from perihelion, must sooner or later command the attention which their importance and bearing deserve. Upon these facts hang "the law and the prophets" of cosmical physics. Their exact values involve no less than these pregnant inductions, viz. that the increasing contiguity of any revolving body to the sun is accompanied by an increasing contiguity and centralization of its constituent molecules; that its minimum distance from the sun is accompanied by the minimum distance of its individual molecules from each other, and from its own centre, involving of course the greatest density and tension of the body; that its elongation is accompanied by separation of its individual molecules, so that at aphelion its conditions of density and tension are the reverse, and always less than at perihelion. This is a simple statement of what all classes of observations exhibit and suggest. The general densities of planets, so far as determined by the ablest astronomers, appear to diminish in the order of their heliocentric arrangement. In this respect there is a correspondence with the internal changes observable in comets which even come from beyond the outmost boundaries of the solar system, and make their perihelic circuit within the orbit of Mercury. Molecular attraction throughout all those displays of its functions which constitute the cohesion, density, and sphericity, that is to say, the creation and continued existence of all these bodies, seems to hold special relations to the length of their radii vectores; in other words, to the general forces of gravitation and repulsion which maintain them at constantly fluctuating distances from the sun, whatever their elementary constitution, whether simple or compound, gaseous or opaque. All appear to bear similar relations to the solar centre that a balloon full of hydrogen, or of common air, bears to its proximity to the earth's centre. All know that if a bladder of common air be transported from the surface of the earth to the summit of a lofty mountain, the density of the gaseous contents will diminish, and the particles separate from each other to such an extent as to burst the vessel. The same process of molecular repulsion and expansion appears to be developed throughout the mass of planets and comets, in direct ratios to their distances from the sun.

Now, since our planet revolves in an elliptical orbit, with the sun in one of its foci, so as to present at aphelion and perihelion a difference of three millions of miles in the length of the radius vector, it becomes manifest that molecular movements must be unremitting throughout the globe, and that its density can never for a single instant be uniform, from its central atom to its outermost gaseous limits. While its tension—that is to say, its super-saturation with the forces of molecular attraction and molecular repulsion—is positive and persistent, and so vast as to originate and develop the terrestrial phenomena of light, heat, electricity, magnetism, organization, eruptive motion, &c., we know that this tension fluctuates in degree from solstice to solstice. The evidences of these changes are so numerous, whatever may be their theoretical explanations, that we need only mention tidal, magnetic, and meteorological observations to show that the intensities of phenomena vary with every point of the planet's orbit. Whether these changes are *all* directly or indirectly connected with and dependent upon molecular motion in the planet's constituent elements, superinduced by solar excitations, is a question which further observations will determine, and which will be more fully developed hereafter.

There is, however, an important class of phenomena next to be considered, the numerical elements of which, in connexion with final causes, are now for the first time presented in their broadest and most definable aspects. Seismic, that is to say, earthquake and volcanic, action—the dynamical expressions of which, whatever their degree, form, or extent, all directly depend upon aggregated molecular movements—is so universal and profound as to be truly cosmical, and must no longer be regarded as "strictly local" or accidental in its phenomenal aspects. Its analogy to the general internal and eruptive agitation, so manifest *in comets at perihelion*, clothes it with peculiar interest; and it thus becomes elevated into a subject of unexpected importance. It is beyond question that the density, internal tension, and eruptive action of comets increase with their contiguity to the sun. The sun appears in some way to superinduce increasing attractions and increasing repulsions within their individual molecules, ending in positive augmentations of density in these bodies and of intensities in central reactions proportionate to the operation of the general law of gravitation exerted upon their masses and modifying their mechanical motion in space. These individual molecular excitements, or fluctuating central and decentral expressions of energy (a *simulation* of inductions, absorptions, and expulsions of varying amounts of force), these molecular increments of excitation, or local translations of force, appear to resolve themselves into each other, and to coalesce and become a unit, upon the principles which have been enunciated and illustrated in preceding chapters. Thus centralizing by their own inherent power of coalescing, they assume cosmical proportions, functions, and values in direct ratios to the sums, or physical nature, of the molecules constituting these spheres. As the two forces, whose functions are always antagonistic, react upon each other, they manifest their character and energies by the motions of atoms; and it is through this explanation that we may comprehend the physical phenomena in comets the analogues of which are repeated in our own globe by the appalling molecular disturbance so demonstrable in every class of plutonic agitations and disruptions, and in every display of subterranean dynamics. The identity of these cosmical conditions and of their results is so manifest, and so abundantly sustained by facts and reasoning, as to require no further treatment or illustration. Nevertheless, if it can be conclusively shown that the tension of our planet, like that of comets, is *augmented at perihelion,* not only will solid evidence be presented of our correct analysis of the remarkable phenomena superinduced in distant worlds at *their* perihelia, but new proofs will also be added in support of the philosophy herein set forth, that all cosmical nature is influenced and absolutely governed by the action of another great force, beside gravitation—its equal and co-ordinate throughout the universe—that of COSMICAL REPULSION.

10.

Disruptions of any body, spherical or otherwise, large or small, in consequence of force excited, arising, or augmented within its own boundaries, manifestly prove tension. If the internal dynamic conditions be the constant, external dependent phenomena will be correspondingly so, all things being equal. Observation shows the earth to be, and to have been through geological time, constantly subject to eruptive action. Hence it is, and has been, in a condition of constant dynamical tension; and I have already shown this to be the result of molecular repulsion aggregated into a cosmical force, and have equally demonstrated its analogy and identity with that so manifestly excited in comets during the perihelic relations into which solar attraction draws them. The earth moves towards the sun in obedience to the same attraction. Not only, however, is this force exerted in such a manner in conjunction with an opposing one, as to define the earth's path and periodic motions within measurable and known limits, but it so fluctuates in degree with this opposing force as to develop an elliptical orbit with the sun in one of its foci, and to permit or create, among other phenomena, a difference of 3,000,000 miles in the length of the radii vectores at the aphelic and perihelic points of this orbit. Now, if facts and numbers show that eruptive phenomena are *uniformly* more frequent during the planet's greater proximity to the sun—that is to say, *in the perihelic arc of its orbit,* than in the aphelic arc—and that they maintain general inverse numerical ratios to the length of the radius vector, it will follow that the sun exerts the same control over the fluctuating molecular forces within our planet that it does over the tension and eruptive phenomena of comets.

It is only of late years that special attention has been devoted to seismic phenomena in order to bring them within the range of severe scientific discussion.

M. Alexis Perrey, Professor of Natural Philosophy at Dijon, conceiving that a molten nucleus must ebb and flow within the earth's crust, in like manner and upon the same theory that ocean tides rise and fall, devoted much time, labour, and expense to collecting data; and has published extensive catalogues of earthquakes and plutonic eruptions, including dates in detail, where he could find them recorded. Encouraged by remarkable numerical results, and sustained by the illustrious Arago, his labours have been persistent, of great value, and in the right direction. He assumed that earthquakes and volcanic eruptions must, like marine tides, be results of lunar attraction; and by discussing all his observations, he found that these phenomena are more frequent near the moon's perigee than apogee, and more frequent at the syzygies than at the quadratures.

These facts are positive sequences of numerical discussions; and it will be immediately perceived that, as far as the simple conclusions extend, they sustain the principle which I have announced, that internal molecular agitation and eruptive or repulsive phenomena are increased in proportion to the proximity of the heavenly bodies to each other. It is this point, *simply and alone*, which I desire to present in this connexion; and its weight is greater, and its importance the more suggestive, since it is derived from the labours of a distinguished authority in support of a theory which future discovery will most probably modify.

11.

A critical glance at the "lunar *theory*," when brought into comparison with the *facts* observable in the internal reactions of the celestial masses, may deprive the former of the value which physicists have commonly attached to it from Newton's time to our own. A plain statement of the "Theory of the Tides"—which is only a plain statement of the law of universal gravitation, as expressed by the Newtonian philosophy—is, that the moon's attraction upon the earth is two and a half times greater than the sun's; that the moon and sun, when in the meridian or otherwise, draw the water immediately under them away from the solid globe with these proportional degrees of force, and at the same time draw the solid globe away from the ocean on the opposite side, so as to make two simultaneous swellings of the sea, or four, as it may be; and that these solar and lunar attractions act, in the strictest manner, inversely as the squares of the distances from the sun and moon of each separate molecule upon and in the planet. On this hypothesis, the molten nucleus of the globe, not subject to atmospheric perturbations like the ocean, would swell gently and gradually against and around the interior of the crust, and would be drawn with gradually diminishing attractive forces in a similar manner from the remotest sides. It must be remembered that, in the prevailing theory, there is no other force at work than attraction; and that molecules as units, not the centres of planetary bodies, are acted upon with constantly diminishing strength according to the squares of their respective distances. The double elliptical *tidal waves of the nucleus*, produced by the dissident positions of the sun and moon, would, upon the Newtonian theory, ever be regular and tranquil in their movements, and would unite, during conjunctions, under the influence of less degrees of force than are at the same time drawing the crust away from the internal molten matter.

Thus it will be perceived that while Professor Perrey's seismical statistics augment and reach their maximum with the meridian positions of the sun and moon, and diminish and reach their minimum at apogees and quadratures, they disprove the very theory to prove which they have been collected, or at least they eliminate themselves from it, since earthquakes are frequent in longitudes which should, upon the theory of the tides, be wholly free from plutonic pressure or any dynamical action at the moment these phenomena occur. But when Professor Perrey's numerical discussions are regarded in the light of this new philosophy which I am endeavouring to unfold, they fall with wonderful precision into natural order, under laws which appear to be universal and which embrace all physical anomalies throughout the planet.

The energies of gravitation, viewed as a dynamic entity, act directly and effectively FROM the *centre* of every sphere UPON the *centre* of every other sphere, inversely as the squares of the distances of these centres, and not upon the nearest circumferential particle first and strongest, and so on less and less strongly upon each successive particle, *straight through* a sphere, as is assumed in the theory of the tides. But the actual mechanical effect of solar gravitation as seen displayed in the diaphanous spheres of comets, is first to excite attractive power in the centre, so that every molecule shall be drawn from the circumference, and so on successively, toward this focus, at the same time that the body as a constituent of the system is drawn towards the sun. The resultant of such action must naturally be *reaction;* and, so far as our own planet may be influenced by these conditions, all phenomenal and periodical movements of fluids, whether within or without the crust, must be subjects of reactionary, that is to say *repulsive,* and not of direct attractive, force. If matter and force be everywhere united upon the same principles, physical changes must also be everywhere the same, and the periodic synchronisms of correlated phenomena must self-evidently be traceable to the same causes. One of these general causes we know to be attraction ; and since Newton has also proved that "*action and re-action are always equal and contrary,*" it follows that the reactionary principle can be no other than a positive force of repulsion. When voluminous combinations of molecules depend upon the sun for sphericity and motion —that is to say, for their physical existence, as comets and planets suspended around it—or upon the earth, as seen in bodies of gas and vapour suspended around the latter —the action of attraction and repulsion upon and within these masses is observed to be similar where their physical constitutions permit measurable fluctuations of density. For instance, if a bag of any size were filled with common air at the base of Chimborazo, and then transported to the level of the sea, the volume of this mass of molecules would diminish, and its density would be increased. When transferred back again, it would expand so as to fill the vessel. When transported to the mountain's summit, it would expand still more, and the molecules would separate from each other with so much dynamical energy as to burst it. All these changes of volume and density are expressions of molecular attraction and repulsion, which proceed from circumference to centre, and *vice versâ.* To say that these changes of volume and density result from simple differences of atmospheric pressure is only an indefinite and incomplete explanation of phenomena whose causative conditions lie in more profound directions. That these changes spring from absolute spontaneous developments of attraction or repulsion existent and latent in all molecules, or from their inherent capacity to draw and imbibe instantly fresh and enormous increments of analogous force from cosmic sources, may be proved by many striking illustrations. Phenomena explicable in this way are not unfrequent where repulsion is so instantly developed in the atmosphere near the level of the sea, that the windows or sides of houses will instantly burst outwards, and entire fabrics be made wrecks. Active repulsion in such storms has suddenly supervened in surrounding space, in consequence of some profound telluric condition not hitherto recognised or understood; and the response of the repulsive force, existing as a lively, energetic principle in the atmospheric molecules within the edifice, being as instantaneously sensitive, inductive, accumulative, and motive as the attractive principle, is so quickly manifested, and accumulates its energies so instantly from the great telluric reservoir, that intervening obstacles fly in various directions, and even beams are burst open as with gunpowder, in order that the equilibrium of the repulsive and gravitative forces, as cosmic powers and quantities, may be re-established. Every change of the barometer is but an influx and efflux of these same ultimate forces, which, in grander developments, produce eruptions, dilatations, and disruptions of comets, and in our own planet earthquakes and volcanoes. That all this category of facts and observations are resolvable into cosmical, and not local phenomena, and that all, furthermore, depend upon solar causations whose secret influences act with energy inversely proportioned to the length of the radius vector, may be demonstrated, since the maximum oscillations of the pendulum and the maximum ranges and oscillations of the barometer—all disturbing elements being eliminated—occur at perihelion, and the contrary at aphelion. Observations also upon tides show the same periodic connexion of distant cosmical forces with the forces of every movable molecule on our planet; and when the endless forms and expressions of seismic phenomena are observed to arrange themselves into similar periodicity, it may be questioned whether gravitation does not so act upon and within our planet as to reduce all dynamical phenomena to reactionary conditions, and correlate them with repulsive agencies; and whether these phenomena may not be subject to *laws of repulsion* which are as capable of numerical determination as those of gravitation. That exact observation is as certain a means of reaching scientific truth as experimentation, there can be no doubt: and when extended observations upon cosmic phenomena lead to the discovery of general principles which experiments and the mathematics coincide in confirming, physicists would do equally wrong to reject either one or the other.

By the study of comets and of our own globe, we do not discover solar gravitation to act in such a manner as directly, and *at first,* to draw or attract their nearest molecules most strongly, and so on successively, away from their centres and remotest sides, thereby making constant ellipsoids whose major axes are persistently directed to the sun, or, as in the case of lunar gravitation and lunar tides, to the moon, when in the meridian. On the contrary, we observe gravitation, as a *solar* force, so to act upon them as to increase the gravitation of their individual molecules toward marked centres, creating rotundity, not ellipticity; and simultaneously exciting them into excessive tension and *active* states of repulsion—reaction and eruption being subsequent, and also in direct numerical ratios with their proximity to the sun. Solar attraction is observed to act upon the sphere as a whole or unit, drawing it by established values; while its individual molecules are excited into more intensely attractive states and concurrently fall toward its centre. And inasmuch as dynamical phenomena are developed as local and periodic manifestations, subsequent in time and fact to meridian positions of sun and moon, thus presenting themselves as indirect rather than direct results of solar or lunar attraction, and being consequently reactionary in character, they conclusively demonstrate the internal activity and universal display of cosmical repulsion in all revolving bodies whose orbits are elliptical, and whose controlling forces are in one of the foci of these orbits.

The physical facts here presented cannot be set aside as of no importance to philosophy. On the contrary, their importance is such that they bear upon every branch of physics, and I cannot doubt they must hereafter profoundly modify the directions of theoretical and solid research.

When the development of seismic phenomena in their greatest activity and intensity is found to result from the action of forces identical with those at work in comets to produce *their* greatest eruptive displays, and when this maximum obtains at perihelion in like manner, as observed in comets, we certainly have reason for inferring that gravitation, as a solar force, acts upon all cosmical masses in a similar manner, and excites their constituent molecules into similar states of attraction and repulsion, centralization and reaction, condensation and eruption.

Such being the causation and the *modus operandi* of the agencies within the earth's crust which produce earthquakes and volcanoes, or, as the illustrious Alexander von Humboldt calls them, "the reaction of the interior upon the exterior of the planet," it will be observed that these phenomena cannot be explained, as the indefatigable philosopher M. Alexis Perrey has endeavoured to explain them, upon the "Newtonian theory" of tidal waves in a molten nucleus. And it may be authoritatively questioned in view of the preceding critical glance at, and comparison of, theory and facts, whether the so-called "theory of the tides" does not require profound modifications in its application to oceanic movements, since the severest analysis of the immortal Laplace failed to bring tidal phenomena within mathematical requirements, and my own personal observations upon tidal waves at Tahiti, and upon the coast of Peru from 3° to 14° south latitude, include amazing anomalies which no applications of the calculus have thus far succeeded in bringing within the lunar hypothesis as at present taught.

12.

The facts collated and discussed by M. Perrey, however, and the numerical results at which he has arrived, are, beyond question, of the utmost value. That they may be presented with full candour and receive their greatest weight, I shall so enlarge upon his labours as to cite the concluding passages of a "Report" made to the French Academy of Sciences, June 12th, 1854, by a special commission upon two of M. Perrey's memoirs, which he had communicated to that body; the first entitled: "On the Connexion which may exist "between the Occurrence of Earthquakes and the Moon's "Age;" the second: "On the Occurrence of Earthquakes in "connexion with the Moon's passing over the Meridian."

"The apparent result from this" (the preceding analysis of M. Perrey's Memoirs) "is, that the difference between the "unequal attraction exercised by the moon at her greatest "and nearest distance has a sensible influence over the "occurrence of earthquakes. In the note on the 'Occurrence "of Earthquakes in connexion with the Passing of the Moon "over the Meridian,' which he presented to the Academy "January 2d, 1854, M. Alexis Perrey discusses the question "whether the division of the shocks of earthquake during a "lunar day is, like the tides, connected with the passage of "the moon over the superior and the inferior meridian. For "this method of investigation, he could only avail himself of "824 shocks felt at Arequipa, which are registered with day "and hour in the above-mentioned table of M. de Castelnau. "By means of proportional calculations, which must have "occupied considerable time, he has calculated to which "hour after the passage of the moon over the meridian each "of these shocks corresponds. He thus formed a first table

" (which he afterwards changed by dividing it into sixteen " equal portions grouped, side by side, to form eighths) con- " taining the 24 hours 50 minutes and a half, of which a " lunar day generally consists.

"By these two methods (notwithstanding some marked " anomalies which could not but exist in so limited a number " of facts as 824), the results obtained in both arrangements " manifest the existence, in the length of a lunar day, of two " periods of *maximum* for the occurrence of shocks, and two " of *minimum*. The two periods of maximum occur at the " hours of the passing of the moon over the superior and " inferior meridians; and the periods of minimum fall about " the middle of the intervals.

"M. Alexis Perrey has thus succeeded, by the simple " analysis of catalogues which he had previously drawn up, " in proving by three different and independent methods the " influence which the moon possesses in the production of " earthquakes.

"1st. That earthquakes occur more frequently at the " syzygies.

"2d. That their frequency increases at the perigee, and " diminishes at the apogee of the moon.

"3d. That the shocks of earthquakes are more frequent " when the moon is near the meridian, than when she is " 90° away from it.

" But the numerical tables from which these three proposi- " tions are derived, present some anomalies; and the author " has omitted nothing in endeavouring to account for them, " and to prove the law which is revealed at their first " inspection. . . .

"By means of the formula thus obtained" (which the Commission have here interpolated and which I omit), " the author was enabled to draw up numerical tables, corre- " sponding to those deduced from observation alone, and " in which the law of the phenomena appears disconnected " from the principal anomalies which tended to obscure it in " the first tables. The numbers contained in these new tables

" are carefully arranged, and form regular curved lines, in " which the law is clearly manifest. These curves have a " marked resemblance to each other, although they are not " entirely alike—which could not be, for they are only " approximative—and each bears the stamp of the group of " figures which it represents. The resemblance of these " curves is essentially increased by the fact that each presents " two principal maxima corresponding to the syzygies, and " two principal minima corresponding to the quadratures. " We are thus brought back to the conclusion so evident by " M. A. Perrey's toil—that, for half a century, earthquakes " have been more frequent at the syzygies than at the " quadratures.

"The Academy fully conceives of the importance of this " conclusion, and appreciates the labour the author has taken " to collect nearly 7,000 observations on the first half of this " century. This number, however, is very small for the solu- " tion of a question of this nature; and it is very desirable " to have it increased, either by collecting all future observa- " tions from year to year, or by going back to past centuries, " as the author has commenced doing."

While M. Perrey was busy in France collecting facts, and elaborating his theory and discussions with such commendable assiduity, encouraged by Arago and the French Academy, another indefatigable physicist was occupied in Great Britain, under the special authority and auspices of the British Association for the Advancement of Science, in studying seismic phenomena from a wholly different point of view, and in classifying or tabulating data without reference to lunar or other cosmical causation. The labours of Mr. Robert Mallet have been intended to embrace this subject in its entire ranges both in fact and literature, and to treat it, moreover, critically and exhaustively in its telluric and dynamical aspects; and they are not the less valuable for having been prosecuted independently of speculations upon final causes. As his elaborate "Reports" to the British Association embrace also candid expositions of Professor Perrey's tables and conclusions, as well as his own numerical summaries, I have availed myself of the statistical data presented by both, in order more strongly to support the conclusions drawn from the physical aspects of comets, and the deductions to which molecular physics developed into the widest ranges has inevitably brought us in our study of such periodic phenomena.

In order to present these numbers in the most exact and definite manner, I shall here introduce M. Perrey's tables embracing his monthly summaries of earthquakes, transferring them from Mr. Mallet's Report to the British Association for the Advancement of Science, made in September 1858, and I shall append to these tables Mr. Mallet's own mensual classification of 6,102 of these phenomena, reported in the same communication. Thus exhibited, the invaluable labours of these investigators will draw the attention of astronomers and physicists to the main point of our discussion. And here I may be permitted to say that an important law of periodicity was deduced by me some years since from an analysis of earthquake phenomena, as they had been observed and collated by myself, independently of their labours and without the knowledge of their pursuits or catalogues. I then announced what accumulating facts steadily confirm, viz. that the intensity of internal telluric agitations—similar to those which exist in comets in the perihelic arc of their path, and ending in disruptions of surface—is, as a general law, dependent more or less upon the planet's proximity to the sun; and that these phenomena hold certain numerical relations to the length of the radius vector.

The tables of M. Perrey are so numerous and lengthy, that I have condensed some and omitted others of no special bearing upon our subject, his immense toil having been based upon a special hypothesis, which, like all simple speculations, however suggestive or temporarily useful, must ultimately vanish like vapour before the sunshine of truth. In reference to the accuracy of these tables, it is proper for me to remark that I have noted no errors which affect the general conclusions which may be drawn from their totals.

"TABLE I.

" *Earthquakes of Scandinavian Peninsula and Iceland.*

Century. A.D.	January.	February.	March.	April.	May.	June.	July.	August.	September.	October.	November.	December.
12th to 18th	16	9	10	6	9	4	9	5	8	7	8	11
19th . .	17	11	11	7	7	6	8	8	10	10	11	6
Totals .	33	20	21	13	16	10	17	13	18	17	19	17
	Winter.			Spring.			Summer.			Autumn.		
	74			39			48			53"		

In connexion with the preceding table, Mr. Mallet makes the following commentary:—"On examining this table, " M. Perrey remarks the same preponderance of earthquakes " in the winter half of the year, that is evident from many of " his other calculations for various regions. Here, for the " six months of winter, there are (including some other facts) " 127 shocks, and but 87 for the summer half of the year.

"The total number of earthquakes given with dates is 252. " Representing by 12 the mean annual number, he tabulates " the proportional number for each month, thus :—

"TABLE II.

" *Scandinavia. Relative Frequency through the Year.*

January.	February.	March.	April.	May.	June.	July.	August.	September.	October.	November.	December.	Proportional Number.
1·85	1·12	1·18	0·75	0·90	0·56	0·95	0·75	1·01	0·95	1·06	0·95	=12
Winter.			Spring.			Summer.			Autumn.			
1·38			0·73			0·90			0·99			

"At the two months of each solstice and equinox—

"March and April . . . . 0·94
"June and July (*Aphelion*) . . . . 0·74
"September and October . . . . . 0·95
"December and January (*Perihelion*). 1·36."

The interpolation of the words *Aphelion* and *Perihelion* is here made by myself, in order to call special attention to these periods, because nothing in connexion with these elements was in the minds or theories of either M. Perrey or Mr. Mallet, during the discussion of these tables.

"TABLE IV.

"*Earthquakes of the British Islands and Northern Isles.*

Century. A.D.	January.	February.	March.	April.	May.	June.	July.	August.	September.	October.	November.	December.
11th to 18th	12	7	9	9	8	4	4	8	12	9	11	15
19th . .	9	9	10	7	8	6	5	11	12	8	11	12
Totals .	21	16	19	16	16	10	9	19	24	17	22	27
	Winter.			Spring.			Summer.			Autumn.		
	56			42			52			66"		

Mr. Mallet here remarks:—"The number occurring in "spring and summer together is but three-fourths that of "autumn and winter united, the relative number for the four "seasons being—

"Winter . . . . 1·03
"Spring . . . . 0·76
"Summer . . . . 0·96
"Autumn . . . . 1·24

" and the two months of the critical epochs (as M. Perrey
" characterises the solstices and equinoxes)—

" Winter solstice (*Perihelion*) . . . 1·28
" Spring equinox . . . . 0·96
" Summer solstice (*Aphelion*) . . . 0·53
" Autumnal equinox . . . . 1·13

"TABLE VI.

" *Earthquakes of the Spanish Peninsula.*

Century. A. D.	January.	February.	March.	April.	May.	June.	July.	August.	September.	October.	November.	December.
11th to 18th	15	9	10	11	5	8	8	11	3	12	15	9
19th . .	10	5	6	7	4	6	10	5	9	11	7	5
Totals .	25	14	16	18	9	14	18	16	12	23	22	14
	Winter.			Spring.			Summer.			Autumn.		
	55			41			46			59		

" Taking the mean monthly frequency = 1, the relative
" monthly frequency, and that according to season, are as
" follows :—

January.	February.	March.	April.	May.	June.	July.	August.	September.	October.	November.	December.
1·49	0·84	0·95	1·07	0·54	0·84	1·07	0·95	0·71	1·37	1·31	0·84
Winter.			Spring.			Summer.			Autumn.		
1·09			0·82			0·91			1·17		

" Or in autumn and winter together, 114 earthquakes against " 87 in the spring and summer.

" Table VII.

" *Earthquakes of France, Belgium, and Holland.*

Century, A.D.	January.	February.	March.	April.	May.	June.	July.	August.	September.	October.	November.	December.
4th to 18th.	56	47	32	42	29	28	32	23	35	31	39	54
19th. .	27	17	21	13	13	8	15	17	15	17	21	25
Totals .	83	64	53	55	42	36	47	40	50	48	60	79
	Winter.			Spring.			Summer.			Autumn.		
	200			133			137			187		

" and for two months at each critical period of the year—

" December and January . . Winter solstice . . 161 (*Perihelion*)
" June and July . . . . Summer solstice . 83 (*Aphelion*)
" March and April . . . . Spring equinox . . 108
" September and October . . Autumnal equinox . 98."

The earth is (as a general statement) in perihelion on the 1st of January, and in aphelion on the 1st of July. The remarkable differences in number of these phenomena at the perihelic and aphelic extremities of the planet's orbit, must arrest the attention of even superficial observers, and they palpably present a problem which demands solution upon broader principles than those embraced in the lunar hypothesis.

"TABLE VIII.

"*Earthquakes of the Basin of the Rhone.*

Century. A.D.	January.	February.	March.	April.	May.	June.	July.	August.	September.	October.	November.	December.
16th to 18th	14	8	8	7	8	9	7	5	13	9	6	10
19th . .	12	12	8	3	3	2	2	4	6	6	8	14
Totals .	26	20	16	10	11	11	9	9	19	15	14	24
	Winter.			Spring.			Summer.			Autumn.		
	62			32			37			53		

"presenting considerable similarity to the results for France "as a whole.

"The following are the proportional numbers for the "months—

January.	February.	March.	April.	May.	June.	July.	August.	September.	October.	November.	December.
1·69	1·31	1·06	0·66	0·71	0·71	0·59	0·59	1·24	0·98	0·92	1·57

"or for—

"Winter . . . . 1·35
"Spring . . . . 0·69
"Summer . . . . 0·81
"Autumn . . . . 1·16

" and for the two months each of—

" Winter solstice (*Perihelion*) . . . 1·53
" Spring equinox . . . . 0·81
" Summer solstice (*Aphelion*) . . . 0·61
" Autumnal equinox . . . . 1·05

" Table IX.

" *Earthquakes of the Basin of the Rhine and Switzerland.*

Century. A.D.	January.	February.	March.	April.	May.	June.	July.	August.	September.	October.	November.	December.
9th to 18th	47	37	31	25	25	24	23	19	26	19	34	46
19th .	15	17	13	12	11	6	12	11	10	17	24	25
Totals .	62	54	44	37	36	30	35	30	36	36	58	71
	Winter.			Spring.			Summer.			Autumn.		
	160			103			101			165		

" The autumn and winter together here present a number " having nearly the same ratio to that of spring and summer " together, as 3 : 2.

" And at the critical periods of the year, of two months " each, we have—

" Winter solstice (*Perihelion*). . . . 133
" Spring equinox 81
" Summer solstice (*Aphelion*) 65
" Autumnal equinox 72

"TABLE X.

"*Earthquakes of the Basin of the Danube.*

Century. A.D.	January.	February.	March.	April.	May.	June.	July.	August.	September.	October.	November.	December.
5th to 18th	17	16	5	8	13	11	10	14	5	7	8	14
19th .	14	15	9	8	12	8	16	11	11	16	10	12
Totals .	31	31	14	16	25	19	26	25	16	23	18	26
	Winter.			Spring.			Summer.			Autumn.		
	76			60			67			67		

"The relative numbers are for—

"Winter solstice (*Perihelion*) . . . 1·33
"Vernal equinox . . . . 0·70
"Summer solstice (*Aphelion*) . . . 1·05
"Autumnal equinox . . . . 0·91

"TABLE XI.

"*Earthquakes of the Italian Peninsula, with Sicily, Sardinia, and Malta.*

Century. A.D.	January.	February.	March.	April.	May.	June.	July.	August.	September.	October.	November.	December.
4th to 18th	64	60	60	49	48	62	30	41	40	51	42	48
19th . .	37	39	38	35	32	24	33	36	23	41	22	29
Totals.	101	99	98	84	80	86	63	77	63	92	64	77
	Winter.			Spring.			Summer.			Autumn.		
	298			250			203			233		

"M. Perrey having obtained access to the work of Mura-"tori and other documents, produced a supplement to this "Memoir, the result of which he has given in

"SUPPLEMENTAL TABLE XII.

"*Italian Peninsula, Sicily, Sardinia, and Malta.*

Century. A.D.	January.	February.	March.	April.	May.	June.	July.	August.	September.	October.	November.	December.
8th to 18th	18	8	13	15	12	11	10	15	15	12	9	12
19th .	7	5	10	8	8	10	8	10	4	4	4	10
Totals.	25	13	23	23	20	21	18	25	19	16	13	22
	Winter.			Spring.			Summer.			Autumn.		
	61			64			62			51		

"In the first of these, the winter and spring earthquakes "together are to the summer and autumn together as 6 : 5.

"In the supplemental table alone, however, the winter "season has lost its preponderance, and autumn shows the "smallest number.

"The number in winter and autumn together, however, "still slightly exceeds that for spring and summer, in the "ratio of 9 : 8.

"While this shows the usual doubtfulness of generaliza-"tions from partial data, the result rather tends to awaken "increased attention to the very prevalent excess of seismic "action in the winter half-year, shown by so many catalogues, "and here sustained, though by a supplement, that, taken "alone, somewhat departs from the principle.

"TABLE XIII.

"*Earthquakes of Algeria and Northern Africa.*

January.	February.	March.	April.	May.	June.	July.	August.	September.	October.	November.	December.
5	2	6	7	3	2	2	5	1	4	8	1
Winter.			Spring.			Summer.			Autumn.		
13			12			8			13		

"TABLE XIV.

"*Earthquakes of the Turco-Hellenic Territory, Syria, the Ægean Islands, and Levant.*

Century. A.D.	January.	February.	March.	April.	May.	June.	July.	August.	September.	October.	November.	December.
4th to 18th	18	15	15	20	21	20	21	18	26	17	21	19
19th .	22	20	16	10	16	15	14	22	14	17	12	14
Totals.	40	35	31	30	37	35	35	40	40	34	33	33
	Winter.			Spring.			Summer.			Autumn.		
	106			102			115			100		

For the four critical periods of the year he finds—

" Winter solstice (*Perihelion*) 73
" Spring equinox 61
" Summer solstice (*Aphelion*) 70
" Autumnal equinox 74

"TABLE XV.

" *Earthquakes of the United States and Canada.*

Century. A.D.	January.	February.	March.	April.	May.	June.	July.	August.	September.	October.	November.	December.
17th and 18th	10	10	9	3	3	4	6	8	5	8	12	12
19th .	4	4	3	3	3	...	4	6	3	2	7	5
Totals.	14	14	12	6	6	4	10	14	8	10	19	17
	Winter.			Spring.			Summer.			Autumn.		
	40			16			32			46		

" Here the number of earthquakes in autumn and winter " are to those of summer and spring as 88 to 49, or nearly " as 2 to 1; and for Perrey's critical periods—

" Winter solstice (*Perihelion*) 31
" Spring equinox 18
" Summer solstice (*Aphelion*) 14
" Autumnal equinox 18

" Perrey wholly disputes the verity of Humboldt's conclu- " sion (*Cosmos*, t. i. p. 519, trad. p. M. FAYE), that earthquakes " are most frequent at the equinoxes, and declares that the " results of all his memoirs prove the contrary.

"TABLE XVI.

"*Earthquakes of Mexico and Central America.*

Century. A.D.	January.	February.	March.	April.	May.	June.	July.	August.	September.	October.	November.	December.
16th to 19th	3	5	8	5	6	5	4	2	4	4	3	3
	Winter.			Spring.			Summer.			Autumn.		
	16			16			10			10		

"TABLE XVII.

"*Earthquakes of the Antilles.*

Century. A.D.	January.	February.	March.	April.	May.	June.	July.	August.	September.	October.	November.	December.
16th to 18th	6	8	4	5	4	6	11	7	10	10	5	3
19th .	9	8	19	12	12	10	9	16	12	10	13	12
Totals.	15	16	23	17	16	16	20	23	22	20	18	15
	Winter.			Spring.			Summer.			Autumn.		
	54			49			65			53		

"Contrary to the result usual for Europe, the number of "shocks in summer here seems to preponderate; and in the "critical periods we have—

"Winter solstice	30
"Spring equinox	40
"Summer solstice	36
"Autumnal equinox	42

"Or, for autumn and winter together, 108; spring and summer, "114;—a result equally contrary to what has been found so "uniformly for Europe, and to the prevalent belief of the "inhabitants of the Islands themselves, who deem the equi-"noxes the dangerous times.

"Representing by unity the mean degree of frequency, and "by twelve the whole number of earthquakes given with "date of month, we find for each month the following "proportional number:—

January.	February.	March.	April.	May.	June.	July.	August.	September.	October.	November.	December.
0·81	0·87	1·25	0·92	0·87	0·87	1·09	1·25	1·19	1·09	0·98	0·81
Winter.			Spring.			Summer.			Autumn.		
0·98			0·89			1·18			0·96		

"The last of Perrey's monographic memoirs is that on Chili "and La Plata. . . .

"The following table contains his numerical results, for a "region, however, in which shocks of greater or less intensity "are almost of daily occurrence:—

"TABLE XIX.

"*Earthquakes of Chili and the Basin of La Plata.*

Century. A.D.	January.	February.	March.	April.	May.	June.	July.	August.	September.	October.	November.	December.
16th to 18th	1	2	2	...	2	1	...	1	...	1	...	1
19th .	14	10	14	8	19	11	16	15	16	9	27	8
Totals.	15	12	16	8	21	12	16	16	16	10	27	9
	Winter.			Spring.			Summer.			Autumn.		
	43			41			48			46		

". . . M. Lambert, mining engineer of Chili, in a memoir "on the causes of earthquakes in Chili and Peru (*Ann. de* "*Chim. et de Phys.* t. xii. pp. 392—405), published in 1829, "mentions that the Chilians vulgarly divide their year into "three seasons or 'temporadas,' and that one of these, the "*first,* composed of January, February, March, and April, is "called 'temporada de los temblores,' or earthquake season. "On comparing the facts of his catalogue with the popular "belief, however, Perrey finds the facts palpably contradict it."

It will be apparent to all familiar with earthquake phenomena in Chili that the above table is very limited in data, and that no reliable conclusion upon the general laws of periodicity can be drawn from it.

"Finally," continues Mr. Mallet, "we come to the last "two of Perrey's memoirs which have been referred to— "those in which he has brought under one view many of "the facts of his monographs, and graphically discussed the "results in tables for all Europe, with the adjacent parts of "Africa and of Asia, and for the north of Europe with the

" north of Asia, viewed as one great boreal band. The results
" of the former are given in the following table :—

" TABLE XX.

" *Résumé of the Earthquakes of Europe, and of the adjacent parts of Asia and of Africa, from* A.D. 306 *to* 1843.

Century. A.D.	Earthquakes with date of day or month.												Date of season only.			
	January.	February.	March.	April.	May.	June.	July.	August.	September.	October.	November.	December.	Winter and Autumn.	Spring and Summer.	With date of year only.	Total.
4th	...	...	...	...	...	...	1	...	...	1	...	2	3	1	12	21*
5th	1	...	...	3	...	2	1	...	2	...	2	...	3	...	11	25
6th	...	1	...	2	2	1	2	1	2	3	2	3	1	...	11	31
7th	...	...	...	1	...	1	...	2	...	...	...	...	...	...	6	10
8th	2	2	1	1	1	...	...	...	...	...	...	...	1	...	3	11
9th	4	2	...	1	1	1	...	1	2	2	...	6	5	1	10	36
10th	1	...	1	1	...	...	...	...	1	2	1	1	1	...	8	17
11th	1	4	5	1	2	1	2	2	4	3	3	3	1	...	19	51
12th	8	2	2	3	3	2	...	3	3	...	1	4	3	...	34	68
13th	3	2	3	1	5	...	2	...	1	...	2	5	4	...	27	55
14th	1	1	3	...	3	4	3	2	4	3	4	4	2	2	22	58
15th	...	1	1	1	2	2	2	2	1	2	2	7	...	1	17	41
16th	10	5	6	8	10	4	2	3	9	3	6	10	3	...	31	110
17th	21	16	15	13	6	9	10	3	14	3	10	17	1	1	41	180
18th	77	53	45	52	36	49	49	49	32	62	55	62	14	4	21	660
19th	99	100	90	59	55	55	74	78	72	92	60	78	6	1	6	925
Totals	228	189	172	147	126	131	148	147	147	176	148	202	48	11	279	2,299*
	Winter.			Spring.			Summer.			Autumn.			* An accidental error of 1 in the addition of the first line of this Table does not affect the general result set forth in it.			
	589			404			442			526						

" Autumn and winter still preponderate thus for entire " Europe. As regards the 'critical periods' of the year, the " results are—

	For 19th century.	For the whole period.
" Winter solstice (*Perihelion*)	177	253
" Spring equinox	151	170
" Summer solstice (*Aphelion*)	129	150
" Autumnal equinox	164	159

" And for the half year, and nineteenth century only—

" Autumn and Winter	527
" Spring and Summer	394

" And for the whole period of nearly fifteen and a half
" centuries—

" Autumn and Winter 1,165
" Spring and Summer 857

" or about as 1 : 0·75.

" TABLE XXI.

" *Earthquakes of the Northern Zone of Europe.*

Century. A.D.	January.	February.	March.	April.	May.	June.	July.	August.	September.	October.	November.	December.
8th to 18th	15	13	5	6	7	3	3	8	6	4	6	7
19th .	12	5	4	5	6	3	2	4	2	9	7	6
Totals.	27	18	9	11	13	6	5	12	8	13	13	13
	Winter.			Spring.			Summer.			Autumn.		
	54			30			25			39		

" TABLE XXII.

" *Earthquakes of the Northern Zone of Asia.*

Century. A.D.	January.	February.	March.	April.	May.	June.	July.	August.	September.	October.	November.	December.
18th .	3	6	2	1	1	...	1	2	2	2	1	3
19th .	4	6	6	4	4	3	5	7	6	3	4	5
Totals.	7	12	8	5	5	3	6	9	8	5	5	8
	Winter.			Spring.			Summer.			Autumn.		
	27			13			23			18		

"TABLE XXIII.

"Earthquakes of the Northern Zone of Europe and of Asia together.

Century. A.D.	January.	February.	March.	April.	May.	June.	July.	August.	September.	October.	November.	December.
8th to 19th	34	30	17	16	19	9	11	21	16	18	18	21
	Winter.			Spring.			Summer.			Autumn.		
	81			44			48			57		

"TABLE XXIV.

"General Result as to Mensual Relative Frequency of Earthquakes.

Regions.	January.	February.	March.	April.	May.	June.	July.	August.	September.	October.	November.	December.
Europe (the whole). .	1·35	1·11	1·07	0·95	0·85	0·81	0·87	0·95	0·89	1·02	0·93	1·21
France and Belgium .	1·52	1·17	0·97	1·01	0·77	0·66	0·86	0·73	0·91	0·88	1·09	1·43
Italy and Savoy . . .	1·16	1·13	1·27	1·05	0·96	0·96	0.94	0·94	0·76	1·13	0·76	0·94
Basin of the Rhone .	1·69	1·31	1·06	0·66	0·71	0·71	0·59	0·59	1·24	0·98	0·92	1·57
Basin of the Danube .	1·38	1·38	0·62	0·71	1·11	0·84	1·16	1·11	0·71	1·02	0·80	1·16
Scandinavia	1·85	1·12	1·18	0·75	0·90	0·56	0·95	0·73	1·01	0·95	1·06	0·95
Europe, Northern Zone	2·19	1·46	0·73	0·89	1·05	0·49	0·43	0·98	0·66	1·05	1·05	1·05
Asia, Northern Zone .	1·04	1·78	1·19	0·74	0·74	0·44	0·89	1·33	1·19	0·74	0·74	1·19
Both Zones, united .	1·78	1·57	0·89	0·84	0·94	0·47	0·58	1·10	0·84	0·94	0·94	1·10

"TABLE XXV.

"Result as to Relative Frequency in Season.

Regions.	Winter.	Spring.	Summer.	Autumn.
Europe (the whole) . .	1·18	0·87	0·90	1·05
France and Belgium . .	1·22	0·81	0·83	1·13
Italy and Savoy . . .	1·19	0·99	0·88	0·94
Basin of the Rhone . .	1·35	0·69	0·81	1·16
Basin of the Danube . .	1·13	0·89	0·99	0·99
Scandinavia	1·38	0·73	0·90	0·99
Europe, Northern Zone .	1·49	0·81	0·69	1·05
Asia, Northern Zone . .	1·33	0·67	1·13	0·89
Both Zones, united . .	1·41	0·75	0·84	0·99

"TABLE XXVI.

"*Results as to Relative Frequency at the Equinoxes and Solstices.*

Region.	Winter Solstice.	Spring Equinox.	Summer Solstice.	Autumnal Equinox.
Europe (the whole) . .	1·25	0·99	0·82	0·93
France and Belgium . .	1·43	0·96	0·73	0·87
Italy and Savoy . . .	1·02	1·13	0·93	0·92
Basin of the Rhone . .	1·53	0·81	0·61	1·05
Basin of the Danube. .	1·33	0·70	1·05	0·91
Scandinavia.	1·36	0·94	0·74	0·95
Europe, Northern Zone .	1·74	0·87	0·48	0·91
Asia, Northern Zone . .	1·20	1·04	0·72	1·04
Both Zones, united . .	1·48	0·96	0·58	0·98
	Perihelion		Aphelion	

"There remains to be noticed of M. Perrey's labours his " discussions of the periodicity of the earthquakes of his " annual catalogues for 1844, 1845, 1846, and 1847, with " reference to the phases of the moon's motions. . . .

"The result he arrives at, as respects these four years, is " that the number of earthquakes occurring at the Perigees " (when the tides are highest and lowest) are to those occur- " ring at the Apogees as 47 : 39—a conclusion which, inde- " pendently of the assumptions by which it is arrived at, " must be as yet accepted with caution upon so narrow a " base of induction, although possessing more than enough " probability, from physical considerations, to induce further " inquiry."

To the preceding tables Mr. Mallet has added the discussion of Mr. David Milne's catalogues of Scottish and English earthquakes, the numerical results of which are compared with M. Perrey's, and comprised as follows :—

"TABLE XXVII.

	Scotland.	England.	
"January	14	11	
"February	14	13	74 Winter.
"March	12	10	
"April	9	10	
"May	8	4	44 Spring.
"June	4	9	
"July	5	5	
"August	12	9	58 Summer.
"September . . .	12	15	
"October	14	11	
"November . . .	20	12	79 Autumn.
"December	15	7	
	139	116."	

It will be observed that during the four months in which our planet was passing through the perihelic arc of its orbit there occurred 106 earthquakes, while during the four months in which it was most remote from the sun there occurred only 56.

Mr. Mallet also introduces in juxtaposition with M. Perrey's summary of the earthquakes of the Antilles, in which a preponderance appears at the summer rather than at the winter solstice, the statistical results of M. Poey's labours upon the earthquakes of Cuba, in which it will be observed the contrary appears, sustaining the general law of greatest plutonic intensity with smallest heliocentric distances.

These results are embraced in the following table, wherefrom it will be perceived that the numbers recorded for the four aphelic and four perihelic months stand relatively as 14 to 21, thus presenting one-third more for the former than the latter period, a proportion coinciding remarkably with the great majority of our tabular summaries.

"TABLE XXVIII.

"*Earthquakes at Cuba.*

Century. A.D.	January.	February.	March.	April.	May.	June.	July.	August.	September.	October.	November.	December.
16th to 18th	...	4	...	...	...	...	...	...	...	...	...	...
19th .	4	3	2	3	3	4	5	2	6	5	6	4
Totals.	4	7	2	3	3	4	5	2	6	5	6	4
	Winter.			Spring.			Summer.			Autumn.		
	13			10			13			15"		

Notwithstanding the inevitable direction of these discussions toward the discovery of *solar causation*, it is proper to state here that neither Professor Perrey nor Mr. Mallet, the two most accomplished investigators of seismic phenomena who have yet lived, recognised in any of their labours and statistics the law which I assert to exist at the foundation of every department of cosmical physics ; and I shall so far cite Mr. Mallet's Report, made in 1858, to the British Association for the Advancement of Science, as to place the statement and the development of the subject beyond all question to future students.

"If it ultimately prove a fact," he says, "that there be a "real relation in epoch between earthquakes and the ocean "tides, or the moon's and sun's position in respect to the "earth, the phenomena will probably be found in relation "only through the intervention of changes in terrestrial "temperature, or in the great circulations upon or within "our planet, of its electrical or magnetic or thermic currents,

" or the conversion of these into each other reciprocally, and " not to the direct action of the variable attractive forces of " our primary and our satellite. To some such conversion of " force into heat, developed at local foci, it would appear " much more probable that all volcanic phenomena are due, " than to a universal ocean of incandescent and molten lava, " beneath our feet, with a thin crust of solid matter covering " it, the present or historical existence of which is not only " not proven, but for which no argument of weighty proba- " bility has been, as I conceive, advanced.

"In the present state of our knowledge of the obscure " relations between the internal mass and actions of our " planet with the cosmical forces that act upon it both within " our own atmosphere and from the abysses of space beyond, " and our comparative ignorance even of the terrestrial phe- " nomena themselves, no speculation, however hazardous or " hardy, that is based upon a natural hypothesis, need be " regretted; such views in the beginning of every separate " road of inductive science are eminently suggestive, and, " although in themselves false, may point towards truth. It " is only in this aspect that a memoir by Dr. C. F. Winslow, " M.D., 'On the Causes of Tides, Earthquakes, Rising of " Continents, and Variations of Magnetic Force,' requires " notice. The communication appears to have been made " to the Academy of Sciences of San Francisco, California, " by the author, in 1854 or 1855. I have met with it only " through a printed copy, for which I believe I am indebted " to the author."

Thus says Mr. Mallet in the course of his able and exhaustive Reports upon the entire literature and physics of seismic phenomena, wholly rejecting M. Perrey's opinions, admitting my Memoir of 1854 to be " eminently suggestive," and presenting no definite view of his own to explain facts which appear from closest annual observations to depend as certainly upon solar causation as do the annual variations of intensity of magnetic force.

But as my object is to present here facts derivable from

seismic studies, in order specially to prove the existence and action of a *force,* the dynamical functions of which are developed and intensified both in comets and in our own globe, in virtue of the reign of a cosmic law of *action and reaction proportionate to distances inverse ;* that is to say, of a force which develops intensities of internal mechanical effects that increase and decrease directly as these bodies approach and recede from the sun. Such being my object, I will confine myself to numerical data, and add such remarks only as are necessary to bind these facts with the principles which have been heretofore, or which may be subsequently, unfolded.

Mr. Mallet has prosecuted these studies in various ways, and has arrived at his numerical and final results by laborious discussions of catalogues, tables, and curves for both the northern and southern hemispheres, embracing the entire planet, as far as data could be collected, or observation extended from the earliest records to the year 1851.

"The total number of earthquakes, classed by months, is " as follows :—

	Northern.	Southern.	Seasons, North.	Seasons, South.
January	627	19		
February . . .	539	14		
March	503	9	1669	42
April	489	17		
May	438	20		
June	428	19	1355	56
July	415	18		
August	488	12		
September . . .	463	17	1366	47
October	516	25		
November . . .	473	32		
December . . .	500	21	1489	78
Totals	5879	223	5879	223

" Total of catalogue for both hemispheres capable of mensual classification	6102
" Total of unclassed, except as to annual date	670
" Total number catalogued	6772

" of which there are recorded by season only—

" Spring	6
" Summer	7
" Autumn	7
" Winter	5
Total	25

" January, February, and March have been taken for the " spring of the Northern Hemisphere, and for the Southern, " July, August, and September."

Thus it will be perceived that Mr. Mallet, in the classification of seismic phenomena presented in his Reports to the British Association, strictly confined his observations, and the direction of his opinions, to "*seasons*" *of local or periodic heat and cold;* overlooking profound cosmical causes, hesitating to interrogate the mobile conditions of a plastic globe, constituted of manifold chemical elements, revolving eccentrically around another mass of chemical elements, 1,200,000 as large and 300,000 times as heavy ; and never for a moment recognising the grand fundamental fact that molecular attraction and molecular repulsion must inevitably develop themselves into cosmical amplitudes, and exert positive mechanical and reactionary effects upon the planet's crust, with intensities always varying inversely, *as a general law,* as the length of the radius vector subtending their respective centres undergoes variation.

The final result of all Mr. Mallet's discussions which bear upon the point in question, is included in the following paragraphs :—

"In Fig. 5, Plate X. a curve has been obtained for the " whole period of the catalogue, and for both hemispheres, " representing graphically all recorded earthquakes occurring " near or at the equinoxes and solstices (the *critical epochs* of

" Perrey and others), within a limit of twenty days, *i.e.* ten " days before and ten days after each equinox and solstice. " The base of induction is moderately large, the catalogue " containing the following numbers :—

" Vernal equinox (March 10—30) . .	310
" Summer solstice (June 11—July 1) .	254 (*Aphelic*).
" Autumnal equinox (Sept. 13—Oct. 3) .	249
" Winter solstice (Dec. 11—31) . . .	318 (*Perihelic*).

"This we may call the equinoctial and solstitial curve " of comparative seismic energy. It indicates a distinct " maximum about the winter solstice, and an equally distinct " minimum rather before the autumnal equinox. Taking the " average of the whole year for any lengthened period, it " may admit of much doubt whether there is a real seismic " paroxysm at the equinoxes and solstices, although a clear " preponderance is shown by our catalogues at two out of the " four annual epochs at which all are recorded ; yet from the " accordance of Perrey's results with those given by this " much larger base of induction, we cannot put aside the " possibility that the fact may have a cosmical basis.

" The most direct connexion in such case that we should " expect to find with other ascertained periodical phenomena " would be with the annual march of the barometer. In " Fig. 4, Plate X. the annual curves of mean mensual " barometric pressure are laid down to the same scale of " ordinate for time as the equinoctial and solstitial seismic " curve below (Fig. 5), giving the variation in atmospheric " pressure for places in several and distant latitudes, Macao, " Havanna, Calcutta, Benares; and in Europe, Halle, St. " Petersburg, Berlin, Paris, and Strasburg ; the curves them- " selves having been reduced from those of MM. Buch, Dove, " and Kaemtz.

" On comparing these barometric curves with the seismic " one, an obvious similarity addresses the eye. . . .

" An earthquake in a non-volcanic region may, in fact, be " viewed as an uncompleted effort to establish a volcano.

"The forces of explosion and impulse are the same in both; "they differ only in degree of energy, or in the varying sorts "and degrees of resistance opposed to them. There is more "than a mere vaguely admitted connexion between them, "as heretofore commonly acknowledged—one so vague that "the earthquake has been often stated to be the *cause* of "the *volcano* (Johnston, 'Phys. Atlas,' Geology, p. 21), and "more commonly the volcano the cause of the earthquake; "neither view being the expression of the truth of nature. "They are not in the relation to each other of cause and "effect; but are both unequal manifestations of a common "force under different conditions."

13.

Such are the theoretical views and the numerical bases upon which seismic phenomena have been discussed in recent times. The numerical results have been attained with scientific precision, and exhaustively treated upon data collected from earliest records down to 1851 by rigorous physicists of France and Great Britain, stimulated and encouraged by the most learned bodies of these enlightened realms. Embracing these with the entire labours of the German physicists also, in their speculative efforts to reach final causes, it may be safely stated that the opinions of Aristotle, based upon earliest observations, the details of which have been lost for ages, were more extended and probably more accurately directed than any which appear in the writings of modern scholars. And this is the more remarkable, because the numerical studies to which I have alluded have persistently pointed to solar causation. More than two thousand years ago the immortal Aristotle recorded the fact—now repeated by M. Perrey, and considered probable by Mr. Mallet—that earthquakes often happen at lunar eclipses. He further stated that the greatest number of these phenomena occur in the night (an important fact to which I shall have occasion to refer hereafter, and point out its bearing upon celestial mechanics); and what is very remarkable, that the sun in some way influences the earth to produce earthquakes; and that the force which produces them (whatever it may be, for it is difficult to understand what the ancient Greek philosophers meant by "exhalations," "winds," "spirit," &c.—their idea of motive or impulsive power within the bowels of the planets) "*received an impulse from the path of the sun.*" Crude, hypothetical, and imperfect as these ideas may appear, nothing more definite has been expressed by M. Perrey, and nothing at all explanatory of interior telluric phenomena has been attempted by Mr. Mallet, notwithstanding the constant indications of all their tabular results. The weight and bearing of their facts, discussions, inductions, and conclusions demand, however, the serious consideration of astronomers, physicists, and mathematicians. So authoritatively sustained and so highly respected have been the investigations of these indefatigable collectors, that the numerical results at which they have arrived may be declared unquestionably as worthy of acceptance and credit as bases of physical induction, as the discussions of Gauss, Sabine, Schwabe, or Bache, in other departments of numerical observation.

14.

Up to this point I have not specially alluded to my own investigations in this wide and pregnant field of seismic philosophy. Mr. Mallet honoured my cosmical views in his last "Report to the British Association," with the brief critical notice which has been already quoted. Appended to the Memoir therein alluded to, were the following note and table:—

"Of 232 earthquakes and volcanic phenomena recorded by " myself as they have occurred from time to time within

" three years, in different parts of the globe, and which have " been collected from different sources of anterior dates " taken at random, the numbers for each month show remark- " ably strong evidence in favour of this theory; and they are " so extraordinary as to stimulate the most active inquiry " into this department of the physics of the solar system. " The statistics copied from my memorandum are as appended: " as the notes of the phenomena were made without any " reference to favouring or discouraging the hypothesis of " molecular repulsion as an earthquake force, the facts afford " almost conclusive proof of the theory."

October.	November.	December.	January.	February.	March.	April.	May.	June.	July.	August.	September.
27	30	25	27	18	15	16	13	12	11	17	21
142						90					
Perihelic Numbers.						Aphelic Numbers.					

This Memoir concisely developed the theory of cosmical repulsion which is here amplified; and it was intended to confirm the data and inferences which I had previously published.

As early as 1843 and 1844 my attention was specially drawn toward earthquake and volcanic phenomena and to their cosmographical bearing, in consequence of explorations of volcanic islands in the Atlantic, Pacific, and Indian Oceans, and of inductions derived from my own observations upon the general outlines of the planet's surface. The names of the French and British *savans* whose labours so largely enrich the preceding pages, did not become known to me until years afterwards. And it is not wholly devoid of interest that so searching an analysis of the same class of phenomena should have been initiated, at nearly the same time, by three persons so widely separated, who conducted investigations from such different points of view. But that which invests this circumstance with much importance is the identity of numerical results attained by all, and the influence this fact must exert upon the future developments of this branch of physics.

My own records and discussions of plutonic phenomena have been brought down to the close of the year 1866. They embrace many authentic and well-marked periods of eruptive activity. Events and time reduced to numbers as exactly as the character and variety of such data permit, must correctly represent the relative dynamic strength or activity of the reactionary force "of the interior upon the exterior" of the entire planet; and they exhibit a periodicity in the movements of this repulsive energy, conspicuously and constantly, but inversely, dependent upon the length of the radius vector, as a general law. By equalizing apparent differences of intensity as far as this can be done, giving unreservedly the fullest weight to the aphelic period, and reducing innumerable or rapid and prolonged series of local concussions to limited quantities, so as to represent facts and days by units, the result of the complete discussion exhibits an aggregate of 11,872 facts distributed as follows: for

January.	February.	March.	April.	May.	June.	July.	August.	September.	October.	November.	December.
1,328	1,172	934	935	807	942	798	823	870	966	1,060	1,237

For the six months during which the earth is most remote from the sun, there stand 5,175 against 6,697 during the six months of nearest proximity of these two bodies, thus presenting a difference of 1,522 *in favour of the perihelic arc.* For the two months immediately preceding and succeeding the aphelion passage, they present 3,370 against 4,797 for the corresponding perihelion period, exhibiting a difference of 1,427 in the same direction. For one month before and after aphelion there stand 1,740 against 2,565 for the same time before and after perihelion, exhibiting a difference of 825, the ratios of difference constantly increasing as these points are approached and passed by our planet in its annual revolutions.

Now, if the tables of Professor Perrey and Mr. Mallet be re-examined, the same general monthly and periodical results will be found to exist. That such remarkable coincidences of numerical data as these different series of facts expose are a mere accident, is beyond the range of probability. On the contrary, they all significantly point to the prevalence of some occult law, and evince the existence of a general cosmical agency, whose *action upon the molecular forces* of the heavenly bodies, from their *centres to their circumference*, is dependent upon, or at least coincident with, the ellipticity of the orbits in which they revolve around each other.

Professor Perrey, connecting his facts and numbers with time, discussed them as results of lunar causation, limiting solar action to a secondary rank, and this of a simple attractive character only, in conjunction with the "Theory of the Tides." In his latest disquisition upon this subject, however, solar influence receives more attention than appears in his earliest memoirs.

Mr. Mallet, discovering the same relative numerical difference of facts in connexion with time, yet discarded every condition of cosmical causation; and, ascribing all classes of seismic phenomena to changes of temperature, and the possible conversion of electrical, magnetic, and thermal currents into each other in local foci, classed them by simple thermal seasons, as is clearly shown by special references to the seasons of the northern and southern hemispheres.

In my discussions of plutonic and seismic phenomena, I reject all considerations of a theoretical character, and all arbitrary divisions of astronomical time, except as an element in numerical determinations. The names of the months, in my treatment of the subject, are only used, and must be considered, as simple terms in the problem, representing definite arcs of an ellipse, along which our planet, as a plastic mass of molecules, moves, in performing its revolutions around the sun; and the numbers, as simple representations of varying intensities of dynamical force, exerted from the centre of the planet upon its exterior, in like manner as similar movements are seen in comets, when in proximity to the sun.

As to the local foci of volcanic or seismic activity, spoken of by Mr. Mallet, Sir Charles Lyell, and other scholars, who reject the view of Leibnitz, that the earth possesses a molten and fluid nucleus, they may, and probably do, exist. But the conditions are wholly reconcilable. Immense caverns and empty canals under all mountain ranges, are not only probable as consequences of geological changes, as I have shown in my discourses upon the "Dynamics of Geology," and in my memoir upon "Volcanoes," read before the American Association for the Advancement of Science, at Albany, in 1856; but subterranean thunderings, the eruptive phenomena exhibited by some volcanoes, and the abundant materials accumulated over many volcanic regions, prove beyond question the existence of immense and profound hollow spaces and channels of communication from one to another, all of which must be immediately filled with the atmosphere, or with mineral gases and sublimates, when the molten matter of the nucleus recedes toward the centre of the planet, after its radial dynamic impulses have subsided. When the eternally fluctuating pulsations of the molten nucleus burst anew, either with spasmodic or with prolonged ejective currents, into these hollow spaces, or "local foci," after longer or shorter periods of repose—during which periods air, water, gases, and vapour have mixed with dust and pumice, thus granulating slag and tufa into sand and powder, producing mud or ashes, and preparing the caverns for new thunderings, or terrific combinations of eruptive conditions—then follow physical events, whose inconceivably vast attendant conditions demand the development of dynamic energy, which the entire heart of the planet, excited into fullest tension of repulsive force, can alone impart.

It is not my design, however, to dilate in these connexions upon this important, interesting, and diversified branch of terrestrial physics. Extensive explorations, nevertheless, of our globe, close observation of various volcanic regions, personal experience of many earthquakes in Mexico, Peru, and Ecuador, together with observations of the volcanoes of Fogo, Kilauea, Isalco, Massaya, and Cotopaxi in greater or less degrees of eruptive action, and of scores of extinct volcanic mountains and smaller craters, have fully convinced me of the truth of what I say.

There is, however, more evidence to be adduced upon the relative differences of seismic and plutonic intensity in the aphelic and perihelic arcs of the earth's orbit; and notwithstanding these discussions tend to enlarge the fields of geology, I can but indulge the hope that the cultivators of that science will turn their attention to this subject and so extend their inquiries and speculations as to lift all similar facts above novelty and doubt.

Von Tschudi ascertained while in Peru, as a general result of experience and observation, that an excess of earthquakes occurred in that country in November, December, and January of each year.

It was stated many years since, as a well-observed fact, that in Chili earthquakes are more frequent from October to March than at the other period of the year. The table presented by Professor Perrey seeming to contradict popular observation and opinion upon this point, I have reduced the facts in my own catalogues to a tabulated form, and give them with this single comment, that if the entire number of earthquake *shocks* which have been hourly and momentarily noted in Chili, at certain epochs of violent and prolonged "reaction of the interior" upon that region of the South American continent, were taken into account, the proportion from November to April would be much greater than appears. Thus it may be declared as established beyond further question that popular opinion in Chili upon this subject is correct; and it will be seen that the facts arrange themselves into the same numerical and natural order so observable and remarkable for other regions and for the entire planet.

The whole number of 546 facts are distributed as follow : for—

January.	February.	March.	April.	May.	June.	July.	August.	September.	October.	November.	December.
78	47	44	40	44	30	26	17	32	62	64	62
Winter.			Spring.			Summer.			Autumn.		
169			114			75			188		

the perihelic period exhibiting 357 against 189, or nearly as 2 : 1.

I find the same notable circumstance to exist at Hawaii, where the great plutonic outlet of Kilauea, being continually open, probably acts in such a manner as to modify considerably the expression and intensity of the seismic phenomena felt upon the Island. Nevertheless, in tabulating a complete and valuable catalogue, which has been kept with very commendable care from June 1833 to May 31, 1867—a period of thirty-four years—by Mrs. Sarah J. Lyman, one of the resident missionaries of the American Board, and which has been kindly furnished to me by that worthy lady, it appears that of the whole number of 173 shocks, distributed through the different months of this long epoch, 45 more occurred during the perihelic than the aphelic ones—presenting the proportion of more than 3 to 2. By reducing the data where, at one period, circumstances prevented their hourly record, and where, for certain days, an interpolation reads, "there were several shocks each day and night;"—reducing this irregularity to a fair ratio (indeed, to the smallest) by substituting the number *two* for the number "several," the summaries of the Hawaiian catalogue are as follow :—

January.	February.	March.	April.	May.	June.	July.	August.	September.	October.	November.	December.
7	10	17	11	7	10	18	10	8	11	36	28
34			28			36			75		
Perihelic Arc.			Aphelic Arc.						Perihelic Arc.		

And it may be here observed, that the great variety of monthly numbers with different ratios appearing in seismic tables, are only the more natural, since all the phenomena under discussion spring from pulsating and irregular perturbations of its molecular forces, as our planet rotates and revolves along an orbit the arcs and axes of which are never for two successive instants the same, governed and influenced as they are by the exhaustless action and reaction of correlated and more voluminous forces, located in one of the foci of this path.

At Petropaulaski, Kamtschatka (a region subject to earthquakes), similar observations have been made. An intelligent observer, Mr. John W. Widdifield, of New York City, who resided there for a considerable period, has informed me that it is a fact well known to the inhabitants, and confirmed by his own observations, that these phenomena occur in greater numbers in the *winter* than in the *summer;* and he further stated that a volcano in sight of his residence was more active during the winter than summer months.

A French gentleman, M. Victor Provost, who resided two years in, and travelled much upon, the Isle of Bourbon, nearly the antipode of the last-named locality, became familiar with the phenomena of its famous volcano. He stated it to be the common observation of the inhabitants, and of his own knowledge, that eruptions were annual and periodical, and that the lava generally flowed with more than ordinary force in December and January.

Dolamieu, Hamilton, and Scrope, names well known to science, long ago stated on the authority of the inhabitants of the Lipari Islands, that the eruptions of Stromboli were far more violent during "*the winter seasons* than in summer."

By careful and extensive inquiries into the observations of natives and foreign residents, during a fortnight spent in Acapulco, Mexico, in 1853, I discovered that a similar preponderance of earthquakes, both as to number and violence, prevailed there also from October to February.

During my journey through the Cordilleras of the Ecuadorean Andes in 1864, I learned, by constant inquiries between Loja and Quito, that the volcano of Sangai, which exhibits unceasing activity, displays the greatest violence from December to March. This is the common observation of the white people and Indians of the regions within a hundred miles of this most terrible of present active volcanoes.

During my residence in Munich in 1866, Doctor Hermann Schlagintweit (the last now living of the three intrepid and accomplished young brothers who recently explored the Himalayas under the auspices of the British East India Company, upon the recommendation of the illustrious Von Humboldt) kindly contributed to me from the proof sheets of his great work on "High Asia" the following fact, which sustains the already abundant evidence upon this important physical question, to which I have long invoked the attention of geologists and astronomers.

"Earthquakes are frequent in Assam chiefly in November, "December, and January, when I, too, had occasion to observe "several shocks along the right bank of the Bramaputra. "Also in the meteorological registers I found them regularly "noted." (Vol. iv. p. 167.)

Dr. Schlagintweit informed me that these results were obtained from the records he found in Assam.

The same excess of seismic phenomena for the perihelion period—or what is commonly called the autumn and winter months—has been observed in California since the settlement of that part of the American continent. For the purpose of establishing the truth of this common opinion, I will append the monthly summaries of earthquake shocks as I find them noted in my records for the entire State.

January.	February.	March.	April.	May.	June.	July.	August.	September.	October.	November.	December.
40	15	9	11	9	6	6	14	16	24	14	22
Winter.			Spring.			Summer.			Autumn.		
64			26			36			60		
Perihelic.			Aphelic.						Perihelic.		

In the whole number of 186 there appear 124 for the perihelic period to 62 for the aphelic, or exactly 2 to 1. But the points nearest aphelion and perihelion present differences still more striking and remarkable.

15.

Thus it will be seen from all the numbers, facts, and statements which precede, that *the activity of eruptive force in our globe is, as in comets, greater in the perihelic points of its orbit, than in the opposite points.*

Any *single* statement in this enormous mass of evidence, whether presented by Professor Perrey, Mr. Mallet, or the writer, would be entitled to no weight upon a physical question of so much importance to the future developments of general science. But the results of the researches of all three, thrown together upon this inquiry, become very striking; and I think they cannot fail to excite attention even among the most strenuous advocates of the prevailing system of celestial mechanics. Indeed, they must convince physicists that there is no such principle or condition in the constitution or in the motions of a planet, or of the universe, as a *vis inertiæ;* but, on the contrary, that all molecules throughout revolving globes are points charged with positive repulsion and attraction which unite into planetary units; and that these latter act and react dynamically to and from the centres of cosmic bodies, and to and from the centres of revolving systems, intensifying as globes approach each other, diminishing as they separate, and always so acting as to guide their movements, shape their orbits, and originate the mechanical and numerical elements involved in their elongations and in their approximations to any focus whatsoever.

The internal action and re-action of cosmic bodies mutually affect, as we shall more clearly see hereafter, the superficial disturbances and the general and relative external motions and perturbations of one another; and these actions and re-actions are dependent upon the reciprocal influences of one upon the other, from centre to circumference, and *vice versâ,* in the course of their translations through space. This is a plain statement of a general law.

Upon these points no further evidence, arguments, or illustrations need be offered in this connexion.

16.

In order that the facts and principles already laid down and proved in relation to the functions and forces of molecules, and in relation to their cosmical developments in comets and the earth, may be seen to have universal expansion, I will briefly allude to phenomena of a kindred character, which are conspicuously displayed in other planetary and in stellar bodies.

The eccentricities of the orbits of the Asteroids are known to be as various as these bodies are numerous and remarkable; and their perihelion and aphelion distances are equally different and striking. Now, while comparatively little or nothing is known of the physical constitution of these bodies, it is a singular and suggestive fact, that (as in comets) the intensity and hues of their light are different, constantly vacillating, and even undergoing hourly or momentary changes; that the magnitudes of some of them have been observed to undergo unmistakeable transitions; that the atmospheres of some are dense and definable, of others diffuse, extensive, and nebulous; that some have no appreciable atmosphere, nor a sign of nebulosity; that some undergo such intense and sudden changes in the appearances of their discs as to lead to the belief that they are not always round: and observations upon these planetoids have been so carefully and accurately made as to convince telescopists that their modifications of physical aspect can neither be attributed to changes of distance from the earth or sun, nor to atmospherical perturbations. No astronomers who have made special observations upon these bodies, agree even approximately on the measurements of those the longest known among them. Herschel declares the diameter of Pallas, for instance, to be 135 miles; Schroeter, 2,295 miles; and Lamont, 738 miles. Allowing for instrumental defects, micrometric errors, and for inaccuracies, accidents, and imperfections of every sort; yet, considering the fact that with the best instruments the same observer never finds the asteroids to present precisely the same aspects under the most favourable atmospheric conditions, the strongest reasons exist to conclude that their molecular constitutions are of the most elastic, sensitive, and heterogeneous character. Measurements of the oldest and largest of these bodies have been often attempted, both in respect to their solid discs and the extent of their envelopes. With such scrutiny, indeed, have these observations been made that Schroeter determined the *atmosphere* of Ceres to be 668 miles high, and that of Pallas not less than 465 miles at certain epochs. The differences of opinion upon the appearances and magnitudes of these bodies, indicate with certainty the fluctuating character of their physical aspects; and we may safely infer that the various action and re-reaction of their constituent molecules and internal forces are the causes of these external changes.

I am aware that some of these statements do not coincide with the calculations and conclusions of Mr. Stone, the eminent Assistant Astronomer of the Greenwich Observatory, relative to the magnitude of the asteroids. But the records of Arago and other astronomers, and a special study of the elements and physical descriptions of these bodies, individually recorded by them, impress me with the conviction that previous telescopic observations cannot have been completely inaccurate, and that future ones, acutely directed to these objects, will only confirm their variable characteristics, as above generally delineated.

Whatever opinion may be entertained of the origin of the asteroids—whether they are fragments of one original planet, or of a planet which had satellites, or bodies eternally separate and numerous as they are now, or portions of a former nebulous ring—it will be perceived that they must have always been in some manner influenced and acted upon, as distinct masses, by a force the opposite of gravitation: for, in the first two conditions, they must have been rent asunder and separated by cosmical repulsion, as has been illustrated by similar divisions of numerous comets; and in the last two they must have been originally endowed with a cosmic principle *repelling union;* otherwise ONE solid sphere, composed of these separate masses, would be rolling between Mars and Jupiter, in strictest accordance with Bode's remarkable and pregnant law.

However their creation was accomplished, whether it be comparatively recent or otherwise, their physical elements and various molecular constitutions, thus anomalous and eccentric, compel us, in the light of present knowledge, to classify them, as an astronomical order, between comets and planets, partaking as they do of the characteristics of both. Now, since repulsion has been demonstrated to augment the tension and internal agitation of both these classes of bodies *at perihelion*, to intensify the light of, and even to split, comets, and constantly to strain the earth into reactionary and eruptive activity with periodic intensities, varying inversely with the aphelic and perihelic length and sweeps of the radius vector, we might expect to find kindred phenomena developed in the asteroids under similar circumstances; that is, we might expect them to contract and expand like comets, and intensify their light by their own internal molecular commotions. So far, no special observations have been made upon this point; and perhaps the small size, great distances, and transient appearances of these bodies may long preclude positive knowledge in these particulars. But it is a remarkable fact, and one which demands the closest scrutiny of telescopic astronomers, that the greater the eccentricity of the orbits of the asteroids, and the greater the differences in the lengths of their respective radii vectores, the greater is the transient variability of their light, and the more their discs, their atmospheres, and nebulosities, when visible—in a word, their general molecular conditions and physical aspects undergo notable modifications.

When the diversified observations which have been recorded in relation to these members of our system (now numbering one hundred) are compared with one another, and closely studied in detail, the points presented in this generalization will appear as striking as those in relation to the periodical phenomena exhibited by comets and our own globe to which I have already called attention.

To deny, then, even with our present limited knowledge, that molecular repulsion, developed into planetary amplitudes, plays its part as a cosmic function in the asteroids, is to deny what the telescope has revealed to a hundred eyes, from Piazzi and Olbers down to present observers.

17.

Grouping the rest of the planetary system into two natural orders, embracing the bodies which roll between the sun and the asteroids in one series, and those exterior to the asteroids in another, and keeping steadily in view the definite point of study, we shall not be slow to discover the same principle of repulsion which has been so fully established as the great phenomenal agent in the earth and in comets, exercised in all of them, although variously exhibited.

To say nothing of evidence deducible from differences of the specific weights of the interior and exterior planets (which fact of itself alone is sufficient to indicate, and even establish, the principle that molecular repulsion is developed into planetary quantities or functions affecting individually the densities of the cosmic masses);—to say nothing of this fact, their variable physical aspects, the eruptive phenomena noted in some, and the atmospheric perturbations and the broken or mountainous surfaces of all, indicate that the same molecular conditions which operate in the earth, reign, as cosmical causations and powers, throughout the volume of every one of them: and since all move in elliptical orbits of different degrees of eccentricity, and since all maintain determined and different aphelion and perihelion distances, it follows that the law of minimum and maximum intensities of molecular and cosmic eruptive action established for comets and the earth must reign also throughout their various masses. If like causes and conditions did not pre-exist, it may be safely concluded that like phenomena could not appear.

The uncommon eccentricity of the orbit of Mercury renders it a peculiarly interesting object of study from our present point of view. That violent and general perturbations of its elastic envelopes occur is well known, since dark bands suddenly form across its face, obscuring its light, and as suddenly change or vanish, producing sensible variations of brightness and sharp and dazzling scintillations.

During the transit of this planet across the sun, in 1799, several astronomers saw a luminous point upon its dark disc which they supposed to be a volcano. As this phenomenon was seen from different observatories, and with different telescopes, and especially by Harding, at Lilienthall, whose acute eye afterwards discovered Juno, it must be considered a decisive fact in its bearing upon this discussion. Observations upon its cusps, moreover, prove its mountainous and broken features as rigorously as these features can be shown in our own satellite.

18.

The hypothesis that all these various appearances and changes in the atmosphere and body of Mercury are produced by different degrees of heat and light received from the surface of the sun, will not satisfactorily explain them when they are compared with similar phenomena which are constantly occurring in, upon, and around our own planet, and which we now know do not depend upon *simple exposure* of the terrestrial surface to the solar surface. But when the mind rises to grand generalizations, and comprehends that the entire series of phenomenal events transpiring in the physical universe rests upon the simplest plan; that all spring from the action of one overruling law; that all flow from one incomprehensible central Being, whose ways are eternal and immutable—whose grasping, quivering arms, for ever stretching out and drawing back by turns, are nerved by the same invisible and immortal elements as our own, but which, shooting from the very central atom of the cosmos, strengthen as they glide and tremble from star to star, spreading asunder and binding together atoms and worlds alike; that the wonderful agents of this universal power are the simple molecular activities of repulsion and attraction, developing from zero into infinite quantities, and assuming isolated cosmic conditions with mutual affinities and antagonisms;—when the mind embraces this simplicity unfolded into its infinite and amazing grandeur, philosophy will have discovered the natural path for a fertile, harmonious, and happy solution of past and present celestial problems.

The *universal ellipticity of the paths* in which the heavenly bodies move, and the other equally wondrous fact that controlling globes never occupy the centre but always one of the foci of these ellipses—these facts, when studied in the light of gravitation alone, are involved in utter obscurity; and such phenomena can never be satisfactorily explained as long as gravitation is maintained by scientific men to be the only active force which controls the motions and sustains the system of the heavens. When, however, repulsion also is considered to be an active principle in celestial dynamics, possessing powers and functions co-ordinate with, but antagonistic to those of attraction, an explanation is at once afforded for every puzzling and obscure mechanical problem in the universe.

As these fundamental forces act and react among molecules, attracting and repelling, drawing them into foci, and disrupting foci again into ultimate elements or prior conditions, so they act externally and at distances as cosmical entities, assuming cosmical integrals. Thus mutually acting and reacting upon each other as masses and spheres of forces; suns and planets, satellites, asteroids, and comets (the densities of which, never equal in any two, are fluctuating from instant to instant in all) are compelled by the influences of their various chemical conditions to impress special degrees of eccentricity upon their individual orbits, and institute their respective elongations and approximations. This development of our discussion I suppose to contain the essence of Dana's idea of "pulsating molecular force," an expression obscure in itself when applied to interstellar space as absolute vacuum containing bodies wholly disunited and widely separated from each other; but pregnant with meaning when cosmic phenomena are demonstrated to result from *ever-changing densities* in the cosmic spheres, and when these changes of density are shown to depend upon molecular forces excited into varying degrees of activity through the mutual dependence of masses upon each other, as all revolve in elliptical orbits, and act and react upon each other from centre to centre, according to the law of inverse distances.

Applying these principles in their widest ranges and bearing upon cosmical physics, it will be discovered that the molecular conditions of, and the phenomena developed in, revolving bodies—that is to say, their power and facility of expansion and contraction, of condensation and eruption—hold definite relations to the ellipticity of their respective orbits and to the varying length of their radii vectores.

19.

Objections to these propositions, based upon geological and mechanical theories—for instance, that the earth is composed throughout of augite, granite, or iron—in other words, that it is solid rock or metal, unyielding, undilating, and uncontracting; that its condensed and crystalline constitution precludes the possibility of continuous active molecular movements from centre to surface, and therefore that it is an individual forceless mass, and can of itself have nothing whatever to do with its own motions, any more than a boulder or cannon ball; that it moves in virtue of its *vis inertiæ*, possessing no energies within itself capable of accelerating, retarding, or modifying its own course and motion; and in consequence of these assumptions, the conclusion that the elliptical form of the earth's orbit, and the location of the sun in one of the foci of this orbit, can have no possible connexion with or relation to its molecular and chemical conditions, or general physical constitutions;—objections and conclusions of this character should, I think, dissolve and vanish before the facts which have been set forth. By a diversity of data, I have endeavoured to show that this planet is not only *not solid* throughout, but that its interior conditions are both fluid and actively mobile and elastic; that these elastic matters are contained as a boiling, surging nucleus within a crystallized shell; that this shell is pierced in many places, and is universally subject to tension, blows, shocks, fissures, faults, cleavages, and fluctuations of outline, in consequence of contractions and *reactions* of its fluid nucleus; that the number and intensity of these reactionary blows and eruptive phenomena vary, as periodic fluctuations of dynamic energy, with inverse distances from the sun. Now, inasmuch as these coincidences establish positive connexions between the internal dynamics, or molecular forces, of the earth, as a cosmical mass, and the length and sweep of its radius vector, I hold that there ought no longer to be question that the ellipticity of its orbit is determined by the mutual interaction of physical forces between itself and the sun; and now, finally, inasmuch as there can be no reaction without repulsion, it follows that the nature of the centrifugal force which unites and fluctuates with gravitation in order to produce these elliptical elements, is no other than an active living *cosmical repulsion,* which plays, in some occult way between the terrestrial centre and the solar centre, in like manner as cosmical attraction itself.

Such is the light which a microscopic study, so to speak, of the ultimate forces of Nature throws upon the most wonderful problems of the universe. A studious and persistent development of this subject will elucidate (as we shall see more clearly in the sequel) the true character and action of all mechanical force, will simplify investigation in every geological, astronomical, and physical direction, and will sooner or later profoundly modify many doctrines in every branch of modern philosophy.

20.

Passing from these considerations, the pregnant value of which, springing from all that precedes, will especially bear upon all that follows, the physical conditions of our instructive satellite will next claim brief attention. All its aspects of surface declare its subjection to the same elements which have been heretofore demonstrated to act as central and conflicting forces in the constitution and physical agitations of comets and of our own globe. The object of this study will not permit descriptive details, and I shall therefore only step from universal facts to universal principles, and *vice versâ,* for the purpose of binding one with the other, and consolidating the general structure upon which we are at work. But it is manifest to all telescopists that the extraordinary volcanic and wildly-broken conditions of the entire lunar surface only indicate a stronger conflict of antagonistic forces within the secondary than apparently has ever agitated its primary. This fact, when studied by the light of preceding demonstrations, will not appear so strange or inexplicable; for the density of the lunar mass being only about half that of the earth, and gravity at the lunar surface being estimated at about one-sixth of our own, its general repulsive force as a rarefying and reactionary agent must be comparatively very great. Whatever its exact relative quantity may be in comparison with the repulsive force of the earth, there is nothing in the physical conditions of the moon which does not fully harmonize with the developments of this discussion, and strengthen its consequences in every direction.

The fact which telescopic discovery is of late so plainly demonstrating—viz. that many aspects of the lunar surface must have been produced by the *falling in* of vast areas of its shell—far from conflicting with principles thus far deduced, actually proves that similar material conditions, similar forces, and similar creative and economical designs have reigned throughout the constitution and being of both the earth and moon. In two discourses long since read before the Boston Society of Natural History, I endeavoured to overthrow the popular geological theory of slow and continental upheavals (a theory which I *formerly* believed sound because Leopold Von Buch taught it), by pointing out the existence of enormous cavities, and their continuous formations within the terrestrial shell, into which vast areas and entire hemispheres have bent or plunged at various epochs in the earth's history, deducing therefrom the important truths which I predict must sooner or later be accepted—that our planet was, during its azoic and palæozoic ages scores of leagues larger than now, in all its diameters; that the linear projections of its features (now improperly called "upheaved mountain systems"), and its watersheds and its deep seas and oceans—in short, its general geographical outlines—have been created by successive sudden and cataclysmal subsidences; and that although upheavals do exist as matters of fact, and as results and in some measure unavoidable adjuncts of subsidences, they are limited, and are not, and never have been, the general agency in dividing the geological—that is to say, the creative or morphological —periods.

In order to sustain this position, I might, were it admissible, adduce here, as collateral proof, an important and hitherto unpublished fact of an archæological character, in addition to my geographical and geological observations made upon the coasts and islands of the Pacific Ocean. The fact is this, brought to my knowledge by an unusually extensive practice of my profession, that a uniform custom of *dorsocision* has existed throughout the various Polynesian Islands from periods unknown, and beyond all tradition, embracing alike New Zealand, Esther Island, Tahiti, the Marquesas, and Hawaii—a rite wholly different from, but similar in its result to, the Jewish one of circumcision; and that this has been performed at the eighth or ninth year in all of them, and transmitted by father to son, with undeviating precision, from generation to generation. A fact of this character so deeply rooted in the moral, social, and traditional life of many peoples thus widely distributed throughout that vast ocean, so remotely separated from each other, and without intercourse, indicates even more strongly than colour, caste, or language, not only the unity of their progenitors, but also the wide-spread existence of a single race, the vestiges of which were left here and there above the waters when the land sank between America and Asia, and received the older seas into a new basin.

Physical changes of this kind and magnitude, sustained by evidences derived from diversified sources, have not only a local importance to the planet in which they may occur, but the bearing of such facts upon the physical appearances of other globes within telescopic scrutiny is of the utmost value, as a means of instruction and inference in regard to their respective dynamic conditions.

The latest investigations of the moon's features with great telescopes are bringing selenographers into clear and decided opinions that many of *its* appearances and irregularities must have been produced by the "falling in" of numerous and vast areas of its surface. Facts and determinations of this character indicate, as in my studies of the earth, that the size of the moon was formerly greater than at present. Whether greater elasticity of matter and forces existed in these bodies during earlier ages, and consequent greater eccentricities of their orbits, or whether these forces have been always employed in embellishing and populating and diversifying their surfaces, leaving their orbits eternally the same, are interesting questions; and although progressive acquirements of scientific knowledge are fast removing them from the realms of speculation, it would be out of place to discuss or further to allude to them in this connexion. It is only the identity of phenomena, of physical aspects, as actually observed, and of the forces which have produced, and continue to produce, these broken and revolutionary features in both the earth and moon—in primary and secondary alike—to which the attention of philosophers is specially invited. The indications of terrific eruptive actions throughout the visible hemisphere of the moon are so remarkable as to convince the commonest observer and the most superficial scholar, that if a central force, coincident in name, nature, and function with cosmical repulsion, can exist in the earth as a planetary quantity, and even perform beyond the earth important functions in solar mechanics, the same force must unite with gravitation in the material structure, internal dynamics, and general motions of its attendant body.

21.

Every fact and argument adduced and brought to bear upon the planet Mercury, the earth, and the moon, will apply to Venus and Mars. The features of the latter are more or less visible to the aided eye. Their physical characteristics are similar; they are embraced in the same system; they exist by, and are subject to, similar general laws.

22.

Reaching the outer members of our solar group, and omitting further allusions to the asteroids, we next encounter the prodigious volumes of matter constituting the systems of Jupiter and Saturn. The small density and specific gravity of these planets, in comparison with similar conditions in the interior ones, confirm at once the wide prevalence of the law of development of molecular forces into planetary ones; and when these physical differences, so well known to exist in the interior and exterior planets, are compared with facts developed by experimental physics, the law of inverse distances will also be found as applicable to cosmical and celestial as to molecular and mundane phenomena. For instance, when balloons charged with air or gases at ocean levels are elevated, their contents expand as the balloon recedes from the centre of gravitation, and so on indefinitely; that is to say, their internal property of molecular attraction, or of condensing energy, diminishes, and that of molecular repulsion, or of rarefying energy, increases, and new series of phenomena become apparent. As the entire volumes of the atmospheric and aqueous envelopes of the earth are only, after all, cosmic conditions of matter, and would be and act much like a comet, were they peeled from the globe and dropped in space as an independent body, it becomes manifest, from all which precedes, that this gaseous and aqueous body would contract, that is to say, increase its own density at perihelion, and expand and become more rare at every successive point of its elongation. It is a *principle*, already perhaps sufficiently referred to, that I wish to elucidate by this illustration, in order the more firmly to establish the theorem, that the further a mass of matter, whatever its physical constitution or dimensions, recedes or is forced from the central point with which it is connected, or about which it revolves, the more its individual molecules will also recede from each other, the rarer it will become throughout its entire volume, and consequently the less will become its density and specific gravity. By comparing exact physical experiment with approximative data attained by astronomical observation and mathematical analysis, we find this principle to prevail as an inflexible law from the centre to the outermost boundaries of our solar system. There can be no doubt upon this point, since Jupiter and Saturn, Earth and Mercury, all in succession proclaim its truth; and since comets entering and vanishing in and out of every latitude and longitude of the heavens—north, south, east, and west—declare it to be one of the universal results of the union of matter and force wheresoever such conditions are seen to exist. Inasmuch, then, as various conditions of density and specific gravity in planetary bodies, and the changes and oscillations of molecular distances incident thereto throughout their respective volumes, cannot take place without manifest developments of corresponding intensities of repulsion, it follows that the force which has always been recognised and taught in elementary and experimental physics by the name of "*molecular* repulsion," does absolutely develop itself in infinite space into "*cosmical* repulsion," and become a central, measurable, and variable power in planetary physics and celestial dynamics. Limited by academic and technical philosophy to the simple name of "molecular repulsion," with corresponding conditions, capacities, functions, and possibilities alone, this force has been heretofore generally ignored, overlooked, or rejected as non-existent, unimportant, or redundant in celestial dynamics; while its molecular co-ordinate of attraction can be demonstrated by Newton's analysis of Kepler's discoveries to expand into gravitation, be traced and amplified into general conditions, and verified as identical in all its molecular and cosmical relations, developments, and aspects. Hereafter, I trust, the magnitude and universality of this element will receive the mathematical development and verification of which I feel sure it is capable.

Applying these considerations to the entirety of phenomena presented in the changing aspects of Jupiter, Saturn, and all their appendages, problems heretofore obscure become illuminated, and the active conflicts of repulsive and attractive forces, as central and cosmic powers, become the demonstrable elements and causes of all their strange and various appearances. The illustrious Arago, full of doubt and prophecy, said well, when treating of the "very variable apparent magnitudes of Jupiter's satellites:" "On voit que ces mondes "éloignés présentent encore bien des problèmes à résoudre à "nos successeurs." (*It is manifest that these distant worlds present many problems for our successors to solve.*)

The general configuration of the planet Jupiter, its constant varying dimensions (independent of terrestrial or solar distances), however delicately micrometric observations upon its diameters are made or accurately discussed, and a remarkable local flatness upon its southern border once clearly defined with different telescopes by Schroeter and his assistants, but which distinct bend in the outline they could not discern in the following revolution ;—all these singular facts denote such plastic conditions of elemental constitution, that, when considered together with its varying belts, shifting cloud-spots, and surprising transitions of luminous intensity in different regions, we can no longer doubt that the orb entire, from its innermost parts to its outmost atmospheric boundaries, is as much under the influence of active reactionary force—pure cosmic repulsion—as our own planet, and even more so. The different degrees of eccentricity so remarkable in the orbits of its satellites establish the elastic character of the action of their central forces : and thus the evidence of the reign of cosmical repulsion throughout both the primary and the secondaries not only becomes probable, but positive.

23.

What we have said of the Jovian system will apply with equal weight and significancy to the Saturnian one. In addition to this, the rings of Saturn, so anomalous, so wonderful, so impossible when studied in view of gravitation alone, seem to loom from the abysms of space in the light of our present knowledge, as if to rebuke analysts for indulging in vagaries, wasting the powers of the calculus upon erroneous theories, and persistently endeavouring to establish final truth upon insufficient causes and doubtful foundations. The earliest opinions upon the physical constitution of these rings were announced by Laplace, who considered them *rigid*, and *proved them to be so* by the severest mathematical analyses. To account for the planet's eccentric position within the rings, and for the maintenance, under these circumstances, of stable equilibrium between the primary and its massive appendages, he declared it essential that the latter should be loaded with a superabundance of matter at certain parts of their circumference. So profound has been the conviction of scientific men since the time of Newton that gravitation is an *unique* principle of celestial energy, that even Laplace felt himself justified (and not in this instance alone) in announcing the results of his equations as decisive physical facts, even where visible conditions of nature did not meet the demands of an hypothesis. I do not say this with any reflection upon the scholarship, integrity, or memory of Laplace, for he is one of the most illustrious men of modern times. But my object is to place this fact in a strong light in order to show how the scientific world, even up to our own days, has been swayed by the weight of authority, and how dangerous and improper it is in mathematicians to warp numbers, or assume quantities, to meet the requirements of hypotheses, when observations are imperfect or adverse to such hypotheses. Imperfect theories rarely embrace anomalies. Anomalies often indicate the way toward subtle principles of truth.

The fact had been long known, and all observation has confirmed it, that Saturn is not centrally situated in relation to his rings—that there is, indeed, a peri-cosmic and an apo-cosmic relation existing between them; and it therefore appears from this, that the same elements of stability are incorporated into and must exist between Saturn and his rings, as exist between him and his moons, between our earth and its satellites, between other primaries and theirs, between all planets and the sun, and between all double and revolving stars. Nevertheless, embraced as this peri-Saturnian mechanism evidently is, within the plan upon which the universal clockwork of the heavens has been established, it has not yielded its anomalies to the dissolving menstrua of mere formulary mathematics.

The history of Astronomy decisively establishes the facts, that the rings of Saturn have separated into many bands, and parts of bands; that they decompose and recompose themselves in like manner as water or quicksilver might do in flowing, breaking up, dividing and uniting again in order to form homogeneous volumes or oceans; and that, contrary to the opinion of Laplace, *they are not rigid.* Moreover, in later times, Mr. George P. Bond, one of the late eminent astronomers of Massachusetts, discovered an inner variable ring not previously known, and detected also such movements in the "old" rings, as to afford unmistakeable proof that voluminous interchanges of force and material currents are constantly oscillating from the inner to the outer, and from the outer to the inner regions of these breaking and still apparently continuous appendages. These immortal discoveries are positive demonstrations that the peri-Saturnian rings are not solid or rigid. Nothing beyond these disclosures is wanting to prove the existence and incessant fluctuation throughout these bodies, of attraction and repulsion as cosmic co-ordinates and co-efficient principles of energy.

The doctrines of Laplace upon the rigid constitution of these rings prevailed in the schools for fifty years, and checked all investigations upon their character, notwithstanding the frequently observed dislocations of their form and substance, until Mr. Bond's discoveries and discussions settled the question, overthrowing the mathematical demonstrations of the "Mécanique Céleste," and establishing, on the contrary, their fluid and elastic constitution and mobile and fluctuating functions. And now, strange to say, since Mr. Bond's important observations have furnished new data for discussion, other mathematicians have carried analysis so far in opposite directions to those of Laplace, as not only to describe (as Pierce of Harvard College has done) the exact degree of fluidity possessed by these peri-Saturnian rings, but so much further also as to announce that they are not fluid at all, but probably continuous currents of meteors, like the showers of shooting-stars periodically visible from our own planet.

Whether this last conclusion is more accurate than the

first announced by Laplace, time will probably determine ; but the point which I desire to set forth conspicuously in this connexion and by these contrasting facts, is the effort made in the present state of cosmo-mechanical philosophy to bridge or fill a vacuum by numerical or mathematical speculation, so to speak, as a result of the incompleteness of our knowledge of universal physical laws.

Every effort to satisfy disturbances of equilibrium, where there never was stability, and where no condition or element of stability exists or ever existed—where gravitation must have a counterpoise of absolute antagonistic energy—where the assumption of the *vis inertiæ* of masses vanishes both from hypothesis and possibility—and where, as in the rings of Saturn, it is impracticable for analytical mechanics to base research exclusively upon current theory without reaching extremes of physical impossibility and scientific absurdity ;—every effort and result of this character exposes more and more plainly not only the imperfect development of celestial dynamics, but also the insufficiency of the system upon which analysis is founded.

Amid all this confusion and uncertainty, however, two simple facts are clear : molecules exist and attraction reigns throughout the Saturnian system, as cosmic entities. If so, then repulsion reigns also, as the co-ordinate of gravitation ; for we have previously seen that, as a fundamental and eternal law, all three are united and inseparable, and that all three are equally essential to the universal harmonies and general stability of nature, whatever combinations or forms molecules may assume, or however masses of matter or volumes of force may be associated with, related to, or separated from one another.

Were any facts or evidence wanting to illustrate the truth of my proposition, or strengthen the soundness of my discussion, they will be found in that wonderful condition of things presented by the planet Saturn and his rings. The current mechanical hypothesis does not apply to their conditions any more than to the phenomena developed by comets, or the interior physics of our own globe. But if the basis of inductive philosophy be enlarged so far as to admit the antagonistic element of repulsion to be a living mechanical agent, capable of reacting with attraction under every molecular and cosmical circumstance and condition, all difficulties vanish in connexion with the elastic character and decompositions of these otherwise anomalous appendages. Unless this fundamental and reactionary principle be recognised as a power capable, like gravitation, of the utmost numerical as well as graphical development, the physical aspects, constitution, and phenomena of the peri-Saturnian rings must for ever remain insoluble problems. Permanent yet flexible creations, looming and fading, gilded and glowing with vacillating currents of light and shadow, revolving in space with stupendous velocity as independent continuous rings of matter, unattached and yet chained to a rolling projectile globe, thus far defying the calculus and setting at nought the most varied conjectures of the greatest analysts, overwhelming astronomers and philosophers alike with wonder, there they are, established in the heavens, impassable and unavoidable barriers to the triumph of prevailing doctrines. The intellect, like the eye, must accept them as new and pregnant facts; and if they embrace or expose elements or conditions of force and motion beyond explanation upon principles announced by our illustrious predecessors, philosophy must certainly expand itself to meet their requisitions, even if a complete reconstruction of its present dogmas be found necessary.

24.

If repulsion, the same potential principle which has hitherto been traced as a radical and voluminous energetic existing and active in and throughout the structure of the various members of our system, be sought after in the central and controlling body itself, abundant observation verifies its existence and activity there, and establishes the fact of its enormous and far-reaching influence as a reactionary and phenomenal agent. For not only do its dynamical functions appear to be correlated with mechanical events occurring constantly in our own globe, as I have endeavoured to show; but also, as is now being demonstrated by De la Rue and Balfour Stewart, with the relative positions of the other planetary bodies which revolve around it. Since gravitation as a cosmic force binds to the solar centre all bodies or spheres of force which revolve around it, by their respective centres, so that there exists, as Newton has proved, an absolute mutual interchange of attractive influences based upon specific laws of quantity and inverse distances, it follows by the closest induction that repulsion must exist and act as a central cosmic force also, and become the great reactionary agent, affecting every molecule bound to the sun, in any manner or relation whatever, and *always in an opposite sense to the force of attraction.*

The fact that the density of the sun is only one quarter that of our planet, shows, in the first place, the pervading reign of repulsion between its molecules, throughout the entire solar mass. This, together with its enormous magnitude (its diameter being 882,000, while the mean diameter of our globe is only 7,912 miles), will, in the next place, give some idea of the prodigious volume of repulsive force existing and active within this central and controlling orb.

Since repulsion has been shown to be a dynamical agent in molecules, in comets, and in planets, the facts just noted are certainly suggestive that the mechanical energy heretofore believed to exist in the sun is not dependent upon and confined to the force of gravitation alone. If gravitation acts in proportion to the magnitude of this body, drawing its surrounding appendages towards its centre with degrees of energy dependent upon inverse distances; then, upon the principles which have been antecedently developed, it follows that repulsion must be a component of the solar energies, that it must also act in the capacity of a magnified co-ordinate, and be a co-efficient principle, but active in a manner contrary in every sense to that of solar gravitation. And such I hold we shall find to be the case.

Besides the points above stated upon the density and specific gravity of the sun, its physical aspects abound with visible proofs of uninterrupted reaction and of unusually active molecular repulsion in and over every part of its surface. Even the appearance and origin of the faculæ and "willow leaves," and of the general mottled appearance of the photosphere, are no longer subjects of doubt to physical observers of large experience. That they are material conditions arising from the breaking up and rolling and surging of luminiferous gases or elastic rarefied fluids of constantly fluctuating densities, of the swelling and puffing up of one point and the subsiding of another, and the mingling of contiguous irregular masses of vapoury matter into one great spherical envelope of mobile molecules, no astronomer or physicist longer doubts.

If observers could view our globe as they examine the sun, enormous areas of its atmosphere would often present appearances similar to those in the solar envelope. Such phenomena can be demonstrated to be cosmical in character, and to depend for their existence as much upon primary repulsion as subsequent gravitation. For I hold that no motion of vapour, or of atmospheric molecules, can take place, unless repulsion first creates a radiating or upward current, and that wind is only the gravitation of particles to fill what would otherwise become an absolute or comparative vacuum. Hence the most violent cyclones, either in the terrestrial or solar envelopes, are only so many positive proofs of the phenomenal energy of repulsion as a force the opposite of gravitation. The recent observations of Lockyer, upon the solar spots, establish the truth of these conclusions in a remarkable manner, and are of the highest importance.

On the morning of July 15, 1844, a rare moment for comparison of terrestrial and solar phenomena of analogous characters, I happened to be upon the summit of Hale-a-ka-la, the vast extinct volcano of East Maui, one of the Hawaiian Islands, and from a height of ten thousand feet above the ocean saw the aspects imparted by the rising sun to a field of vapour which, from three to five thousand feet below me, spread to the horizon in all directions. Here and there the black summit of an island pierced the resplendent picture, and a vast irregular fleecy mass or a deep cavity broke conspicuously its general uniformity. Otherwise, it was uninterrupted and apparently endless. The magnificence and impressiveness of the view exceeded everything I have yet seen upon the earth; and I never look at the sun through any telescope without beholding the counterpart of that extraordinary vision stamped upon its disc. The only observable difference was that the snowy aspects of the terrestrial vapours presented on that occasion are replaced by yellowish tints in the solar ones. All below, and wherever I looked, was endless confusion of soft, fleecy, brilliant inequalities; rounded and projecting points gleaming with silver, and transmitting light with different shades; swelling masses divided by immense and diversified shallows, with here and there a bursting snowberg, looming in gigantic boldness above all the rest; and as the rays of the advancing sun gleamed from point to point and changed the aspects of drift and hollow, this endless, undulating mantle presented not only every aspect of luminous beauty, but also all that mottled appearance so characteristic of the solar photosphere. Beside all these general identities of aspect, "a spot" was observable here and there—a rare object, but a veritable hole—piercing the mottled luminous field, in and beneath which were only blackness and darkness. The common dark colour of land, and the blacker hue of ocean, were alike undefinable.

Such was the exterior surface of this mottled vapoury mantle suspended for several hundred miles over a tropical region of our planet; and which, under my limited range of observation, was to all intents and purposes as much an envelope of the earth as the solar photosphere, seen through our telescopes, is an envelope of the sun. It was separated, moreover, from our globe as the photosphere appears to be separated from the sun, and as the rings of Saturn are separated from that planet, by a force the opposite of gravitation. To say that clouds of dust, and that such vast fields of dense and glittering vapour "only float," or "are sustained by their *vis inertiæ*," is only to employ meaningless parlance for the profound and pregnant scientific term of repulsion. It is no difficult task to show that this repulsion, which thus acts throughout and around the orb of the sun, is a cosmical force; inasmuch as the telescopic observations and discussions of astronomers, from Fabricius and Galileo to those of our own day, including especially those of Wilson, Wolf, Schwabe, Carrington, Dawes, and Lockyer, the polariscopic studies of physicists since Arago initiated them in 1811, and the late spectral discoveries of Frauenhofer, Kirchhoff, Bunsen, and others;—all these varied observations assert and confirm such physical conditions to exist in the photosphere as are inseparable from the play of potential energies of a reactionary, uplifting, or radiating character pervading the entire mass embraced within it. So palpable, indeed, to every philosopher is the energetic action of some central outbreaking element, that no theory upon the causation of the solar spots has been advanced which does not involve the necessity of the presence and agency of force the opposite of gravitation. Even the theory of my eminent friend Professor Kirchhoff of Heidelberg, so exceptional and probably quite wrong upon this particular point, is nevertheless more decided than all others upon the absolute incandescent and molten constitution of the solar sphere itself.

The same statements may be made in relation to every other physical aspect of the sun. The convoluted masses of dense luminous vapour which envelope and swell the central body to its vast dimensions are continually surging, bursting, driving hither and thither, flashing and fading in their lights and shadows, and exhibiting such constant molecular changes and voluminous movements as to demonstrate conflicts of force within and pervading the entire globe, which are wholly irreconcilable with the theory that gravitation is the only element of potential energy at the base of its constitution. Attempts to explain these phenomena by magnetic and electrical hypotheses fail to meet the demands of exact science, for these must all fall to the ground when it is considered that attraction of gravitation *alone* is not electricity, nor magnetism, nor heat, nor light, and cannot produce them, nor be substituted for or converted into them. But when repulsion is recognised as a conflicting and reactionary agent in the central and controlling mass of our system, swelling from molecular into stellar potentials, and from limited into voluminous manifestations, as an active central force fluctuating with gravitation through every molecule in the sun and in every globe,—it is then that philosophy will command a solid basis for theory, and a link will appear and become established between gravitation and magnetism, and every other expression of polar force. For the antagonism of the molecular forces of attraction and repulsion is the very fundamental principle of all polarity. When this is traced—as will subsequently appear—into its solar developments and stellar amplitudes, and when the manifold members of the cosmos are shown to sway and hang between these two elements of Omnipotent Power which, first creating, still generates and governs all other physical principles and unfoldings, the puzzling problems of magnetism, electricity, heat, and light, as planetary phenomena, will receive immediate solution, and their polar characteristics will appear legitimately dependent upon the duality of their primaries and originators, and become palpably correlated to gravitation through the great link heretofore wanting to perfect modern physics into a harmonious system of natural philosophy. When the reign of repulsion shall be admitted to be universal, and the principle itself to be a co-efficient element of energy with gravitation in all its varied manifestations of local and planetary activity, as I have endeavoured to indicate and illustrate, the secondary phenomena of magnetism, &c. proceeding out of their functions as a sort of secretion or new creation and birth, will no longer be objects of greater wonder or more difficult explanation than their congeners which are produced in the workroom of the chemist or physicist in his daily experiments.

25.

To prove by another order of data—by a sort of inductive or synthetical method of investigation—that repulsion exists and acts as a central solar force, I will again summon the earth for testimony, and so interrogate it that its responses, although indirect and reflective, shall leave no doubt upon the question, if the contributions of Lamont, Schwabe, Wolf, and Sabine to cosmical science be facts, and not delusions. That they are facts and solid elements of inductive knowledge, continued observations most certainly attest. The connexion ascertained by Sabine to exist between the decennial period of magnetic fluctuations discovered by Lamont, and the periodic character of the solar spots observed by Wolf and Schwabe, evinces ceaseless contests between potential forces of a gravitative and *reactionary* character, which not only rend enormous areas of the solar surface and create visible agitations and convoluting alterations of form and substance throughout the solar disc, but also simultaneously act beyond and across immense depths of space in such a manner as to engender dynamic disturbances within other bodies, which disturbances also appear to act and react, from their centres to their surfaces, and *vice versâ*, with intensest power. Heretofore the *molecular conditions* and *forces* of attraction and reaction—that is to say, repulsion—existing and acting as causal or phenomenal entities *within* the spheres revolving round the sun, in *all*—whether primaries, secondaries, comets, or asteroids—have been considered wholly independent of the central forces and molecular conditions of attraction and of reaction existing and acting within the solar globe as an individual body of matter. Nevertheless, so great and positive is the phenomenal action of this dynamic element in the central body, and so mutually responsive is its cognate principle in the primaries of our system, that I am led to conclude, as the result of prolonged observation and study, that not a throb of molecular force or an eruptive disturbance occurs in the sun that is not instantly felt and responded to within the bowels and from the centre of our own planet. Indeed, such mutual and definite relations, and such delicate adjustments, do these molecular and central forces appear to hold towards each other in the two bodies (perturbations from the action of third bodies proving with greater certainty the accuracy of my conclusions and the universality of this law), that reactionary, upheaving, and radiating phenomena occurring in the sun are simultaneously felt in the earth, and not less positively indicated and proved by variations of intensity in the magnetic force than by associated disturbances in the molecular and massive structure of the planet; for the extensive researches of Dr. Emil Kluge exhibit numerous synchronisms between volcanic eruptions and variations of terrestrial magnetic force and the solar spots. These valuable contributions to cosmical physics, together with various other authentic coincidences of remarkable deflections of the magnetic needle with earthquakes, recorded by different observers in widely separated parts of the globe, and the frequent notes existing among my data of the synchronism of seismic disturbance with the aurora borealis and with magnetic storms, are profoundly suggestive of the wealth of physical discovery lying within the range of the forces which react with such activity between the centres of the sun and of the earth, and of the sun and of every other body which revolves around it. An interesting observation, personally communicated by Professor Lamont, during one of my visits to the Astronomical Observatory near Munich, is not without the broadest significancy. I present the statement in substantial form as this distinguished astronomer permitted me to use it. Descending one morning into a subterranean observatory, where, some years since, he was conducting a special series of magnetic observations, he found his instruments in the most tumultuous agitation. He was filled with surprise, having at no time seen them in such a state. The observatory was too deep and remote from roads to be affected by vehicles or passing loads of iron. The anomaly was so striking, that he noted the time and peculiarities of the fact in his record, by reference to which the date was found to be 1842, April 18, 9 h. 10 m. A.M. He thought nothing more about this until he noticed, some days afterwards, the occurrence of a violent earthquake in the Grecian Archipelago; and on examination of his register, he ascertained that the magnetic perturbations were simultaneous with the seismic phenomenon. Upon the same authority, I will also state that Professor Colla made a similar observation upon his magnetic instruments at Parma, Italy, the same morning.

I shall take occasion in this connexion to invite in an especial manner the attention of physicists to another remarkable synchronism—a fact presenting a still wider range in its bearings, and possessing a very wonderful and impressive character. For a detailed account of the phenomena I will refer to "A Description of a Singular Appearance in the Sun on September 1, 1859," noted by Mr. Carrington, and to another observation by Mr. Hodgson, noted at a different observatory, and published in the monthly notices of the Royal Astronomical Society, Nov. 11, 1859.

Mr. Carrington describes the sudden, violent, and dazzling commotion as he was observing a great sun spot, and goes on to say: "The instant of the first outburst was not 15 seconds "different from 11h. 18m. Greenwich mean time, and 11h. "23m. was taken for the time of disappearance. In this "lapse of five minutes, the two patches of light traversed a "space of about 35,000 miles," and they were computed to be as large in area as the earth itself.

Mr. Hodgson, who fortunately was observing the sun at the same time, at another and distant place, says that this sudden and surprising outbreak was "*much brighter than the* "*sun's surface, and most dazzling to the protected eye,*" and that "*the centre might be compared to the dazzling brilliancy of the* "*bright star Alpha Lyra, when seen in a large telescope with a* "*low power.* It lasted for some five minutes, and disappeared "instantly about 11h. 25m. A.M."

"Mr. Carrington," remarks the Secretary of the Astronomical Society, "exhibited at the November meeting of the Society "a complete diagram of the disc of the sun at the time, and "copies of the photographic records of the variations of the "three magnetic elements as observed at Kew; and pointed "out that a moderate but very marked disturbance took place "at about 11h. 20m. Sept. 1, of short duration; and that "towards four hours after midnight there commenced a great "magnetic storm, which subsequent accounts established to "have been as considerable in the southern as in the northern "hemisphere."

In order to establish the connexion of these solar and terrestrial phenomena, and to fix the attention of physicists upon a point so interesting as the intercosmic relationship of matter and force; in order, indeed, to bring, as I design to do hereafter, these reciprocities of phenomena within range of explanation on principles recently made known by the observations and experiments of Mayer and Joule upon the correlations, conservation, and equivalencies of the terrestrial forces; I will here present an *exact coincidence of events*—a fact which to logical minds must be as conclusive of a unity of dynamic action and a connexion of final causative agencies (although the masses were 90,000,000 miles asunder), as if Lisbon and London were simultaneously shaken by an earthquake, that the events proceeded from the same dynamic impulse, and were in some hidden way connected with the same preceding conditions and the same final sources of mechanical power.

A note addressed to my eminent friend Dr. Balfour Stewart,

the Superintendent of the Kew Observatory, has enabled me to present the time-scale readings of the photographic record of the magnetic disturbance noted at the epoch mentioned. The following memorandum was communicated, in the absence of Dr. Stewart, by Mr. G. M. Whipple, his able Assistant:—

"As near as I can determine, it appears to have commenced "on Sept. 1, 1859, at 11h. 16m. A.M. The culmination was "at 11h. 22m. A.M. The magnet gradually came to rest, so "that no definite time can be stated for the end. The original "curve appears to be resumed in the horizontal force at 11h. "52m. A.M.—in the declination at noon—and in the vertical "force at 12h. 15m. P.M.

"The greatest extent of movement in the declination "was 17′ 15″."

During my visit to the Royal Observatory at Greenwich, on June 6th of the current year, I was furnished with the amplest opportunities for examining the records, as well as the instruments in use and the methods of observation adopted at that renowned establishment. Placed by Mr. Glaisher in the hands of his accomplished first assistant, Mr. Wm. Nash, I was enabled to inspect the hourly photographic curve and the sudden and extraordinary magnetic disturbance recorded by the Greenwich instruments at the date referred to. Upon the application of the time-scale to the photographic sheet, Mr. Nash found the time of the commencement of this sudden and short storm to be (Sept. 1, 1859) 11h. 15m. A.M., and its end to be 11h. 43m. A.M. Its culminating point, as indicated by the greatest deflection of the magnet, was at 11h. 27m. A.M. *The movement of the magnet, as traceable in this record, was as sudden as the flashes in the solar spot observed at the same moments by both Carrington and Hodgson.*

Mr. Nash informed me that on the morning of the 2d Sept. 1859, the photographic sheets indicated the greatest magnetic disturbance which the instruments have ever recorded.

With regard to this point, Mr. Whipple writes from the Kew Observatory as follows: "A much larger disturbance "occurred at 5 A.M. on the 2d Sept., which lasted for 11 hours, "the movements of which were in several instances too rapid "and too large in extent to be recorded."

The same phenomenon, I am informed by Dr. Stewart, was recorded at Lisbon at the same time.

Mr. Nash informs me as the result of his unintermitted observations for many years, that "more great magnetical "storms and general magnetic disturbance take place in "*winter* than in *summer*;" preferring to use the names of the cold and warm *seasons*, rather than the names of the winter months, December and January, and summer months, June and July.

The general results of the magnetic storm observations made at the Greenwich Observatory correspond with those made at all other stations on the globe, whether upon the surface or deep in the bowels of the earth. As periodical phenomena studied in the light of simple meteorological data, scientific men have heretofore regarded them as dependent on seasons of heat and cold—that is to say, the exposure of the *surface* of the planet to the solar *rays*; and observers allow themselves to speak of them only in connexion with the *seasons* of the year. But when they are referred to in this work I wish to give them a broader and deeper meaning, and I use them in connexion with the names of *the months,* as December and January, June and July, because these terms are symbols of the earth's position in its orbit, because they measure the length and sweep of the radius vector, because they mark the position of the perihelion and aphelion, and bring the general dynamical phenomena observable in the globe into correlation with the dynamical movements in the central body seen by the eye of Schwabe and Wolf, and photographed and confirmed by De la Rue and Stewart at Kew.

So positively do all these synchronisms (differing in intensity and in the form of their manifestation as they may) exist between visible physical disturbances in the earth, and even between the earth and sun, and so numerous are they, that when considered in connexion with the general fact that *maxima* and *minima* intensities of seismic phenomena and of magnetic phenomena are constantly periodic and coincident with perihelic and aphelic distances, few will disallow the correctness of my positions and conclusions.

As all these diversities and alliances of phenomena, so interesting and so pregnant with useful results, lie as yet upon the verge of a comparatively unexplored field of inquiry, they can but incite scholars to enlarge their researches, and even stimulate governments to contribute unsparing facilities to extend this branch of knowledge. And it is the more promising of fruitful discovery, since all experimental researches in molecular physics bring out only such conditions and results as are entirely harmonious with those observable in cosmical phenomena, and the former appear to be only diminutives of the latter, when all are viewed from the cosmographical point of view in which I have endeavoured to unfold them. No luminous, thermic, magnetic, or electrical element appears, or can be generated in physical experiments of any sort without motion—that is to say, molecular compression and reaction—or those constant disturbances of the molecular forces of repulsion and attraction called friction. And I may be permitted to repeat just here, at the risk of being tedious to the reader, that heat, magnetism, electricity, and light viewed in their cosmic amplitudes are not molecular attraction nor attraction of gravitation, and that they are not convertible into, nor the equivalents of, attraction or gravity. This has been conclusively established by Faraday and others. That they do appear, however, as correlative entities with repulsion, when repulsion is in any manner cognizable or demonstrable, is most certain ; and the intensities of their development appear to be in direct proportion to the action and *reaction* of the fundamental forces of matter, molecular *repulsion* and molecular attraction. In so far as the relations of cause and effect are considered, the molecular forces maintain the conditions of primaries, out of the vibrating functions or decompositions of which proceed a series of *secondary* forces—magnetism, light, heat, &c.: in other words, *motion* and dynamical phenomena in every form and amplitude, the partial discovery of the laws, correlations, and mechanical equivalencies of which by Mayer and Joule constitutes the highest intellectual achievement of the present age. Every department of experimental physics declares and confirms the correctness of these principles.

Attraction or gravity, as a pure abstract force, is not magnetism, nor light, nor heat, nor electricity; and, as a unique fundamental principle, without an active opponent inherent in cosmical masses, would be powerless to vary the motions of these masses in space, or to create differences of ellipticity in their orbits, or develop seismic and other dynamic and correlative phenomena within their respective circumferences. But REPULSION as a *polar* antagonist, capable of conversion also into a dynamical associate of attraction, *is the principle which links all phenomenal expressions of the* SECONDARIES *with the ultimate conditions of nature,* and harmonizes experimental facts with every display of light, heat, electricity, and magnetism, and with every form of motion and mechanical action observable in the cosmos. When celestial physics shall be more generally studied from this point of view, and in the light of exact knowledge derived from experiments in molecular physics, discarding all hypotheses and speculative doctrines, the central relations and mutual attractions and reactions existing between the sun, and the earth, moon, comets, and all other bodies revolving in orbits of different eccentricities, will be better understood. So harmoniously do observations upon the varying intensities of terrestrial electricity and magnetism correspond with the same elements in seismic and eruptive phenomena, that it may be clearly inferred that magnetic and thermic forms of energy are generated, on a terrestrial and solar scale, by general molecular disturbances, as we see their diminutives generated in the experimental workshop of the physicist and chemist. Whatever may be the physical constitution of any object, its molecular status cannot be disturbed in the slightest degree unless *secondary* force, in one of its forms of electricity, heat, light, magnetism, becomes manifest. Now, since Kepler has shown that our planet moves in an elliptical orbit with the sun in one of its foci, and since I have shown that the *maxima* and *minima* intensities of the molecular disturbances in our planet coincide with the perihelic and aphelic arcs of this ellipse, it follows that no particle of matter or iota of force in the globe is ever at rest, and that the density of its mass is never for an instant the same, any more than that of comets; and, inasmuch as Sabine has shown that the intensities of the total magnetic force are also greatest when the earth is in the perihelic part of its orbit, and least in aphelion, it follows with equal certainty that this total force is in direct proportion to the molecular agitation, motion, and friction of particles; that is to say, to the oscillations of repulsion and attraction on a vast scale—the very energetics which I have already demonstrated to be the active agents in producing seismic and volcanic phenomena in the earth, and the spots and other physical evolutions in the sun.

Thus the various forms of experimental research, the results of general observation at different points of the planet's surface, and the tendencies to unity of all scientific inquiry lead at last to a generalization of great interest and importance. The data derived from these sources do more than suggest that in worlds, as in molecules, *attraction* and *repulsion* are the root of all other forces and phenomena. They compel us to believe, if motion, heat, light, magnetism, electricity, actinism, or life, are correlated in any manner or degree with the force of gravitation, that it is through the agency of REPULSION that this connexion takes place. So broad and vast a function does this latter force, as a pure and ulterior principle, perform in the universe, considered as a field of creation and continuous existence, that I hold it to be more essential to the present system of being than any other force or principle within the range of human thought, or of experimental and inductive philosophy.

26.

Having pursued the study of this force into the very centre of our planetary system, and grappled with it in its relations to the various spheres and forms of matter which, clinging to and revolving round this centre, are nevertheless not permitted to unite with it, however closely approximating (as proved by the great comet of 1843, which traversed the luminous atmosphere, or photosphere), we will examine its action in still more distant regions of space.

To name phenomena known to exist in connexion with multiple and variable stars (and I here direct attention to them in order that they may be elsewhere studied in detail) is sufficient to present a hundred problems which defy all solution within the hypothesis that gravitation is the sole active central force in cosmical bodies and in celestial mechanics. The assumption that heat and its correlatives of electricity, light, &c. are the motive powers, either in combination with or in opposition to gravity, within and beyond the respective circumference of these stars, is, in the present state of physical knowledge, inadmissible. The wondrous fancy that all the stars of heaven were pelted one by one from the hand of God, as was assumed by Newton in regard to the moon and planets, whereby to claim a suitable momentum and inertia for gravitation to play upon (through which conception—splendid in him and fortunate for the world—his immortal discovery was happily achieved), fails to resolve the mighty problem which lies at the base of creation, or grasp the force which imparts continuous impulse to universal nature.

All this class of distant worlds whose motions have been determined, travel, like the bodies of our own system, in elliptical orbits, which differ only in this particular, that many of them are enormously elongated, thereby resembling cometary orbits. This fact, it will be perceived in view of the preceding discussions, is pregnant with the most striking significancy. So vast indeed are the periods in which some of these multiple stars perform their revolutions around each other, that "the great year of Alcor, Mizar, and the small "star near them, cannot be less than 180,000 of our years, "while that of the quadruple ϵ Lyra must exceed 500,000." "The orbits of the planets and their satellites are," says Nichol, "almost circular, or ellipses of very small eccen- "tricity; while the curves of the double stars have every "degree of elongation—challenging the freedom of the "comets. The orbits in the case of α Centauri and γ Vir- "ginis are especially eccentric, so that at opposite periods "of their cycles the two connected suns must be very close "to each other and very far apart. If," he continues, "each "of these suns is attended by planets, how extraordinary "the physical condition of these planets, and how inextricable "their mechanical relations!" The luminous intensity and general variability of physical aspect observable in these classes of orbs appear to be directly proportional to their motion, internal agitations, and approximation to each other, thus establishing the *universality* of a law of correlation between intensities of phenomena—intensities in the developments, I mean, of *secondary* forces, as of light, heat, &c. which always spring from and coincide with intensities of action between the primary forces of attraction and repulsion in and upon molecules. The manifestation of this law is equally observable, and coincident with the play of repulsion and attraction, in the molecular constitution of the most distant stars; in that of comets, and of the earth in their radial and orbital connexions with the sun; and in that of terrestrial matter subjected to experiment in every department of physical investigation. While phenomena developed by the latter depend upon compression and reaction, composition and decomposition, absorption and dispersion, &c. and are induced by human inventions; phenomena of a similar kind, developed by molecular agitations in cosmical masses, result from and depend always upon *incessant fluctuations of their density.* Upon this point we have already sufficiently enlarged.

27.

The course of study developed by the illustrious Francis Arago upon the polarization of light, reduced the general physical constitution of the stellar bodies and the play of the molecular forces within them to such positive knowledge, that many years before his death he declared himself prepared to announce the following propositions without hesitation :—

"Si une étoile changeante, examinée avec la lunette polari- " scope, reste parfaitement blanche dans toutes ses phases, on " peut donc assurer que sa partie extérieure ou incandescente " n'est pas liquide, et que la lumière émane d'une substance " analogue à nos nuages ou à nos gaz enflammés. . . .

" La constitution physique des photosphères des millions " d'étoiles dont le firmament est parsemé, est identique à la " constitution physique de la photosphère solaire." (Arago, " Ast. Pop." liv. xiv. chap. 25.)

" If a variable star examined with the polariscope keep " perfectly white through all its changes, we may be sure " that its external or incandescent part is not liquid, and that " the light emanates from a substance analogous to our " clouds, or burning gases. . . .

" The physical constitution of the photospheres of the " millions of stars with which the heavens are studded, is " identical with that of the solar photosphere."

The spectrum observations of Messrs. Huggins and Miller confirm and extend the discoveries of Arago. Extracts from an important paper of theirs on the new star in Corona Borealis, communicated to the Royal Society of London in 1866 ("Proceedings," &c. No. 84), will exhibit our field of inquiry in a new and striking light, illustrate still more clearly the power and extent of cosmical repulsion, and fortify every point of the preceding discussions.

"'A new star has suddenly burst forth in Corona,' wrote " Mr. Baxendell, of Manchester, to one of us. . . . Last "night, May 16th, we observed this remarkable object. . . . "*In the telescope it was surrounded by a faint nebulous haze* "*extending to a considerable distance, and gradually fading* "*away at the boundary. A comparative examination of* "*neighbouring stars showed that this nebulosity really existed* "*about the star. . . . On the 17th this nebulosity was suspected* "*only; on the 19th and 21st it was not seen. . . .* When "the spectroscope was placed on the telescope, the light "of this new star formed a spectrum unlike that of any "celestial body which we have hitherto examined. The "light of the star is compound, and has emanated from two "different sources. Each light forms its own spectrum. In "the instrument these spectra appear superposed. The prin- "cipal spectrum is analogous to that of the sun, and is "evidently formed by the light of an incandescent solid or "liquid photosphere, which has suffered absorption by the "vapours of an envelope cooler than itself. The second "spectrum consists of a few bright lines, which indicate that "the light by which it is formed was emitted by matter in "the state of luminous gas. . . . It is difficult to imagine "the present physical constitution of this remarkable object. "There must be a photosphere of matter in the solid or liquid "state emitting light of all refrangibilities. Surrounding this "must exist also an atmosphere of cooler vapours. . . . Be- "sides this constitution, which it possesses in common with "the sun and stars, there must exist the source of the gaseous "spectrum. . . . The gaseous mass from which this light "emanates must be at a much higher temperature than the "photosphere of the star. . . . The position of two of the "bright lines suggests that this gas may consist chiefly of "hydrogen. . . . The character of the spectrum of this star, "*taken together with its sudden outburst in brilliancy and its* "*rapid decline in brightness, suggests to us the rather bold* "*speculation that in consequence of some vast convulsion* "*taking place in this object large quantities of gas have been* "*evolved from it;* that the hydrogen present is burning by "combination with some other element, and furnishes the light

" represented by the bright lines; also that the flaming gas " has heated to vivid incandescence the solid matter of the " photosphere. As the hydrogen becomes exhausted, all the " phenomena diminish in intensity, and the star rapidly " wanes. . . . The whole class of white stars are distin-" guished by having hydrogen lines of extraordinary force. " It may also be mentioned here that we have found that the " spectra of several of the more remarkable of the variable " stars, namely those distinguished by an orange or ruddy " tint, possess a close general accordance with those of " *α* Orionis, *β* Pegasi, and the absorption spectrum of the " remarkable object described in this paper. The purely " speculative idea presents itself from these observations, that " hydrogen probably plays an important part in the differences " of physical constitution which apparently *separate the stars* " *into groups*, and possibly also in the *changes by which these* " *differences may be brought about.*"

All *purely speculative ideas* touching the physical constitution, or the influence of this upon the groupings, or positions of the heavenly bodies, are rejected, in this branch of analysis; but it will be noticed at the same time, that this speculation of Mr. Huggins favours the positions to which the entire drift of the preceding discussions tend. My only point is to prove the existence in nature of a purely abstract force of repulsion, whose amplitude of dynamical action ascends with that of attraction from molecules into worlds, so as to become everywhere, in planets and stars, a central force with decentral functions, and be the associate, co-ordinate, and coefficient of gravitation, wherever the latter can be shown to exist and act.

28.

Spectral analysis, as a means of chemical discovery of the most subtle character, both on a cosmical and molecular scale, is now placed beyond question.

If, then, the discoveries made by Arago, Frauenhofer, Kirchhoff, Bunsen, Huggins, and Miller, positively prove that the chemical elements and the mechanical (or thermodynamic) actions of matter existing in distant orbs are similar, and in many cases identical, with those existing in this planet, it follows with equal certainty that the radical forces of attraction and repulsion, and therefore, as a consequence, the molecules themselves, whatever may be their ultimate forms or other peculiarities, are identical in all respects throughout the universe, at least as far as this sort of analysis has been carried: for the discovery of chemical and *thermodynamic* identities upon a cosmical and universal scale involves all the other physical conditions in question. But the most profound and important discovery of all embraced in this prolific department of physics, is that which demonstrates the identities and reciprocal influences of the molecular forces as *chemical energetics* existing and acting between all the bodies which roll through space, however widely these bodies may be separated from each other. The wonderful delicacy of the spectrum-analyses exhibits evidence of these physical intercommunications with far greater certainty and by more varied means and facts, than are afforded by synchronisms and periodicities of the magnetic disturbances and plutonic agitations with the solar spots. The two series of facts—the latter connected with the cosmical forces developed from centres, and the former with those developed around the surfaces, of the celestial masses of molecules—strengthen the conclusion derived from each separately. But the amazing subtlety of spectroscopic analysis strikes at once at the *forces* of *molecules* in their atomic and residual divisibilities (so to speak), and acts in such a manner as to bring out in each ray of light the combined action of all the varieties of matter in these masses, and to correlate and connect both the molecules and their forces, wherever they exist, with their analogues in distant cosmical accumulations. For I hold that light itself is only a compound resultant of molecular attraction and molecular repulsion; and that it is generated by, and is an equatorial product or function of, thermic and other moleculo-chemical vibrations, in the same manner as magnetism is an equatorial product or function of electricity and *vice versâ;* all dynamic and sensible effects of the transmissions of antithetic forces depending on the characters of the ultimate elements and of their compounds. For it is plain, that if the sun or any star were composed of iron alone, or of any single chemical element known, the characteristic lines exhibited by that substance, whether bright or dark, would be the only ones revealed by the definite capacities for vibration affixed to its molecules. As many different elements as enter into its constitution, so many will appear when its light—that is, the *sum of the vibrations* combined of the entire compound mass—is analysed. Thus much experiment, observation, and induction teach us. Now, when light is reduced to its ultimate elements—that is to say, to its essential components of force—it is only found to consist of different conditions and degrees of repulsive and attractive principles, inasmuch as refraction and reflection are only other terms for repulsion, partial or complete, and absorption another for attraction, all expressing the specialized properties of molecules, the chemical secret in their ultimate conditions of divisibility, whatever they may be. And it is furthermore plain that only elements of the same sort, and consequently vibrations of the same wave-lengths, can combine through laws of perfect affinity, and so become homogeneous masses. Similar laws govern the harmonies of musical vibrations and of all matter. The association of heterogeneous elements immediately induces successive series of physical conflicts, the chemico-dynamic laws governing which are more or less known; and the violence and persistency of these conflicts are proportional to quantities and diversities of elementary substances. So, if the earth, or a comet, heave from its centre, and repel portions of its interior beyond its periphery as a responsive and synchronous action and reaction with those occurring in the sun (between which and the various component bodies of our system all dynamic forms of force are most delicately adjusted), it will not hereafter appear strange that the attractive and repulsive action so manifest in the photospheres is felt and responded to by the minutest molecule of cognate constituents in our planet. The dynamic properties of magnetism astonish us when they play between distant points, and push and pull the impulses of human *thought* around the earth; but we know they cannot manifest these powers unless suitable connexions between molecules exist, by which perfect compensations or balances of force shall be restored and maintained throughout the entire volumes of cosmic repulsion and cosmic attraction which, composed (as before proved) of molecular increments, charge the plastic matter of the globe. Indeed, it seems to be the effort of nature to preserve this equilibrium between the molecular forces of its different chemical elements, which constitutes these currents in opposing directions, and which thus develops their dynamic peculiarities. If we are astonished at these phenomena which have become so well known by experiment and observation, how much more shall we wonder at beholding in every line of the solar and stellar spectra only so many electrometers, so to speak, which with fingers of light and darkness point to an alphabet of chemical elements, and spell out the physical history of the heavens and of past time with as much certainty as the magnetic telegraph whispers the events of the present from nation to nation! While the simple conditions of the magnetic telegraph demonstrate that molecular attraction and molecular repulsion, as developed in chemical compositions and decompositions, are the bases of electric and magnetic force, and that they all—both primaries and secondaries—are integrants of cosmical forces of the same characters; the marvellous revelations of the spectrum lead to yet more profound mysteries, and indicate that rays of light are local phenomena, and are capable of solution and separation, like the elements and molecules themselves, in such a way as to show that they are simply microscopic impulses, dynamic vibrations, of molecules depending upon the capacities of matter in its elementary specialities to transmit the antithetic principles of attraction and repulsion with varying and specific degrees of velocity. In their massive conditions as forces saturating worlds to constant tension, they appear to flow and ebb, reach out and recede from their great central fountains with strength or elective subtlety, as the capacities of the molecules through which they are transmitted admit or require. Thus playing from atom to atom and from element to element, affecting all combinations of matter and all worlds alike, the relative positions of which are mutually modified by these forces as aggregates, the abstruse truth is disclosed that chemical attractions and chemical repulsions are but so many modifications of general cosmic polarities mathematically measured and infused into atoms, and that crystallization, the basis of all morphological development, depends as much upon impulses generated in some occult way by the most distant cosmical bodies, as it does upon the molecules of the one wherein they aggregate and assume forms possessing geometrical outlines and internal principles of proportion and being.

Thus by chemical as well as by astronomical and mathematical researches, we reach the conclusion that the entire universe combines together to establish laws for the government of a single molecule; and that, *vice versâ,* every molecule unites its force of repulsion with that of attraction in order to develop the sum of cosmical potential energy, and the general phenomenal and working conditions and capacities of nature.

29.

But reaching still more profoundly than all this, facts conduct us beyond the mere discovery of the fundamental elements of nature, and bring us face to face with their mutual relations upon the grandest scale, and with their methods of correlation and universal interaction; and by glancing for a moment at the latter, we shall be able to corroborate the principles heretofore presented. The remarkable bands in the solar spectrum, whose scientific import was first suspected by Frauenhofer, and whose wonderful chemical mysteries were subsequently brought to light by my eminent friends Bunsen and Kirchhoff, are not only telegraphic signals between distant spheres, but they are also the most accurate indices of the methods by which attraction and repulsion chain the universe together and unfold all the dynamical phenomena whose problems are so puzzling to thinking minds.

Every ray of sensible or insensible force, thermic, luminous, or actinic, generated in, and appearing to be emitted by the sun (however the impulse may reach our brain), *must be considered, as it evidently is, an expression of the combined forces of the combined sums of the elementary substances constituting the solar mass.* Hence the ray, although infinitesimal in itself, is a compound ray, and a ray not only constituted of the vibrations incident to the generation of simple heat or simple light or simple actinism, but of the heat and of the light and of the actinic qualities which each and any of the individual elements is capable of generating if acted upon alone by the radical forces of attraction and repulsion. This is the general principle which I wish to convey in this axiom, and which, while it will be immediately comprehended as a self-evident truth, will be more fully understood by discussion in the sequel.

The *molcculo-actinic* force of every *ultimate* ELEMENT *constituting* the enormous volume and energies of the central body, is therefore represented in every ray which, flaming upon its disc, impinges upon the earth viewed also as a mass of mixed elements subject to the same laws of combination and decomposition. When the prism receives and acts upon this ray (whatever the mechanism by which it makes its way through space), it falls apart, as if it were a compound salt undergoing solution in some dissolving menstruum. But it is not the matter—the molecules—of a prism or of a crystal which disintegrates this invisible spectre of compound solar force. It is the repulsive and attractive forces themselves pervading terrestrial molecules, which receive or reject, transmit or refract, this pregnant impression of solar energy. The prism is suspended between two mighty sources of antagonistic energy, each combining special chemico-cosmic currents the vibratory impulses of which are alike everywhere, in sun, stars, comets, and planets ; and the individual forces and properties of the prism are exposed to the reception, intermingling, transformation, separation, and transmission of all the component principles of these currents. Two conditions or opposing principles of force, springing from the combined chemico-molecular vibrations of the two globes, arrest and grapple with each other within the crystal; one penetrating from above and derived from the sun, the other antagonistic, penetrating from below, and derived from the earth. Both these principles appear at their origin and in their essence to be repulsive forms of energy, and are so proven, indeed, by their projective characteristics when viewed as individual rays and soluble currents of active force.

However, or by whatsoever method, these great forces of attraction and repulsion interact between the surfaces and centres of worlds, we will not stop here to investigate; but it is apparent that their reciprocal action and reaction upon each other, in virtue of their union with the chemical elements which are aggregated into globes, not only constitute the invisible network of nature, interlinking celestial masses, binding everything harmoniously into a single system from which no particle of matter can escape nor an iota of force be lost, but that they also decompose themselves into and generate those luminous and spectral phenomena which become sensible only, so far as we can detect, in virtue of the *local* vibrations of molecules, and of atoms considered as elements of molecules, variable conditions engendered by the variable motions and exposures of one body within specific distances of other bodies.

In order to show more certainly the connexion of this principle of repulsion now under treatment, in its universal phases and bearings, with the development of cosmic and universal light and heat, and thereby to prove that its activity is as constant and important a phenomenal agent in creation as that of attraction; and to show indeed that it is a constant associate and co-ordinate of the latter in the generation of the widest forms of secondary force, I shall pursue this topic somewhat further, notwithstanding that I shall have to refer to it again in a subsequent chapter for the purpose of confirming its accuracy by experimental data.

In like manner as a solar ray is a microscopic expression of the molecular action of all the chemical elements in the sun, so every ray of terrestrial repulsion is an impulse, or a succession of impulses, derived from similar conditions existing in our own globe: and, thus constituted, these impulses encounter at their emission similar compound principles which, as before stated, proceeding from the vast central orb, impinge upon the molecules of our own. The incessant encounter and conflict of these two opposing currents of primal force at the surface of all material forms produce those dynamic and lateral agitations known to optical theory as vibrations and undulations, which, observed both by Newton and Descartes to resemble mechanical impulses upon matter, led them to adopt physical hypotheses wholly different and yet both containing dynamic elements of truth. The former asserted that light is composed of attenuated and imponderable particles *emitted* from glowing objects with such prodigious intensities that, traversing space as if it were a vacuum, at the rate (as we now know) of 185,000 miles in a second of time, they impinge upon the retina with varying velocities, and thus create the sensations of sight and colour. The latter also assumed the universe to be filled with an elastic fluid, called ether, of extremest subtlety and imponderable tenuity, which penetrates all bodies and surrounds their molecules; and this according to his view receives and transmits vibrations from one thing to another, thus propagating light by mechanical undulations that impinge upon the retina and create the sensations of sight and colour, in like manner with Newton's emission theory. Newton imagined his luminous matter to be emitted directly from glowing bodies, and shot like arrows into the eye. Descartes supposed his ether to be luminiferous, and a medium by which mechanical impulses are communicated in an undulatory manner, as sound is transmitted by air, and waves by water. The comparative merits of these theories are not to be discussed or adjudged here. It is sufficient to say that Newton's views long ago yielded to the progress of physics; and that those of Descartes, however popular they may have been, are encountering obstacles with every new step in molecular and analytical discovery. It is at this point that cosmical repulsion, as a polar principle—a positive dynamical antagonist to gravitation—a principle entirely immaterial, but inherent and conjunct with matter, and vibrating with attraction between molecules and from the centre to the circumference of spheres by methods resembling or *identical with convection;*—it is at this point, that cosmical repulsion presents itself as an agent to assist in the solution of scientific problems which have hitherto resisted inquiry.

The constant impingement and conflict of the opposing forces of attraction and repulsion upon the surfaces and throughout the masses of cosmical bodies, as these agents shoot in and shoot out, can *primarily* produce no other effect than vibrations of molecules, varying in degrees of intensity. The necessary consequences of these vibrations are instantaneous developments of light, heat, electricity, magnetism, actinism, and, as matter becomes more extensively and continuously agitated, of sound and mechanical motion.

Space being, as I hold, absolute vacuum, no ether, either ponderable or imponderable, exists there. No impediment, therefore, to the transmission of principles as subtle as the human mind can be there encountered. Necessarily, no vibrations or undulations of a *material* character can possibly occur there: no light, nor heat, nor magnetism, nor electricity, can be generated there. It follows that no force can be employed or wasted there. Moving bodies must consequently act and react reciprocally upon each other, as if in actual contact, and in this way generate local phenomena in one another, as if by mutual impact. Worlds, as special chemical entities, thus appear to be strictly economical in the use of the forces with which they are charged. Space thus appears more clearly as an illimitable wild, where suns and planets have been ordained to push and pull, and rush and roll around each other, in order to unfold each other's *innate* capacities; and, for the accomplishment of these wonderful ends, they incessantly retain the primal forces of attraction and repulsion, transforming them into local work in their individual spheres, and into the secondary forces of light and heat, and into the general mechanical motion by which the unstable equilibrium of the cosmos is sustained and perpetuated. And since force and matter alike abhor vacuum, inertia, and inactivity, it follows that they must for ever cling together; and that space is only so much the more a vacuum, inasmuch as it is constituted with negative characteristics in regard to time, force, matter, creation, and every principle involved in the developments of nature. The present state of experimental knowledge, when brought to bear upon this point, seems to sustain these inductions as positive truths.

It follows naturally from these conditions that not an iota of force of any sort—neither molecular force, mechanical force, nor heat, nor light, nor motion in any of its "modes"—can be wasted or lost in space. Consequently, while observation and experiment both show that a dynamical element is inseparable from cosmic light, as seen in the polarization of the so-called "solar ray" (a phenomenon in which the equatorial law affecting the decomposition of energy observable in impact, already spoken of in chap. viii. sec. 17, is delicately displayed), and while all theories relative to the origin and transmission of cosmic light and heat demand and incorporate this dynamic element in some form, it is clear that the hypothesis of Descartes can be no more sound at its root than that of Newton, for the reason that the former, like the latter, *assumes* an incessant and enormous emission of the absolute working energy of the heavenly bodies, the very energy which sustains their motions and on which the very existence of creation depends. Helmholtz and other eminent authorities have shown mathematically that all the motions of the heavens must finally cease, and the present system of the universe sooner or later come to an end, if the mechanical forces of the sun and stars be radiated as waste heat into the frigid wilds of space. Fortunately, at this point the science of chemistry steps in, and shows us, on the contrary, that the necessities of matter cannot spare force, and that the economy of nature will not permit its energies to rest for an instant latent or unemployed. This science furthermore shows that every molecule which exists claims and owns and uses a special quantum of attraction and repulsion, and that it gives away neither, and only suffers their fluctuating exchanges of potency, and their decompositions and conversions into correlated principles, in order to elaborate and perfect the whole machine of which it and they constitute indestructible parts. In his treatise on Heat, Dr. Balfour Stewart has remarked with great sagacity that " the *test of a theory* consists even more in the new facts " which it brings to light than in the facts already known " which it serves to explain."

30.

Now, having laid down and somewhat developed these important principles, we shall be the better enabled to comprehend and decipher the truly cosmical character of the hieroglyphics inscribed by the lines of Frauenhofer upon any spectrum.

Crystals being designed and built by nature and art upon

exact mechanical principles, and sustained in specific geometrical conditions by the forces of molecules—attraction acting within their outlines from planes to centres, and repulsion reacting from centres to angles—a series of perfect, homogeneous, transparent prisms may be regarded as a specially effective instrument of discovery, and the most suitable one for receiving, dissolving, and reducing to ultimate analysis the solar and terrestrial forces which meet, coalesce with, or reject one another within its range of action. While naturalists possess microscopes, astronomers telescopes, and chemists agents and re-agents for exploring and determining otherwise hidden conditions of *matter*, physicists now command instruments also for exploring and reaching those otherwise obscure principles of action and reaction constituting *force*—the motive *energies* of matter. By extending the limits and nomenclature of the Newtonian philosophy, so as to recognise and embrace repulsion as a cosmical power, this abstruse department of research will become much simplified. Absorption, so manifest and determined in the black lines of the spectrum, receives its true physical expression and profoundest analytical meaning—indeed, its absolute solution—in the word *attraction;* for the solar ray, combining in itself the actinic essences—the microscopic and infinitesimal elements, so to speak—of all the forces embodied in the sun as dynamical principles, impinges upon and penetrates the molecules and forces of our globe as an invisible unit of force. In contact with terrestrial matter and force the ray splits and dissolves. Portions of its energies at least are attracted by and penetrate into, are absorbed by, assimilate with, and apparently replace, other amounts of force which simultaneously radiate from the earth as *repulsion*, and which become sensible in the spectrum as bright lines or *points of reflection.* And I may be permitted here to say, in order to set forth my views with clearness, that I think, could an observer in the sun analyse a ray of terrestrial light by a prism, these bright lines would be there replaced in the spectrum by dark ones, the solar molecules becoming in turn excited into similar but opposite absorptive and refractive functions. We have reasons to infer this the more strongly from what we ourselves observe in the sun in connexion with the dark spots; for their number and magnitude as a whole, are now being conclusively shown by the Kew observations, as well as by universal magnetical data, to hold special relations to the attractive and repulsive forces (the agents which generate light) in other planets besides those of the earth. Herein, I hold, lies the subtle mechanism, the occult essence, of electrolytic and actinic vibrations, and dynamical motion: for matter being saturated with force can hold no more than given sums. Repulsion and attraction, as radical energetics, are quantitatively equal in the whole universe; and although the several elementary substances themselves are variously distributed in different globes and unequally charged with these two forces, yet a globe full of force must always be like a cask full of water; and if a quantity of *ab extra* force were thrown into it no larger than that in a ray of light, a similar quantity must gush out. In this functional action of force lies not only the wonderful chemical mystery of the Frauenhofer lines, but also the secret mechanism through which both local material motion and eruptive and projectile dynamics are initiated, and through which the universal motions and phenomena of the cosmos are sustained;—all depending, whether molecular or stellar, upon the mutual influences of these two opposite forms of force upon each other. A ray of light, therefore, emanating from any body contains, so to speak, the veritable image and expression of its internal and elemental constitution of matter and force. Every chemico-spectral observation and experiment upon nebulæ, suns, planets, comets, and terrestrial elements only confirms the truth and universality of this final language of the prism, and of these principles which I have endeavoured to set forth; and they prove the correlation of all celestial mechanical force with terrestrial force and with the forces of molecules, and demonstrate that every molecule and every form of force is linked with universal gravitation through the equally universal principle of *cosmical repulsion.* New and abstruse as this subject may now appear to be as I have attempted to treat it, I feel quite sure that when globes are studied as masses of chemical elements constantly acting and interacting as if in contact, and exciting each other's individual forces as if by direct impact—by "striking fire," so to speak, against or with each other; thus viewed, the present obscurity of this subject will, I think, greatly diminish. But, as it will be referred to again in a subsequent chapter, further remarks upon it will be here omitted.

31.

There is not a fact or law in optics or physics, so far as I now know, which will not happily coalesce with, and consolidate, this final generalization. Even that fundamental law of Kepler, termed "the first law of motion," immediately gravitates within the range of this demonstration; and the principle of the "sufficient reason," as an explanation of the fundamental phenomena of motion, becomes intuitively apparent to common sense independently of logic or the mathematics: for force, playing *between the centres* of the *sun* and *earth,* creates and causes "all motion to be naturally rectilinear and uniform," inasmuch as molecular and mechanical motions are embraced within, and determined by, cosmical principles which act and react as definite units of direct energy. Moreover, the great second law of Newton, declaring "all action and reaction to be equal and opposite," will be here found to stretch beyond simple rational mechanics, and regulate the development of forms and functions of the subtlest and sublimest character.

Even the entire range of physical discussion and mathematical development, so acutely bestowed upon the latter law by D'Alembert and Lagrange, and the profound system of principles so prophetically initiated by Daniel Bernouilli, fall legitimately within this generalization; all tending to confirm the final truths that a pure, abstract, radical force of *cosmical repulsion* exists, and is associated with cosmical attraction, i.e. *gravitation,* and that the fluctuation of these two universal principles with each other in molecules and in suns is necessary to maintain the present mechanical and phenomenal constitution of nature.

Thus, at last, is this part of our sublime and laborious task completed, and the corner-stone of a new and all-embracing philosophy is laid. Rather might it be said that the corner-stone of a new tower is here planted in order to assist and excite scholars to press forward the completion of an imperfect temple the fashion of which was foreshadowed in the abstractions of Thales and Anaximander, outlined in the principles of Euclid and Archimedes, founded in the facts and observations of Copernicus, Galileo, Tycho Brahé, and Kepler, reared in the demonstrations of Newton, and the inside and the outside of whose mighty walls have been overlaid with pearl, and ivory, and precious stones by an immortal host of mathematicians and astronomers, from the youthful Horrocks to Le Verrier, John Herschel, and Frederick Struve.

In order to indicate the existence and potential functions of *cosmical repulsion* as a pure and positive *universal* reactionary force—as much a pure and fundamental principle as *molecular* repulsion—and to prove it to be the same principle, augmenting like attraction of gravitation in proportion to masses and the inherent capacities of matter, we have found it necessary to discuss this force in its relations to matter in every sense and form; otherwise it could only have been treated as a metaphysical abstraction: whereas it is no more a metaphysical abstraction than molecular attraction molecular repulsion, or gravitation, neither of which can be demonstrated or clearly apprehended to be forms of potential force, except through their specific action upon matter. Inasmuch as these various conditions of fundamental force interchange their action with each other by fixed laws, combining, decomposing, and reacting to constitute central forces and dynamic energy, to bind molecules and worlds together, and inspire nature with mechanical and universal activity, to develop motion and all other expressions and modes of force, even to engender morphological energy and govern developments of life, they form diversified subjects of study, and will be progressively unfolded in subsequent chapters.

CHAPTER X.

MECHANICAL FORCE.

1.

THERE is a difference between *mechanical force* and "mechanical power," as this latter has been heretofore understood and treated in theoretical and rational mechanics.

The term "mechanical power" is somewhat deceptive, inasmuch as it is very general in its meaning, and is inseparable from the idea of masses of matter in motion impelled by some external and applied impact. This motion once communicated, and called *vis viva,* or living force, sets forth an idea of mechanical power involving strictly local conditions, and material elements of action.

But MECHANICAL FORCE—*abstract, immaterial, absolute* ENERGY—the life of the world—is different from and more than this, and is not comprehended in the ordinary terms or expositions of modern natural philosophy. As our discussions have developed this subject, it consists in the compositions and decompositions of those potential principles which, existing in all cosmical bodies as the integrals of their molecular forces, generate and control their respective movements of translation and rotation, all their mutual relations, and all their internal eruptive, contractive, and periodic phenomena; and which, so far as our planet is individually considered, play convectively and incessantly between its centre and space, and *vice versâ* (as in all other bodies), effecting all creations and changes upon its surface, and all developments and fluctuations of heat, magnetism, and electricity in its interior and exterior parts, involving crystallization and every chemical phenomenon.

I propose to devote the discussions of the present chapter to the demonstration of mechanical force, as working and phenomenal energy, and of its modes of action in and through matter, in order to generate motion, induce molecules to arrange themselves into geometrical forms, effect cohesion and solution, and excite the secondary forces into being.

If our attempts to discover the origin and mechanism of natural phenomena be at first discursive, appearing doubtful of fruitful results, and somewhat wearisome, reminding the reader of an unsteady search after some precious thing in the dark, the importance and profundity of the points sought after will afford sufficient apologies for fitful and heavy progress. But, ever careful to step upon facts alone, rejecting, as heretofore, all physical theories, we shall be sure to arrive finally at promising conclusions and absolute truth.

A brief survey of the field already traversed, and of the important points established, is here necessary to refresh the mind and prepare it for other equally interesting investigations.

2.

In the preceding chapters the elements or primal conditions of nature have been noticed or elaborately discussed; and they have been traced from their manifestations as molecular quantities, through successive amplifications, into the grand cosmical forms which, representing and containing so many isolated volumes or units of attraction and repulsion, are distributed through space, bound together, and maintained apart, by fixed laws of action and reaction upon each other.

We have been brought clearly to comprehend that not an atom of matter can be annihilated or lost, or can escape the influence of all the rest; that every molecule has inherent capacities for receiving, holding, and exchanging definite quantities of attraction and repulsion, as pure abstract forces; and hence that these forces are in the aggregate also, like matter, given, fixed, and unchangeable quantities established for accomplishing the multifarious economies of nature, as the universe was planned and adjusted in the beginning, and as the Creative Spirit of all things is continuously unfolding and perfecting it.

Beyond these truths follow also the positive deductions that not an iota of attraction or repulsion, as final principles of potential energy, can be annihilated or lost, any more than an atom of matter; that all physical phenomena, whatever their character or method of manifestation, are simple effects of these forces associated with molecules, and of the modifications of these same forces in degree or quantity, associated and exchanging relations with each other among molecules; that all phenomena are definite and positive *resultants* of motions between atoms and between masses of atoms; that both the latter and former, as cause and effect, are dependent upon swift translations of forces antithetically through matter, thereby generating the vibrations of molecules; and that the oscillating actions of these three primal elements with and upon each other, in varying proportions, produce effects always depending directly upon quantities and inversely upon distances.

By this careful tracing of nature from its primordial conditions, we discover the first dawnings of mechanical force, and of mechanical motion; for mechanical force, as virtual power, is really and only the measurable effect of pure abstract potential energy (in other words, of those two forces whose existence and character I have heretofore essayed to demonstrate), accumulated to saturation and tension in, and propagated through, masses of molecules, which masses are brought to bear upon other masses, in such a manner as to produce physical results within the ranges of observation and numbers, and whose compositions and decompositions create the science of Rational Mechanics.

It is not my design, however, to discuss the subject of "rational mechanics," either in its statical or dynamical aspects, as it is considered and treated in ordinary disquisitions upon natural philosophy. Its mathematical elements, and their practical developments and applications, are well understood. They have been elaborately discussed and advantageously employed, as the works of antiquity teach, by every people from periods of earliest enlightenment to our own times; and during the last two centuries the laws of mechanical motion have been extended, by lofty inspirations of human genius, to the most distant regions of space, in order to explain in rational ways the origin, architecture, and stability of the cosmos itself. If mathematical inquiries have thus far failed to solve the profoundest questions in physics, instead of wonder at the incompleteness of theory or at erroneous results of any sort, our amazement must be only the greater at the present stretch and acquirements of the human mind. The origin of every science is accidental, and its progress to a certain point necessarily empirical. While the springs of the various branches of knowledge appear to be distinct, and their courses to be individual and divergent from each other; while indeed we had not formerly sufficient data through which to suspect even their correlation; experiment, observation, and numbers leave us now no room to doubt that all the sciences not only run parallel with one another, but that, however diversified their aspects and currents, all at last plunge into the same fountain of central forces, and receive energy, life, and development therefrom. It is in its profoundest and *radical* directions alone, as before stated, that I propose to discuss this important subject.

3.

Absolute Energy.

Mechanical Force may be developed, rendered tangible, and applied, under the name of "mechanical power," in several different ways; and in such ways, that all the relations of absolute energy to matter and motion—that is to say, to material phenomena; in other words, to "work"—may be easily reduced to numerical quantities, and brought within the practical necessities of society. Its source and its nature, however, as an abstract and as a compound principle, are imperfectly understood. Its modes of action are only known by certain effects developed and imparted by one form of matter to another. In nature these forms are called elements or worlds. When considered as human inventions they are called machines, and their variety is endless. The earth is full of them. Nations live by them. Civilization is unfolded by their perfection. Barbarism depends upon ignorance of their capacities or existence. *"Mechanical Force" itself, however, as abstract energy, acting upon and propagated through these machines, is a part of, and springs immediately out of, the accumulated stores of energy which constitute the motive and hidden forces of the planet's centre and mass.*

4.

Machines.

The machines which manifest and develop this energy are the lever, the wheel and axle, the toothed wheel, the pulley, the wedge, the screw, and the inclined plane. Although these instruments are called "mechanical powers," it is now well known that neither of them create power out of nothing; that is to say, generate force within themselves, and of themselves, independent of a preceding and externally applied force, equal always to the result which may proceed out of them. Their motive energy, then, is derived, as a finality, from *foreign* sources; and we shall hereafter discover that it proceeds out of the earth's store, and returns into it.

Whatever the forms, functions, involutions, capacities, or powers of these instruments, all of them are reducible to two simple conditions or forms for receiving, accumulating, and transmitting force. These are the *inclined plane* and the *lever*. With this simplification and finality philosophers and mathematicians appear to have been for a long time satisfied; and public education rested here, until recently, as if matter itself were force, and as if an acquaintance with final principles were unattainable.

The *principles* of *energy* involved in the development of physical phenomena through the agencies of the *inclined plane* and the *lever* are discovered, when reduced to their final essences, to be respectively *attraction* and *repulsion*.

This theorem, which I hold to be almost self-evident, may however demand some demonstration and illustration, in order clearly to connect established dynamical laws with the operations and real expressions of nature.

Now, inasmuch as I have heretofore shown that this planet possesses definite volumes of these two fundamental forces—simple integrals of the forces of the individual molecules constituting the mass—and that not an iota of attraction or of repulsion can be annihilated or dissipated, it follows, in order for any act or phenomenon to appear, or for any physical work to be effected, that these forces must vary their relations with each other, changing place, generating temporary tension here, and laxity or traction there, so that matter in bulk (and necessarily as separate molecules) may be transiently deprived of *a portion* of its natural quantum of attractive force—only, however, to have this absence *instantly* supplied by an equal quantum of repulsive force; and *vice versâ*. Thus, for instance, wherever successive amounts of *either of these forces* may be locally accumulated (so to speak) in any part of the globe, creating tension or contraction at limited points, this surplus is temporarily abstracted from the rest, and must sooner or later be restored to it, either gradually and tranquilly or suddenly and violently. This occurs in the movements of the various parts of all machines, and is the secret of universal motion.

Inasmuch, moreover, as these forces possess affinity for, and are absorbable anywhere and everywhere each respectively into, its own special principle, and are antagonistic each to the other; inasmuch as these two principles for ever remain distinct and conflicting *radical agents,* even in *all other forms of force,* and in *all forms of matter,* however different their elementary characters (chemical functions seeming to act only in such a manner as to modify the inherent capacities of matter for differentiating and excluding, or for absorbing and holding, more or less of one or the other of these fundamental principles of molecules, or atoms, as will be better comprehended hereafter, or may be understood from Kekulé's idea of saturation); inasmuch as they never lose their respective identities, nor simple ultimate properties of drawing molecules together and pushing them apart, whether they be associated with and act in electric currents, luminous rays, flowing sap, organic corpuscles, flying projectiles, or falling bodies; and inasmuch as they are never in equilibrium, and never at rest, it follows that all mechanical motion, all secondary forces, and all phenomena in nature must spring from quantitative fluctuations of these two forces, one with the other, in their relations to molecules and masses; and that these two forces are the absolute and ultimate elements of work, and of morphological and mechanical development everywhere, whether in the creation of crystals and globes, in the origin and growth of plants and animals, in the lifting and falling of weights, or in the distribution and movements of cosmical bodies in space.

As this planet is endowed with its own specific quantum of working energy, according to its peculiar chemical and general physical constitution, so is every other cosmical body, according to *its* own special composition: and what is said of one will equally apply to all the rest, for there is only just so much matter and just so much force in nature, both being limited, and both being continually employed, acting and reacting, to evolve the local phenomena which take place in comets, planets, suns, and systems throughout the universe, which phenomena in turn become active agencies in, and the mutual excitors of, the celestial motions.

Such is the field, such are the conditions, such the circumstances under which mechanical energy exists in nature—such the fountains from which all mechanical power is drawn, and into which all again in turn falls; so that not a shuttle is driven or hammer lifted by steam, or waterfall, or human arm, or foot of beast, not an arrow shot from bow nor ball from gun, the force to execute which act is not derived from those exhaustless volumes of attraction and repulsion that constitute the central forces of the globe.

5.

Affinities of Matter and Force.

The laws which preside over the *affinities* of matter with forces, and of forces with matter, involve inquiries of exceeding delicacy. Thus far in our discussions we have obtained more or less clear and definite ideas of space, of matter, of molecular attraction and molecular repulsion, of cosmical attraction and cosmical repulsion; and of the actual relations of each of these entities to the other, as *crude and available* elements (so to speak) lying at the base of physical nature and constituting the materials used and directed by some Intelligent and Commanding Agency to execute work, develop motion, heat, light, and other secondary forces. These elements, passing from regions of metaphysics into those of geometrical design and of solid mechanics, become the phenomenal agents of creation, and subjects of experimental study and of academic teaching.

From this point of view it will be discovered that these elementary materials of nature, whatever their origin, lie midway between the Commanding Intelligence which employs and directs them and the designs unfolded and the work effected by them. Efforts to acquire an accurate knowledge of the correlations of the first with the last constitute the highest mission of the physical sciences; while to trace the connexion of the complex and manifold phenomena springing from such correlations with the Great Metaphysical Agency which originated and overrules them all must always be the final aim of the purest philosophy. The former is a strictly positive and inductive study. To aim here, however, at such a development as the latter would be both premature and out of place. I therefore only allude in this connexion to that Supreme Intellectual Principle as a universal presence in nature, reserving a general treatment of this subject for a final chapter. Confining myself in the present chapter to strictly physical investigations, I shall proceed to elucidate the mode and mechanism by which attraction and repulsion, those abstract immaterial entities which link mind with matter, and infuse or transform the spirit and power of the Creator into dynamical, geometrical, morphological, and vital functions, act through atoms, molecules, and masses, in order to initiate vibration and oscillation, evolve mechanical motion, originate the secondary forces of heat, light, electricity, and magnetism, with *their* polarities, and bring forth unending successions of other phenomena from apparent chaos.

6.

Motion and Polarity.

Motion of molecules is the *first sensible* or *perceptible* effect of the influence of force upon matter, or of the relations of matter and force; that is to say, of the subjugation of crude physical material to metaphysical and phenomenal working power.

We see and learn this radical dynamic fact by microscopic aid in acts of crystallization; and in these acts the two fundamental forces which have heretofore been described clearly display opposite and definite functions. These forces invariably direct their energies in the same manner, to similar ends, and always develop identical mechanical results. It matters not whether the results be symmetrical or amorphous, complete or abortive. The motive act is unique. The initial design of nature is the same in all. The final and finished work in crystals is the sum and expression of all the molecular forces of attraction and repulsion expended upon them, whether their construction be perfect or imperfect. Geometrical form, metaphysical preconception, intellectual design, morphological and mechanical activity—in a word, all the principles of creation as we trace and behold its unfoldings—coexist and combine with the action of these forces everywhere, wherever their energies or influences can be detected; and observation traces their uniformity of action upon molecules from the infinitesimal motions revealed by microscopes to the grandest mechanical results unfolded in the motions of the celestial masses. Molecules, according to principles definitely specialized and unchangeably inherent in each of the known elements of nature, demand, accept, and absorb these radical forces in variable volumes. And in virtue of some inherent directive principle, of certain numerical laws of union and dynamic conflict between these two forces emanating from a central molecule; in virtue of such inherent laws of action and reaction of force, molecules arrange themselves into amplifying accretions until the basic elements are exhausted upon which attraction and repulsion exert their respective action and reaction, as the mechanical and working radicals of creation.

It is here that we catch the first glimpse of the influence of force upon matter, and receive the first clear conception and solid knowledge of the positive and indissoluble connexion of one with the other. It is here, too, that we discover the origin of polarity, and obtain the first clear and definite idea of *its* character. It is here that we begin indeed to comprehend that "Polarity" is only a *name* for a LAW which governs the cosmical distributions and the telluric and local translations of force; it being a constant condition—the basis of all phenomena whatsoever—dependent *incidentally* upon matter, but *directly* upon antipodal and antithetical transmissions of the two co-ordinate and coefficient forces of attraction and repulsion, from surface to centre, and from centre to surface, alternately and incessantly, axially and equatorially, in all matter, whatever its form, quantity, place, or relative position to other matter; a law inseparable, as a necessary consequence of the existence of opposing forces, from the fundamental nature of things. Were only one molecule the subject of observation and experiment, the subtle effect of these forces, and of their functions in generating the *initial* phenomena of polarity and motion, could not be discovered. But since molecules are not separately subjects of observation and experiment, and can never be isolated, being on the contrary infinite in number, the polarization of each —that is to say, the *directive action* of force, or, to speak more precisely, the antipodal and antithetic play of the attractive and repulsive forces in and through each—becomes not only observable in definite lines of motion and parallelism, but traceable also to *dependencies* upon the combined forces and motions of all the rest in the globe, *as voluminous sources of power,* and upon the united energies and general principle of polarity, existing as a unit or massive quantity, so to speak, and constituting the directive law in the dynamic functions of the cosmos. All this exists in virtue of the cosmos being a complex and oscillating yet perfected and self-regulating machine; and in virtue of an arrangement of spheres *compressed by attraction* from the galactic poles, and *projected by repulsion* through the plane of the Milky Way—specialized physical conditions permitting no equilibrium and no rest in any world, force, molecule, or atomic principle which has been created.

POLARITY, then, the *principle of dual law* everywhere

pervading motive force, is evidently a concomitant of the first mechanical act in nature, and must have been simultaneously infused into the constitution of molecules at their primal union or endowment with force; and, as a necessary corollary, every one of its phenomena, whether observable in matter under the influence of, or in connexion with, the secondary forces of heat, light, magnetism, &c., is, and must be, inseparable from molecular motion.

Now, since microscopic and atomic motion cannot take place, be seen, or be imagined, except by the action of antithetic forces upon several molecules, it follows that *molecular attraction and molecular repulsion must be the fundamental elements of potential energy;* that neither alone and by itself can create, induce, or awaken successive and continuous phenomena out of nothing or *vis inertiæ;* and hence that both as ultimate principles are the mysterious powers which, springing from some inexhaustible source, glide into, grasp, and move molecules in opposite directions, and mould them into crystalline forms.

This source we have already indicated in preceding chapters to be the massive sums of energy stored in each world for its own special uses, which sums in turn are the simple aggregates of the forces of the individual molecules that are associated as chemical elements, and that constitute the entire masses of these globes. The manner in which these forces act and re-act, are apparently decomposed and re-united, and consecutively exchange their functions so as to perpetuate motion, develop multifarious phenomena, and still never to dissipate their energies, will appear with ample distinctness in the sequel.

7.

This initiative act of the creative forces, so demonstrable by microscopy, to which I have referred, and which accompanies the first appearance of matter and solid form in their sudden birth from chaos, is a purely mechanical phenomenon. It occurs in the same manner in every one of the sixty-four elements of nature which can be reduced by any experimental means to solid conditions. Force suddenly entering and energizing a given point, an instantaneous elective influence darts from this central molecule, which influence is inoculated more or less rapidly into all other molecules possessing similar chemical properties; and the general fundamental and initial forces of attraction and repulsion pass through polarization into cohesion and crystallization, geometrical development, and substantial creation. All elemental materials are acted upon more or less alike by these fundamental forces; and thus the latter appear to develop themselves into agents of cohesive and formative energy on one hand, and of projectile and destructive energy on the other. For at the same instant when a crystal springs into being, a radical and repulsive force—not heat, nor cold, nor magnetism *per se*—projects molecules from its centre into its angles and pyramids, and determines its cleavages; while attraction smooths its planes, determines its density, complicates its capacities for refrangibility, and binds its shortest axes to that same central molecule which first inoculated the entire crystalline mass with mechanical energy. As a *result* of this process, as a concomitant of this primal molecular motion, heat, cold, magnetism, electricity, or light may and do appear so sensibly as to suggest the possibility of the sudden transformation of our radical forces into other and new forms of energy.

The fundamental forces which initiate motion among molecules under the microscope are, therefore, no other than those which initiate and propagate motion everywhere; which originated, or rather developed, themselves into mechanical energy in the beginning; and which perpetually create and maintain it as the veritable working power of nature, giving birth to form and life, and laying the foundations of science, art, and civilization.

Now, therefore, since we know by experimental study and

general observation how broadly and profoundly these principles of mechanical force and motion extend—an extent, indeed, so great that the entire concrete globe (at least, the solid parts thus far explored) appears to be only a conglomerate of crystals either perfect or amorphous, including every element in some combination or other;—since all this is known, we discover that the crystalline structure of our planet is of itself one vast storehouse of composite, mechanical, and convertible energy.

8.

It is here, while desirous to avoid repetition, and at the risk of appearing desultory in this branch of my discussions that I am compelled to reiterate certain data, in order to correlate them with the *compound* action of those agencies which have been heretofore treated of in their broadest aspects and in a separate sense, as distinct entities with potential functions linked with matter, and both of which are ready, so to speak, to burst into activity, and to unfold through their conflicts the varied phenomena of the cosmos.

Comprehending with clearness the important facts stated in the last section as definite steps in the development of our analysis, and looking still further into the interior of the earth in the light of any and every theory of its physical constitution, we discover its central energies to be allied to its molecular conditions, whatever these conditions may be, whether solid or fluid. If solid, we may know the laws and quantities of force involved in their solidification, as certainly as we know those exhibited in the condensation of molecules under the microscope. If fluid, we know with equal certainty the crystalline changes of which this fluid is capable, and which it may sooner or later undergo, whatever the name or nature of its ultimate elements. The facts developed in chemistry and mineralogy indicate that the potential energy entering into these physical processes, and ending in concrete and crystalline creations, whatever they may be, is measurable, definitively exact in quantity, and absolutely dynamical in character.

By the same method of inductive research, adhering rigorously to laws and evidence discovered by physical, chemical, and mathematical analysis—purely experimental and exact inquiry—we know that every gaseous element, whether embodied in air or sea, or expelled from the bowels of the planet, is subject to the action of the same mechanical energy. Oxygen, for instance, an element constituting more than half of the measurable and ponderable matter of the globe, combines and crystallizes, or is capable of doing so, in some manner, with its multifarious associates. Nitrogen, hydrogen, and carbon submit to similar action, as nitre, ice, and diamonds testify. Now the pure force—the *vis viva*—the mechanical energy which all these elements are capable of absorbing, holding, appropriating, employing, discharging, and exchanging with each other, as they pass from one state to another, as they are held in expansion or rendered solid, is, as a cosmical volume, neither more nor less than a given quantity to which no others can be permanently added, and from which none can be permanently abstracted; so that their share of the terrestrial aggregate is ever the same, and equally liable to vanish and reappear, shift and fluctuate with all the rest, in order to constitute the working energy of the planet as an isolated unit or volume in space.

If we turn from the gaseous and aqueous envelopes, and, penetrating the earth's crust, search for the secrets within its bowels, we are no less interested and instructed by similar revelations from these quarters. Not an elemental particle, either of silicate, sulphur, or metallic vapour, cast forth by plutonic eruptions at any period in our planetary history has escaped the grasp of these pre-established polar forces. For, more omnipotent and general than special chemical affinities, they dart through all matter with equal speed, guide all motion with uniform certainty, and stamp mechanical activity, geometrical symmetry, intellectual design, and manifold possibilities, upon every molecule which even apparent chaotic confusion may hurl from the centre of our sphere.

Hence we discover and conclude that there is not a molecule, whatever its chemical properties, from the outermost boundaries of the gaseous envelope to the centre of this globe, which is not the subjugated and menial servant, so to speak, of the cosmo-mechanical forces; which does not accept the influences of these agents with instant celerity; and which, vibrating in virtue of their rapidly varying tensions, does not yield them with similar promptitude, when the equilibrium of nature elsewhere requires such resignations. Thus, one element in this varied storehouse of forms and creations is ever ready to pour, and is incessantly at work in pouring, its overflowing energies into another, in consequence of their conditions of physical connexion and dependence and of their mutual relations of demand and supply—all *minor conditions and relations* appearing as eternally shifting necessities in virtue of *cosmical conditions and relations.* These latter, being universal, are consequently supremely potential; and the varying functions of the former thus become linked with the ellipticity of all planetary orbits, and with the *excentric* position of controlling bodies in those orbits. The functions of attraction and repulsion viewed as molecular entities or attachments, and as subjects of incessant quantitative changes, are, through the influence of these outlying and overruling cosmico-physical circumstances, necessarily and unavoidably made to become causes and affections of polarity—that is to say, of combining and decomposing activities possessing simultaneously opposite directive characters ; and such polarities and activities, studied in the light of general observation and of experimental and applied physics, become in turn the exhaustless fountains and currents of energy (exhaustless because incessantly transformed into one another) which not only push and pull molecules under the microscope, thus initiating mechanical motion and the work of crystallization, but also push and pull masses of these molecules and crystals, multiplying power as the latter aggregate in bulk and receptive capacity, and running through every grade of matter and motion, from the tiniest wheel and axle to the mightiest stars and the universal firmament of heaven.

9.

Identity of Cause and Origin of Molecular and Cosmical Motion.

Of the final cause and certain origin throughout space and matter of the phenomena of polarity and motion as the dual and inseparable antecedents of all other phenomena, and of all mechanical and morphological developments in nature—the fundamental points specified and involved in this vast generalization—there can scarcely be room for doubt in the mind of any physicist who has thoroughly examined the subject. Observation and study in various physical directions are steadily tending to place these truths beyond question. In order, however, to establish them positively, and to render incontrovertible every successive point gained in our analysis, I present in this connexion a physical fact, the discovery of which is due to Laplace, which fact (even if somewhat modified by recent analyses) may be justly ranked among the most important which mathematical research has added to our present stores of knowledge. The weight and bearing of this discovery will be seen in its manifold consequences both in molecular and cosmical results, as we proceed to search into the sources and delineate the character and action of mechanical power. And even should the error exist in the deductions of Laplace which has been asserted to exist by Hansen and Adams (a point upon which mathematicians do not agree), the general result of the calculations and theories will not alter the bearing of the fact as I wish to apply it, inasmuch as I hold that the interior forces of the planet execute the work effected upon its surface, and that no force, notwithstanding all its changes and fluctuations of function, can be radiated into space, and dissipated or alienated from the *matter* of the globe.

It appears from certain deductions of the immortal author of the "Mécanique Céleste," founded upon the lunar observations of Hipparchus (B.C. 150), and upon those of modern astronomers—observations made at an interval of 2,000 years, furnishing consequently as a basis of computation 2,000 revolutions of the earth around the sun, and more than 730,500 revolutions upon its own axis;—it appears by the deductions of Laplace from these data, that only the slightest changes occur from year to year in the length of the sidereal day; that while the former, the orbital motions developing the sidereal year, never change (a point on which all agree), the latter, that is the diurnal rotations, show a possible loss of $\frac{1}{300}$th of a second of time in the course of twenty centuries. Hansen of Gotha, Adams of England, and Delaunay of France, however, in revising the analysis of Laplace upon the lunar influences, have reached the conclusion that the earth loses a larger fraction of a second in each sidereal day; so that they, and those who rely upon their accuracy rather than upon that of Laplace, suppose that there has been an absolute retardation in the earth's movements of about the $\frac{1}{84}$th of a second at most during the last twenty-five hundred years. But instead of supposing the force represented in this loss of rotatory motion to have radiated from the planet, and to have been absolutely dissipated in the interstellar spaces as waste heat, some of them at least, adopting Mayer's hypothesis, imagine it to have been consumed within the planet's own boundaries in the friction of the tidal currents upon the bottoms of the seas. In order to set forth clearly and fairly the uncertain results of modern calculations upon this point, and to show indeed that the calculations of Laplace may not finally be found to be more inaccurate than those of his successors, I will cite the closing passages of a treatise on Astronomy just published by my friend Mr. J. Norman Lockyer, which may be considered to express the latest and most mature scientific opinion upon this subject. Lockyer says: "As the tidal "undulation does not move so rapidly as the earth does, as "it is regulated by the moon, it appears to move westward "while the earth is moving eastward; and it has been sug-"gested that this apparent backward movement *acts as a* "*brake* on the earth's rotation, and that, owing to the "effects of tidal action, the diurnal rotation is, and has "been, constantly decreasing in velocity to an extremely "minute extent. *At all events, if the sidereal day be assumed* "*to be invariable, it is impossible to represent the moon's true* "*place, at intervals of* 2,000 *years apart, by the theory of* "*gravitation.* On this assumption, the moon, looked upon "as a timepiece, is *too fast* by 6″ or 12 seconds (nearly) at "the end of each century. This may be due to the fact "that our standard of measurement of the sidereal day is "*too slow;* and it has been calculated that this part of the "apparent acceleration of the moon's mean motion may be "accounted for by supposing that the sidereal day is shorten-"ing, in consequence of tidal action, at the rate of $\frac{1}{66}$th part "of a second in 2,500 years."

In the address delivered February 9, 1866, by Warren De la Rue, Esq., as President of the Royal Astronomical Society, on the occasion of the presentation of the Gold Medal to Professor Adams, Director of the Cambridge Observatory, he says: "After the thorough investigation which the problem "of the secular equation of the moon's motion has received "at the hands of so many distinguished mathematicians, there "can no longer be any question as to the correctness of the "value of the coefficient determined by Mr. Adams and M. "Delaunay. But here comes a difficulty, which, however, in "no way casts any doubt on Mr. Adams' theory; it is this: "that the coefficient 5·7″, or 6·11″, as M. Delaunay has made "it by further developing the series, does not account for the "amount of the secular acceleration as shown by observation. "It is perfectly true, notwithstanding some uncertainty as to "the actual date and place of the ancient solar eclipses, and also

" as to the correctness of the records of certain lunar eclipses, " that Professor Hansen's value of the secular acceleration, " namely, 12·18″, accords better with these phenomena than " the smaller number 6″. . . .

" We thus see," continues the President in the course of this able address, " that the re-examination of Laplace's theory " of the secular equation by Mr. Adams has indirectly led to " a more formal investigation of how far the effect of the sun " and moon on the protuberances of the great tidal wave may " produce a retardation of the earth's rotation equivalent to " the difference between Hansen's coefficient, 12·557″, and " Adams' and Delaunay's value, 6·11″."

The accuracy or the inaccuracy of the calculations or opinions of either one or the other of these distinguished authorities is not a point to be questioned or introduced at length into this discussion. The bearing of these differing analyses and of the opinions of their authors, however they may conflict with each other, is the same upon my position and my arguments: for while Laplace found that no motion was absolutely lost or force wasted in space, Delaunay and others conclude that whatever *appears* to have been lost, has not really been so; but, on the contrary, that its equivalent, as heat, has been converted into friction and consumed within the envelopes of our planet.

Such physical discoveries, based upon the observations of Hipparchus, are indeed among the noblest fruits of recent research; and I introduce them in this connexion in order to unite the larger and smaller points of my discussion with each other, to determine as the subject is developed the identity of both the origin and the causes of molecular and cosmical motion, and to show the means employed by nature to conserve the potency of her energies.

That *motion*—the embodiment or representative of work—the phenomenal expression of mechanical power—can be reduced to exact quantity, definable in numbers, after such expenditures of energy upon a ponderous mass of molecules like our globe during a period of 2,000 years, the results of which can be seen through the calculus with one eye, so to speak; while the working of the selfsame energy upon the selfsame molecules can be detected and defined with the microscope by the other eye, and this work be measured and computed also;—these facts are a marvel as pregnant with future developments of knowledge, as they are at first startling to human comprehension. Discoveries and results so wonderful, so opposite, and still so cognate, phenomena so universal yet united and inseparable, bring us face to face, in the present state of scientific inquiry, with every "*dynamical theory*" which treats of "heat" as an *ultimate* "cause" or "mode of motion," or as the radical principle of mechanical energy; or which treats of motion and radiation, or of the dissipation of energy, anywhere or everywhere. But as my object is strictly to avoid the discussion of hypotheses, and only to seize upon well-observed and controlling natural phenomena and upon authentic experimental and mathematical facts in order to bind them indissolubly with *the final causes* of their origin and identity, I allude to theories in general only for the purpose of expressing my dissent from all which have been hitherto announced.

10.

Up to this point, we have reached *two positive positions;* one of a microscopic, the other of a cosmical character. Both concern motion, as primal and initiative, or as final and accumulative, functions and effects of force. In the first, the ultimate action of attraction and repulsion as final and radical molecular entities is not only distinguishable, but also *seen* to swell gradually, and more or less rapidly, into measurable quantities. In the last, the sum of all the molecules constituting the planet itself, and the sum of the individual forces of these molecules, constituting on one hand cosmical attraction, and on the other hand cosmical repulsion (the latter constituting the mighty problem for the solution of whose secrets the theories of phlogiston, caloric, latent heat, elastic force, have in turn been originated), become so vast, and enter into such relations to space, and time, and the complicated mechanism of the cosmos, as to be subjects of plain observation and of exact measurement and calculation also. They become, furthermore, I hold, the initial and co-efficient principles of a solid system of mechanical philosophy ;— not disjointed negations in hypotheses founded upon gravitation as a unique force counteracting inertia, or exciting it into activity and modifying or annulling some foreign projectile impulse; nor upon heat filling space and saturating a universal luminiferous ether; but co-ordinate elements in a system of ultimate truth, neither of which elements alone, and unrelated to the other, is capable of producing or explaining any of the constant or periodic phenomena that methodical observation and experimental study are successively exposing, embracing chemical affinity, electricity, magnetism, light, and even heat itself. The single positive term in the present theories of celestial mechanics, and its negative, *vis inertiæ*, have been sufficiently noticed already. The assumption that attraction in any degree, in conflict with *vis inertiæ*, or nothing, should or can initiate or create and *perpetuate* motion, heat, &c. indefinitely, and develop all known microscopic or cosmical phenomena, is not only open to gravest doubts, but at this stage of our discussions begins to take the shape of an absurdity. Indeed, the information we now possess, founded upon observation and experiment, and upon the application of the mathematics to positive data, plainly indicate that some other radical principle of energy besides attraction must be at work in matter to produce the changes which are incessantly taking place around us and in the distant regions of space.

11.

Identity of Cosmical and of Mechanical Energy and Motion.

It will not be disputed in the present state of knowledge that cosmical motion is mechanical motion, or, *vice versâ*, that mechanical motion, in whatever form observed, is cosmical motion subdivided and distributed wherever one mass of matter is dislocated and locomotive upon or around another mass, however small or large these masses may be. Neither will it be denied that cosmical motion is effected by the activity of voluminous energies inherent in and circulating through and saturating cosmical masses as the sum of their atomic life, so to speak, or of their molecular forces, acting and reacting upon each other, as a general system of accurately adjusted but oscillating powers. Hence, minor quantities of dynamical motion, and all phenomena developed upon the surface of any cosmical body—the earth for instance—whether observable by microscopes, or in tides, waterfalls, wheels, levers, complicated constructions, or in animal muscles, are clearly effects of inductions and absorptions of so much cosmical force, drawn out of the general mundane reservoir, as may be required to perform definite local work; and this cosmical force is hereby converted into the local phenomena of creation. The results of motion are equivalent, I hold, to the sum of the two radical forces that initiate and create it, whether in the obedient molecules which in microscopic fields spring to, and arrange themselves around, crystalline centres, converging to and diverging from such centres upon exact geometrical and mechanical principles of action and reaction; or in planets, simple aggregated masses of these same molecules, that spring to and arrange themselves around cosmical centres, converging to and diverging from such centres also upon the same exact principles of action and reaction, so demonstrable by mathematics in all geometrical forms or conditions of matter and force. That these principles operate in their cosmical amplitudes in a similar manner as in their microscopic infinitesimals is proved by the approximative computations of Laplace, Hansen, and Adams: wherefrom it becomes apparent that the forces which cause the earth's rotations upon its axis and its revolutions around the sun are so definite in quantity and effect, as dynamical energetics, that this vast globe of heterogeneous elements has only lost, the $\frac{1}{300}$th part of a second of time, according to the former, or the $\frac{1}{66}$th of a second according to the latter, in the sum of all its motions for 2,000 years. During this period the planet itself has travelled through space, under the potencies of its own attraction and repulsion, reacted upon by solar attraction and repulsion, not far from one trillion, one hundred and ninety-three billion, eight hundred and four million, nine hundred and sixty thousand miles (1,193,804,960,000), while every point upon its equator has rolled additionally, during the same period, about 5,332,000,000 miles in the course of 730,500 revolutions upon its axis. Such is only a fragmentary part of the work performed by the earth viewed as a wheel in the machinery of the solar system. The periods of axial rotation having been comparatively unchangeable for 2,000 years, and those of heliocentric revolution wholly so, it becomes self-evident that the inherent forces and absolute gravity of the terrestrial mass engaged in these motions has not changed to any perceptible degree—that is to say, that no measurable amount of matter or of force has been dissipated nor added, from year to year, during this long epoch.

Now, since the absolute weight and forces of the planet's mass are only the absolute weight and forces of all the different elements combined and calculated in the aggregate, as a unit and mean quantity; and since Newton's second "Law of Motion," viz. that "action and reaction are always equal and opposite," is incontrovertible and of universal application; it follows, if the quantity of attractive force has suffered no dissipation, that the quantity of repulsive force has not altered either. Since we know that molecules and their radical forces are allied and inseparable, that atoms are points specifically saturated with attractive and repulsive forces, and that under all circumstances they maintain the same general quantitative relations to each other, this important fact is the more positively established. Moreover, every experiment, observation, and deduction assists to demonstrate that these two orders of motion—one of the molecule, the other of the planet (a simple aggregation, as we all know, of the same molecules)—are identical, and resolve themselves into attraction and repulsion; which, differing only quantitatively, compose and decompose their functions, so as to become the very mechanical energy which is transmitted through the lever and inclined plane from one mass of matter upon the earth's surface to another, like a fluid or like the diffusive properties of gases, and which everywhere acts in manifold ways and with instantaneous effect to develop the countless phenomena of physics, and execute the entire work of nature, of animal necessity, and of human ingenuity.

Lest this statement of a physical truth may not be immediately comprehended by those who study nature from standpoints different from those which I have endeavoured to foreshadow, it may be here remarked that I regard the action and the reaction of pure abstract force through matter and machines, in virtue whereof the movements of both the latter are effected, to be similar to the flowing of liquids; and I hold that either of the elements of force is capable of transmission through matter by processes of the subtlest delicacy, in like manner as fluids are capable of transmission through tubes, or the magnetic element through conductors; and that general material motion is a simple result of the translation of these two forces in bulk, so to speak, in one direction or another, up or down, in the aspect of a push or of a pull, as one or the other of these antagonistic forces predominates in its tensions and in its specialities of function. Either one or the other, when it actively seeks to unite with its cognate, or to separate from its antithetic, constitutes the cause of motion, because it sweeps matter and masses along with it in its course. This is the general expression of a truth which has been stated in antecedent chapters. But as we are now searching into the occult causes of motion and dynamical laws, both molecular and universal, and are endeavouring to reach the collateral secrets of the conservation and convertibilities of force, the fact may bear repetition in order to bring more distinctly into view another and a correlative fact, which will more fully appear hereafter, that the transmission of these forces in opposite directions constitutes those compound conditions of motion which resolve simple mechanics into phenomenal physics.

Now, since the cosmos is constituted of separate bodies, each of which is composed of chemical elements differing more or less in quantity and character, and charged with forces of its own; and since all these bodies are simultaneously rushing either to combine or to separate their forces, darting from molecule to molecule, from hemisphere to hemisphere, and from centre to surface and *vice versâ*, with subtlest delicacy and instantaneous rapidity, as they approach and recede from one another; it follows that rotation and mechanical motion, as general or simple phenomena, are the results of the subtle action and reaction (a compound function) of the three general elements of nature,—matter, repulsion, and attraction. The interferences and conflicts of these three conditions of nature, incident to general and molecular motion, end in their universal decomposition, and in their transformation and development into other principles, as heat, light, form, &c., the elemental laws of which I trust to be able to unfold as these discussions advance.

12.

Thus, from all which precedes, it clearly follows that a fixed quantity of potential energy exists in this planet—a capacity for performing a definite amount of motion; and that, for the last 2,000 years at least, the capacity of the planet for work, and the mechanical work really effected by its contained energies as an individual wheel in the mechanism of the cosmos, have as a whole undergone little or no alteration. It would not indeed be too much to say that the conclusions derived from the lunar theory, under the analyses of Laplace, and of Adams Hansen, in no way favour any hypothesis which assumes for a basis, or admits as a consequence, the dissipation or the exhaustion of the actual working energies of the planet or of the universe. If I do not here discuss exhaustively the views of Sir William Thomson upon the "dissipation of energy"—views which have many able advocates—it is because I wish, for the present at least, to avoid this discussion; since it would be out of place, and since the main object of this work is to reach points from which such doctrines may be examined in a clearer light than the present state of physical knowledge admits of.

From what has preceded, then, it will be seen that all the work performed in the earth, as an isolated cosmic mass and as a limited reservoir of energy, from the motion, arrangement, and cohesion of individual molecules in the crystallization of various elements to the motion and products of machines and waterfalls, from the motions of matter within, upon, and around the planet to the motions of the planet itself upon its axis and around the sun, must be executed by forces embraced within itself. And since no physicist will admit that a single molecule has ever dropped out of the globe or been annihilated, the preceding deductions may be considered a self-evident theorem; for every one of the sixty-four or more elements constituting the mass of the planet is specifically endowed with its own combining and resolving powers: and since their crystallizing characters exhibit their ultimate capacities for holding mechanical energy, as an entity and fixed quantity, we may clearly conclude that the earth cannot spare, and will not alienate, any of its appropriated energy. The simple manner in which this prodigious volume of energy is distributed and employed upon the surface of the planet, and the progressive developments of matter into form effected by its oscillating activities, we shall comprehend hereafter.

This truth, therefore, becomes apparent, that as motion is the first perceptible physical effect of the influence of force upon matter—that is to say, of attraction and repulsion upon molecules, by which influence each becomes subjugated to and affiliated with the other (a phenomenon seen to unfold under the microscope, from inchoate into measurable terms)—so also motion accumulates and assumes sublime developments when molecules and their respective forces have combined and cohered into cosmical quantities, thus presenting endless themes of study to chemists, geologists, astronomers, and philosophers.

13.

Up to this stage of our inquiry, the elements of nature have been treated as ultimate and chaotic conditions of being, as abstract states and capacities of matter and force, out of which were to spring subsequent developments of a more definite and varied character. Space, matter, and conflicting forces only have appeared; the three latter springing seemingly from nothing, and progressively amplifying from infinitesimal molecules with their individual capacities and endowments into vast expansions and multitudinous forms, the bounds of which, though clearly fixed and limited, cannot be measured or positively known. It is in this Space, with this Matter, and by these Forces, that MOTION and TIME have been initiated, in order to afford opportunity for those mechanical, organic, vital, and intellectual designs, conditions, and events, which have been established and successively unfolded by the Author of Nature.

This is creation—the *beginning*, when the heavens and the earth were made, and when the earth was without form and void. All geological observation establishes the fact that there was a period when our planet was devoid of vital and locomotive being; but that was prior to the granitic ages. Matter and force existing (how and when we will not here inquire), motion and time began.

14.

The Conversion of Mechanical Energy into the Secondary Forces through Motion.

Material motion having been initiated by the action of our two radical forces, and as a consequence of their tendencies to act *antithetically* in their relations to one another upon and through molecules, and in virtue especially of the varied characteristics of the different chemical elements—that is to say, in virtue of the different points of saturation to which elementary atoms can absorb and hold these forces;—these entities and their specific laws of action and re-action having been once initiated, their transformations into heat, light, magnetism, and electricity would inevitably follow; and we find that these latter types of energy do instantly appear as *secondary* phenomena. All experimental and inductive inquiries illustrate and determine this truth; and they furthermore expose the remarkable fact that as motion initiates, or transforms itself into, these secondary forces and phenomena, these latter in turn transform themselves back into motion, so that the quantity of one being known the quantity of the others can be mathematically determined, and the wonderful discovery be finally and surely reached, that the mechanical energy appropriated to this planet (and to every other cosmical body) is a fixed working quantity, which will find its radical in cosmical repulsion, and thereby become correlated with gravitation, the principle of stability upon which the axis of creation, so to speak, turns.

Thus are all the forces and all the elements of nature shown to be linked inseparably together, and to be also exact quantities, limited by weight and measure, meted out and squared or cubed one to the other in smallest or greatest degrees, by an architect and geometer whose abiding-place is everywhere, and whose future plans we can learn only by what is past.

The initial conditions of molecular attraction and repulsion as here considered to be allied to, and to depend for successive activities upon, the vast stores of their cosmic congeners—stores which are infused antagonistically and simultaneously, under cosmic pressure, through molecules, at first establishing molecular and atomic polarities and *subtlest* motion;—such conditions, when initiated, would rapidly develop vibrations in molecules and universal oscillations among masses; and thus the dynamic laws and restless functions of matter and force would be established. The *laws* of action and reaction as quantitative and relative phenomena have been numerically determined. The mysterious mechanical energies of nature thus resolve themselves into motion—an unique and equally inexplicable phenomenon which exists and unfolds itself in matter and masses in proportion to the quantities of repulsion and attraction with which various forms of matter can be charged. These forces—both; not one alone, but both, although in unequal quantities—hereby manifest themselves demonstrably as the absolute radical *vires vivæ.* Motion being thus the product of two antecedent radical agents, is, therefore, a *compound* phenomenon or function of matter; that is to say, the two heretofore known elements of motion, action and reaction, now become reducible to separate living principles which can not only be decomposed into attraction and repulsion, whence motion originates, but also be shown to link all phenomena of every class with matter and motion. These dual principles of motion, furthermore, as expressed by action and reaction between molecules—that is to say, molecular attraction and molecular repulsion, the profoundest radicals of motion—are everywhere and in all forms of matter either converted into forces wholly unlike themselves, yet with similar polarities, as heat, magnetism, &c.; or they generate, induce, control, direct, and modify the development and action of the latter, if these exist independently of the former as definite quantitative entities and distinct imponderable elements slumbering between molecules when undisturbed, and constituting atmospheres around them, as some theorists imagine. In either case, the latter forces universally appear as *sequences of the former*, and are therefore, strictly speaking, *secondary forces.* But Faraday has determined by experiment that these secondary forces are not convertible into gravity, as an abstract attractive force. Yet, since they spring from motion, are equivalents of motion, and convertible back again into motion, as all dynamical phenomena illustrate; and since there can be no motion or mechanical energy without repulsion, which is the essential principle of reaction, elasticity, &c., they therefore become convertible into equivalents of this force; and since this force is the co-ordinate and coefficient of attraction, they finally disclose their special quantitative relations to gravitation itself. Thus, we at last discover that all the forces of nature are linked together; and that they appear and vanish, blend, separate, and assume equivalents as the economy and mechanism of material being demand throughout the universe of things.

Whatever views may be entertained of the *ultimate identity* of the various forms of force (an important point not yet in question, but which this development will subsequently elucidate), if motion be closely studied in its initiation and immediate sequences, one remarkable principle of force, that of polarity, will be perceived to exist as a phenomenon involved particularly in attraction. For, in the first demonstrable act of motion, polarity manifests itself in directing molecules to impinge upon each other from exactly opposite points, thereby creating material poles, and straight lines or axes of molecules, so to speak, extending between these poles. The reactionary result of the impingement of these molecules upon each other, *when they are wholly free to act*, is repulsion at right angles to these axes, or equatorial motion. While attractive force possesses the capacity to impress molecules with equal energy both right and left of its initial action, and to pull them into more intimate relationship from opposite or polar directions, it also appears to possess the capacity to decompose or neutralize itself, and instantly vanish, in order to allow an equally positive force of repulsion to replace it as a radial and equatorial function acting also in opposite directions, and generating a system of opposite, crucial, or antipolar currents.

When these conditions of force and matter are studied experimentally, and when they are more distinctly comprehended in their cosmical developments also, a decided step will have been reached in physical knowledge, from which it may be predicated that investigations of a solid character may be safely extended further backward and downward into the past and occult states of nature.

Thus the *most recondite* forces have passed from their elementary and molecular functions into centripetal and centrifugal phenomena, and motion is transformed into material developments and mechanical results.

15.

The emergence of order and form from chaos—that is to say, the method of action by which the primal divisions of molecules or of masses, the first processes of segmentation, were effected, and whereby these molecules or masses were originally placed in conditions for attraction to act upon and bring them together, in order to generate polar, crucial, radial, and equatorial phenomena—is among the profoundest secrets of creation. Nevertheless, the fact or assumption of their separation equally involves the co-existence of a positive separating force, which could not necessarily be other than a principle of pure, abstract, universal repulsion. This is the force which I insist is as radical as attraction, and co-existent, co-equal, and co-extensive with it; and which, exchanging functions with it by unceasing processes of convective propagations through matter (an inexplicable system of compositions and decompositions of currents of pure physical energy), maintains perpetual motion among molecules and masses throughout spheres and throughout space. That this separating force—that is to say, *repulsion*—was not alone and *per se* originally "heat" or "fire," in the modern or ancient signification of these words, may be legitimately and positively inferred; since, knowing that matter and physical forces have been eternally the same and unchangeable, all experimental research and observation prove that expansion may coincide and co-exist with intensest cold as well as with intensest heat. Therefore neither *heat* nor *cold* is original, radical, initial force—the true *vis viva* of mechanical energy. As we shall see more clearly further on, they are proximate results of profounder forces antecedently at work upon and between molecules. Moreover, since experiments and observation prove that both heat and cold are, in one sense, subordinate to polarity and motion, or at most only co-ordinate and coincident with them, presenting phenomenal results which, when numerically computed, are found to be equivalent to expenditures of motion, and *vice versâ*, it follows that heat is in reality dependent upon motion, or at most an equivalent of motion, and therefore strictly and only a *secondary* principle. For motion and polarity have already been demonstrated to depend upon the pre-existence and propagative action of antithetic principles of energy, as molecular adjuncts, and upon molecules themselves as plastic materials previously differentiated and specialized, so to speak, as elementary entities, with given atomicities, or functions for receiving and rejecting these two radical forces in special quantities.

Furthermore, if the sixty-four known elements, more or less, constituting the earth, were created as diffuse and widely-separated molecules, "*heat*" could not have simultaneously existed as an active phenomenal and mechanical force, according to recent determinations of laws in relation to thermic causes and effects; for thermism only becomes an active motive agent *after* the juxtaposition of molecules has been effected by attraction and compression, and *after* the vibratory relations and reactions of molecules with each other have commenced. And we know by experiment and universal observation that this principle is only generated by the positive impact and friction of molecules one upon the other, whether in larger or smaller quantities, as they vibrate between, or rather under, the action of the two radical powers of attraction and repulsion. Heat, therefore, must have been only a species of secondary force, even in the beginning, and, as now, a concomitant and equivalent of motion, intensifying and diminishing with the latter, and reducible again into its original elements of attraction and repulsion. Heat, as motive force, is consequently a phenomenon whose development and properties are immediate sequences of the action of attractive force on molecules or masses in order to overcome repulsion; and, *vice versâ*, of the action of repulsive force on the same molecules or masses, in order to overcome attraction; in other words, heat is a sequence and result of the propagation of two radical forces, dissimilar in character, through different forms of matter.

However molecules or atoms might have been created and endowed in the beginning—whether, as I hold, in one vast mass of heterogeneous elements, weighed, measured, squared, cubed, mixed, and united in definite proportions, and actuated by pantheistic intelligence (so to speak), which mass was subsequently broken up and projected in fragments through space, as the stars and planets are now distributed (a hypothesis rendered more tenable by spectral and other recent discoveries than any yet suggested)—or, filling space like chaotic, accidental, formless, forceless, inert dust, vapour, or nebulous fluid, as the first Herschel and Laplace conjectured, and as most physicists now think;—*in either condition, repulsion must have co-existed with attraction and molecules as a component of motive force,* in like manner as we may, by experimental and inductive evidence, demonstrate it to exist now.

We cannot penetrate at present beyond this fundamental physical fact; and, rejecting all hypotheses from our discussions, we are compelled to accept it as final truth. We may, moreover, safely build upon it as we advance, inasmuch as we were constrained also by an exhaustive analysis of the subject, at the opening of this philosophy, to found all doctrine upon it.

As attraction pulls molecules axially toward a common centre from opposite or polar points where repulsion had previously left them, or, on the contrary, as repulsion pushes molcules radially and equatorially from a common centre where attraction had previously left them,—in whichever light we study the action and reaction of these binary elements of mechanical energy, or *vis viva*, we observe all phenomenal results in crystals, plants, or animals to develop themselves upon crucial and bilateral principles, and all the *secondary* forces to follow the same axial and equatorial lines of propagated molecular force and motion.

Motion being thus initiated, mechanical phenomena immediately succeed, and the entire machinery of nature becomes inspired with eternal activity, and bound indissolubly together. The mysteries of motion lie concealed in the fact that equal, original, and specific amounts of attraction and repulsion—as living, imponderable, and spiritual entities—were unequally subdivided in the beginning of creation between the ultimate elements out of which nature was constituted. However high this truth may reach, or however profoundly it may cause inquiry to extend, it is nevertheless demonstrable by evidence drawn from every department of exact science.

16

The origin of things is not here in question; only the origin, elements, and character of mechanical force, and the method by which physical phenomena are unfolded, after potential energy and motion have successively commenced their work. I have already stated, and it must be evident to every unprejudiced physicist, that motion never could, and never can, be generated and continued between molecules and masses which are absolutely inert and affected by attraction alone; for most certainly if inertia be not antagonistic force, and be nothing, attraction would be a solitary principle, encountering nothing whereupon to act or react, except itself. This is an unintelligible and absurd idea. Even if molecules fell together by attraction alone, of course there could be no reaction; for matter, upon existing mechanical theories, being inert, there could not be even the slightest "elasticity"—a term substituted for one of the *functions* of repulsion. In order, therefore, to create, generate, or *perpetuate* motion, there must be two positive co-ordinate and coefficient elements, or radicals, in what is now called the "*vis viva.*" There are, then, TWO "living forces" in physical nature, separate and distinct from matter, which reign supremely and equally over every molecule and every mass; while quantitative inequalities only in the oscillation of these forces, together with their power of composition and decomposition, in their relations with molecules and masses as chemical and cosmical elements, constitute the mysteries and phenomena of mechanical energy and material motion. Accepting these fundamental axioms, which microscopy and common observation, experimental physics, cosmical study, and severest induction, all unite in declaring incontrovertible—admitting these elements of force and matter as simple and, in the present state of knowledge, as final physical facts, and then rigorously following axioms and elements alike into all their possibilities and amplifications—science is no longer at a loss to discover the proximate principles of motion and of mechanical action and reaction everywhere, from the centre to the surface of the earth, and throughout the heavens. Nor can physicists reject these elements and deductions, however metaphysical, mysterious, or startling they may at first appear, as primary causes and established quantities in nature, and as bases of development for all secondary phenomena—heat, light, electricity, and magnetism—in their various terrestrial and celestial aspects, nor longer doubt their secret agency in unfolding and perpetuating the diversified forms of symmetry and life which address our senses throughout the surface of the globe. These elements of mechanical energy, however they may be considered, disclose themselves always as attraction and repulsion—the same subtle, active, and opposite principles of motion which constitute the working power transfused through, and developed by, the "lever" and the "inclined plane." That the former of these powers, which is the positive expression and equivalent of repulsion, could be developed numerically to so great a degree as to move a world, was proved by the demonstrations of Archimedes; but he lacked the fulcrum which Newton found.

17.

Such is the inception, such are the causes, such the processes and secrets of motion, in *its fundamental* conditions and mechanical possibilities, both in finite and infinite degrees. Coming and going amid these mysteries, man unconsciously imbibes, through all generations, these identical forces; and thus his material being is sustained, his mind unfolded, his civilization advanced and ennobled, from springs of immaterial power, as hidden and impenetrable as the interior of the planet he inhabits, and as vast as the universe that rolls around him.

Up to this point, our discussions have been firmly founded upon facts. No theory nor assumption of any sort, however popular its form or eminent its authorship, has diverted us into uncertain paths, or has been permitted to break the chain of inquiry, or weaken the foundations upon which we at present rest. With this presentation of the principles, origin, and results of motion it is quite impossible for the present spirit of speculation to be content. But however earnestly the premises may be cultivated or combated, I cannot withhold the prediction that with the extension of experimental and numerical inquiry, based upon these principles, a system of physics will be finally unfolded which will reveal such indications of constant intellectual action and conservative tendencies in nature, that no philosopher will for a moment conjecture that any element in its existence can be dissipated, or that created things display evidences of its internal defects, or signs of its general decay, or of an universal end.

18.

The Correlations of Cosmo-dynamic and Chemical Energy.

The springs of motion being detected and reached, the wheels of creation begin to turn intelligibly. Manifold phenomena successively strike the senses, and excite the intellect to appreciative study; and the history of nature unfolds its past and profound mysteries to the chemist and physicist, like an endless panorama, whose varieties and complexities of form, beauty, colouring, and combination astonish and interest them as much as the substantial present charms and instructs the common observer.

As we have observed force and motion reveal themselves through molecules in crystallization, and traced them through amplifying phenomena into cosmical functions of infinite magnitudes, we have successively been introduced to those forms of matter embraced in the natural realms of chemistry, mineralogy, geology, and cosmology. Without entering the domains of these specialities further than necessary to observe the operations of force and motion in their general development and connexion as an unique system of creation, it is absolutely necessary to glance at the fundamental conditions of matter as nature presents them to us, in order to unite legitimately and definitely the beginning with the end of our study.

The immediate mechanical result of motion, as effective work, among molecules, so far as their individual relations with each other are considered, appears to be crystallization. That heat (the secondary force which can be generated everywhere, even between pieces of ice, by compression and friction) is not the ultimate active element of "repulsion," nor the proximate principle and absolute property of repulsive force as mechanical energy traceable in this basic mechanical phenomenon, we know by the crystallization of water as this fluid passes from 40° Fahr. into ice. Repulsion, as absolute potential energy—that is to say, direct dynamic force—and as an effective phenomenal agency displayed in this special development of form, increases with the *abstraction* of heat: and this fact demonstrates that while repulsion, vibrating with attraction, is positively necessary to generate heat, "heat," on the contrary, is not necessary to generate repulsive force viewed *per se* as mechanical energy; and consequently that thermism, as motive power, and the principle termed "heat," "caloric," "a mode of motion," &c., is not, as some physicists have recently conjectured, the co-ordinate and coefficient of attraction and gravity in the economy of nature. Repulsion is not, then, heat, any more than attraction is heat. But we shall soon discover that heat as motion, or any form of equivalent, cannot be generated independently of repulsion, the identical cosmic agent which has been so much discussed in preceding chapters. That repulsion exists as an abstract and absolute power possessed of properties and functions more profound and energetic than heat, is proved by phenomena in crystallization. "Heat," therefore, examined in the light of experiment and close analysis, appears to lose some of the importance heretofore ascribed to it in various physical theories; and, instead of manifesting the character of a FUNDAMENTAL *dynamic principle, per se,* is clearly seen to assume a secondary range of action, and to depend upon some more profound force for its energetic functions: and thus crystallization, as a phenomenal act or result of motion, involving displays of cohesive force on one hand and of cleavage, segregative, expansive, or wave-force on the other, eliminates its conditions from dependencies upon heat or cold as radical agents, and reduces the origin of its geometric mysteries to pure attractive and repulsive energy, to combinations and reactions of dual agencies in every element, presenting new and important aspects of study in every sense, experimental, theoretical, or practical.

Now, inasmuch as the experimental and mathematical studies of crystallization, by many eminent chemical philosophers, prove the ultimate atoms of the sixty-four material elements constituting our planet to have special endowments and different essential characters and capacities of absorbing and becoming saturated with attractive and repulsive forces; inasmuch as isomorphism and dimorphism—in fact, all phenomena developed in crystallogeny—both indicate and prove these fundamental endowments and capacities to consist in definite and different, yet absolute and uniform, amounts of attractive and repulsive force (properties which render it certain that molecules in their ultimate states are sensitized with a species of morphological, intelligent, and internal creative essence, that forces rather than molecules resolve themselves into combining quantities or crystallizing powers and volumes with affinities for each other, and that sensitive, plastic, crude, chaotic points or molecules are swept by these combining forces into definite geometrical forms); inasmuch as the specific weights and densities of these various chemical elements, thus created, not only indicate specific measures of pure attraction and repulsion to exist in the fundamental constitution of each of them, but also demonstrate their equally specific and unchangeable endowments and capacities for absorbing and holding similar specialities of force, both in their atomic and massive aspects —a fact substantiated by all experimental chemical research; it follows that force—*force alone, and matter in no wise*—produces the chemical conditions of simple and compound bodies, and that the actinic properties and sensibilities of molecules are only specific capacities for absorbing and temporarily or durably holding different and definite quantities or elements of force. It will thus be seen that Kekulé's chemical theory is based upon principles of philosophy capable of the widest extension; and that while chemical experiments are leading to discoveries traceable from atoms into the broadest fields of mineralogy, and the combinations of metallic and gaseous substances into worlds; the course of the preceding studies has descended from these great cosmic masses saturated with simple but opposite forces to the atoms themselves, the universal decompositions and recompositions of which constitute the direct activities of each separate world, and establish the conditions whereupon all the celestial motions depend.

It is at this point of our abstruse and almost metaphysical inquiry that the mystery of *cohesion* begins to appear, and that the birth of *cohesive attraction* in its manifold modifications, as a fundamental element in applied mechanics and embraced in the treatment of "strength of materials," commences, and presents intelligible explanations of its variable and extraordinary phenomena.

Thus the universal forces of attraction and repulsion heretofore delineated as molecular and cosmical generalities, as simple, inchoate, chaotic powers of nature, are now seen to segregate, or segmentate, so to speak, and enter into matter in such manners as to atomize and endow it with specific powers for accumulating, moulding, and building crude, diffuse materials into geometrical forms—forms at first of an infinitesimal character, but pregnant with innate assimilative and germinal energies everywhere capable of vast amplifications, and with capacities for those wondrous developments into vital and morphological, segmentative and locomotive, being which spring up throughout the surface of our own planet.

Crystallogeny, then, in every aspect of chemical and dynamical philosophy or theory, presents the ultimate elements of the material world (so far as they can be traced and discovered, either directly or inductively) in the shape of solids, and as ultimate atoms of unchanging but varied geometrical forms. The very essence of *geometrical* science consists in specific quantities of the binary powers of attraction and repulsion measured out upon molecules; the first so acting as to shape the sides and ends, and the last the angles, pyramids, and cleavages, of all crystals. Such conditional relations of matter and force necessarily reduce the phenomena of cohesion—that is to say, of *cohesive attraction*—into divisible and local functions of the same radical principle which pervades the general constitution of nature as static energy; and they involve the secret of the principle and causation of specific weights; and inasmuch as all elementary bodies display different degrees of density—a property also affecting their specific weights—it may be seen at the same glance that tenacity of cohesion, *i.e.* "strength of material," is equally due to the force of repulsion also, which, associated with attraction, is constantly present in every simple and compound substance. Every simple body then possesses specific capacities for so much attraction, no more and no less (speaking of this now as a pure abstract force and formative agent). In like manner every simple body possesses specific capacities for so much repulsion, and for no more and no less (speaking of this also now as a pure abstract force and disintegrating agent): and by these two forces the molecules of these bodies are held in such special relations as to be differentiated and specialized one from the other as ultimate elements of nature. This is the basis of *chemical* science, involving the mysteries of affinity, and of solution and decomposition—that is to say, of chemical attraction and chemical repulsion. The combinations of elementary bodies forming compounds are plainly, then, only combinations of radical forces in varying and exact quantities measured out by molecules, which quantities affecting each other differently, decomposing one from the other, or being absorbed one into the other, create new conditions, and multiply the working capacities of nature. Thus new materials are added to the original limited stock; and the binary volumes of original chaotic potential energy are divided, and subdivided, and compounded into multifarious generic quantities and specific distinctions, neither of which destroys the differentiated functions of the other, and all of which, working together and oscillating with one another, end in creating all things that vary, and at the same time in holding each equally subject and subordinate to the two primal developing powers, which, as aggregate quantities and units, for ever reign ascendantly and universally over all. By these means, and in this manner, all nature, however diversified its forms, conditions, and phenomena, is nevertheless one; and thus all science and philosophy begin and end with those plastic molecules and their combining and resolving energies, which, at first weighed and measured in exact quantities by an Infinite Creator and Geometer, now eloquently proclaim that harmony and order are eternal, and that a secret ceaseless providence pervades all space, and guides the relations of matter and force throughout their manifold mutations, and in all their developments.

19.

In order to comprehend clearly the modes of action and connexion of attractive and repulsive forces upon, through, and with matter, and of their transformation into mechanical and phenomenal power, it will be necessary at this stage of our study to pass from facts embraced in crystallization and chemical affinity to those of a somewhat different and mixed character.

The conditions existing in the associations of forces and molecules, and the chemical affinities existing in crystalline forms of the grandest dimensions, are identical with those existing in the minutest atoms of any of the elementary bodies or of their compounds, and *vice versâ*.

The *modifications of attraction* involved in the property of cohesion, embracing "elasticity," "strain," "stress," and "the ultimate strength of solids," are very evidently due to the *quantity* of *repulsion* specifically appropriated to and mingled in the capacities of molecules, or in the structure of all ultimate atoms, as final chemical elements. For example, a crystal, or an amorphous mass—consequently each ultimate atom of the elementary body called Gold—is so constituted and endowed as to be susceptible of and chargeable with *attraction* almost alone, comparatively speaking; while carbon is, on the contrary, so constituted in its ultimate atomic structure and capacity as to absorb and hold or induce and conduct *repulsive force*, as pure abstract cosmic energy, with remarkable celerity and exclusiveness. Diamond, for instance, cuts glass and crystals of silex; and the separation of particles thus effected arises from the active and subtle infusion into the points touched of a principle of energy the opposite of cohesive attraction. In order clearly to present natural laws deducible from facts and analysis regarding the action and movements of forces in their relations with matter and worlds, so that this system of philosophy may be fully comprehended, brief demonstrations may be here required, illustrated by facts drawn from the free, rapid, slow, or violent movements of certain elastic gases, and from their capabilities of diffusion, absorption, condensation, vibration, and expansion in connexion with certain other substances of a more solid character. I solicit close attention to this branch of our discussions.

In former parts of this work I have endeavoured to expound the nature and action of force, and to show that similar principles instantly mix and combine into one volume; that force, not matter, is the cause of motion everywhere; that an apple falls, for instance, only because the force of attraction in *it* is precisely like the force of attraction which fills the globe, in virtue of which fact a tendency in both is manifested to integrate and form an unique volume of molecular and cosmic energy; that the force in the apple restlessly strives to rush into and combine with its planetary cognate in the same manner and with like results as one drop of water mixes and coalesces with another drop, or with a lake or ocean, and as one globule of quicksilver falling into, instantly combines with and vanishes in, its cognate element, for ever afterwards to be undistinguishable as an individual drop. It is this wonderful property possessed by pure forces to translate themselves, and act upon their cognates, to create tension, and become transformed into or replaced by each other, that I wish by various illustrations to present in the clearest light possible. These illustrations will in turn throw new light upon the generation of heat and the mechanical phenomenon of combustion. They will at the same time define the cosmic origin of potential energy, explain more clearly the processes of its induction from cosmic sources, and expose the modes of its propagation, general action, and even of its transformation into mechanical motion and phenomenal creation.

Certain substances, either simple or compound, such as carbon, or compounds of nitrogen and carbon, nitro-glycerine, gunpowder, gun-cotton, &c., possess the remarkable attributes of instantly inducing and abstracting from cosmic sources, and of absorbing and accumulating within themselves, pure potential energy, and also of instantly discharging it under certain circumstances with more or less violent mechanical results. The forces expended in explosive phenomena have always heretofore been considered as *latent* principles wholly local, isolated and locked up in explosive substances, and only disengaged at the moment of explosion. It has never been considered that they have any direct relation or connexion with *interior* telluric forces, or that they are instantaneous and positive inductions from cosmic sources of energy. Yet such is really the fact; and a circumspect search into the mechanism of explosive phenomena will not only establish this extraordinary truth, but lead us, moreover, to comprehend the origin and nature of cohesive attraction on one hand, and of solution, expansion, and radiation on the other. When we have studied the manner and swiftness with which certain gases, charged with repulsive forces, move and act in relation to more cohesive substances, we shall be the better prepared to comprehend the conclusions which I present.

Charcoal, for example, absorbs and accumulates within its interstices varying volumes of different gases, even as much as ninety times its volume of muriatic acid gas, and of ammoniacal gas. Spongy platinum instantly selects and absorbs from the atmosphere eight hundred times its volume of oxygen, and holds this gas locked up and unperceived until hydrogen is poured upon it. These examples, and others of a similar character which might be cited, pertinently illustrate a fact and principle connected with the behaviour and inductive tendencies of attraction and repulsion involving the absolute transfusions of the radical elements of energy through matter and into masses, in virtue of which even that state acquired by matter, and called by Sir William Thomson "energy of position," may be legitimately explained. They will elucidate also certain subtle actions of force which have not heretofore excited the attention which they deserve, and the phenomena of which it is desirable to consider, in order to connect the principles and phenomena of "rational mechanics" with their cosmic sources of power, and thus divest physical laws of theoretical complications, lift mechanical science from regions of materialism, and reduce all the facts, laws, and doctrines of dynamics into narrow limits and positive system.

Now, when this spongy platinum above referred to, a pure metal, equally pure and simple in its subdivided and atomic as in its compact and cohesive state, is exposed to oxygen, physical properties appear in both which they did not manifest previously to the molecular subdivision of the metal. The platinum instantly—so quickly indeed, that no thought can conceive of the rapidity—absorbs and accumulates, from centre to circumference of its mass, eight hundred times its volume of this gas.

It is the physical facts and conditions to which I desire to call attention in this connexion, more than to the chemical ones. The latter of course lie at the foundation of the former, and are pregnant with wondrous consequences in physico-chemical directions. But it is the simple facts of the translation of matter under the direct influence of attraction, and of the exclusion of repulsion, so to speak, from the attracted and moving gas, to which I confine attention; this repulsion now being considered as the active agent in expanding the oxygen molecules into eight hundred times the bulk of the platinum mass, when the gas is in its normal condition. It is the swiftness with which these physical translations are effected that I specially cite, in order to illustrate the celerity with which mechanical energy itself, as pure force, can be transferred from one substance or locality to another; repulsion replacing attraction, and attraction repulsion, as surrounding circumstances determine.

Since force clings more or less to molecules, and molecules to force; since forces assimilate with forces of the same kind, and exchange expressions of energy with their opposites, according to special laws of quantity, affinity, saturation, and rejection existing in elementary forms of matter; since even *tension* on one hand, tending to explosion and projection, and *partial* exhaustion of forces on the other, tending to traction, may temporarily and locally exist in all bodies and all parts of the globe; it is evident that the repulsive forces normally inherent in, and separating so widely, these eight hundred volumes of oxygen particles, have either been absorbed and condensed into the interstices of the spongy mass, there resting as inert or latent force (an irrational conjecture, as we shall see more clearly hereafter), or that only the particles of oxygen have been absorbed and condensed, to the temporary exclusion, so to speak, of the repulsive forces of these particles—that is to say, of the positive potential energy which holds them normally asunder. If excluded, this exclusion has been *instantaneous,* and the quantity of repulsive force existing in the oxygen to prevent molecular juxtaposition has been thrown into the general fountain of cosmical repulsion with which the planet is charged. Or, if, on the contrary, the repulsive principle has instantly vanished, though in such a manner as to exist still in the metal as a latent power, a sort of condensed *vis inertiæ*, it must have been as *instantly* assimilated to and absorbed into some form or property of repulsion which the platinum acquired by disintegration. That it becomes stored and lies latent in the spongy metal, *as a definite entity*, as a condensed active and potential mechanical agent, is by no means proved or demonstrable. That this repulsion is instantly induced and pulled, so to speak, by the globe out of the gas as the latter rushes into the sponge, and that repulsive force is subsequently induced and drawn into the expanding gas out of the gross sum stored in the planet, into which it had previously vanished, is demonstrable and proved beyond doubt by the phenomenon developed when a current of hydrogen is poured upon the sponge. In either case, the velocity with which radical energy as a motive principle translates itself anywhere and in any quantity, vanishing and reappearing with violence as it enters matter and effects explosion, or generates continuous motion, is infinite and inconceivable, and so transcendentally subtle as to appear more than metaphysical, if it be possible to conceive of an entity more subtle than thought. The fact exists, however, whatever its explanation. And it is this fact which I wish specially to present in the boldest relief, fortified by plain and unequivocal proof. The ultimate nature of physical energy as a living and active principle in matter we shall probably never know; but that the radical agencies upon which heat, light, magnetism, and electricity depend for their existence and manifestation cannot and may not travel as fast and with as much subtlety as their offspring, is a point that no experimental philosopher will venture to question, in the present state of our knowledge.

The wonderful physico-chemical phenomenon just presented, and shortly to be alluded to again for other purposes, exhibits the connecting link between molecular and cosmical physics, and will, when closely studied, lead us to comprehend the agency of molecular mechanics in initiating and unfolding phenomena which, generated from statical conditions of matter and force, swell from their zero, so to speak, into dynamical potencies, involving every mechanical law and principle in all their graphical and numerical issues. This phenomenon strikingly illustrates the *modus agendi,* at this point or that point, of the instantaneous disappearances, inductions, translations, and reappearances of force, as a motive or energetic principle, and as it exhibits itself in nature or in experiments—I mean of absolute potential energy, such as is contributed from the earth's central quantities, and which, as I hold, is transmitted everywhere through all matter, alike in the motion induced through the "lever" and "inclined plane," and so back again to the earth's centre, as water is transmitted through tubes, or electricity through conductors. Such are the processes, the modes of action, in the transmissions of energy, by which I regard mechanical work of all sorts to be effected. This phenomenon also illustrates the mechanism of transmutation of mechanical force; as, for instance, the way in which attraction existing in cohesion as an attribute of *chemical* law may be supplanted by repulsion, a principle of solution, of projection, and of *mechanical* law. And it illustrates the mode by which matter is excited from static into dynamic conditions, as these radical forces, diffusing themselves in all directions antithetically through molecules, generate successively atomic motion and general vibration, thus transforming themselves verily into heat and combustion, and finally developing "*mechanical power*" and projectile and cosmical motion.

Thus atomic functions—chemical forces and affinities—appear to be directly affiliated with, and convertible into, rational and economical mechanics, and so on into celestial dynamics. And from this point of view it must be plainly seen again, that all manifestations of physical energy spring from the same radical sources, and return into the same, this energy producing everywhere similar or equivalent results.

A critical consideration of the remarkable phenomena

developed by the contact of spongy platinum with oxygen and hydrogen will be fruitful also in demonstrating the origin and truly evanescent character—indeed, I may say, the positive instability and unsubstantiality—of heat. And I will here interpolate, that we must never permit ourselves for a moment to believe that heat, because it is a sensible agent, is a positive entity, or that it is the abstract force of "repulsion," any more than that a galvanic current is abstract repulsive force because it also creates mechanical expansion in platinum, iron, and some other metals. For this separation of molecules galvanism will, indeed, effect independently of heat, thus showing that "repulsion," however or wherever it may appear, is a more abstract principle than heat; that it is virtually the mechanical element in heat and other secondary forces; that heat, electricity, &c. are not repulsive force *per se*, but only the resultants of conflicts of the radical forces, in which conflicts repulsion plays an active part; and that all secondary forces are simple incidents and equivalents of motion. These general remarks upon mechanical force, apparently discursive, are not aimless, but collateral and necessary. For the bearings of our discussion and generalizations are very broad, and it is important to connect theorems already established and the developments of prior discussions with the multifarious phenomena dependent upon heat and other secondary forces which, in one form or another, saturate the earth and all other cosmical bodies, inasmuch as the discussion of these in their largest aspects will soon follow: and it is furthermore necessary, in order thus to unite experimental knowledge and mechanical inquiry with the cosmical forces of which I have heretofore treated, and which open an infinite field for future exploration and discovery.

I am well aware that to minds wholly imbued with mathematical or mechanical ideas, based upon existing dynamical theories, this attempt to enunciate *principles of motion* of a mixed metaphysical and material character, and to reach their origin through inquiries connected with chemical philosophy, may appear futile, and perhaps absurd. To the chemical philosopher they will probably appear less so. But since all branches of physical science are indissolubly connected by threads so subtle as only to be discovered by their effects and mutual influences upon every form of matter, I trust that, as the subject unfolds, our facts and inductions may be clear and convincing to students of nature of every class. If I fail to present them to others as clearly as they exist in my own mind, the defect will arise rather from a lack of ability on my part to state points, than from the points themselves, which are too numerous and important to be long overlooked hereafter.

20.

A few historical remarks at this juncture will also be necessary, in order to complete the elements of our subject; and its grandeur will not diminish, in the end, when we discover that the common, crude, working powers of water, steam, and animal bone and muscle are drawn directly from the central forces of the globe, and that they return again to the same fountain when their action has been exhausted upon the planet's surface in general, chemical, mechanical, and projectile functions.

Two centuries ago, the acute philosopher John Locke stated that "Heat is a very brisk agitation of the insensible "parts of the object, which produces in us that sensation "from whence we denominate the object hot; so that what "in our sensation is *heat*, in the object is nothing but motion." No definition could be more exactly stated, and no subsequent studies or discoveries have changed the correctness of the idea at the base of that plain theorem.

Rumford, in 1798, instituted a series of well-known experiments upon the generation and laws of heat in the boring of cannon at Munich, and he declared, what subsequent experimental inquiry has only confirmed, in these terms: "It is "hardly necessary to add that anything which any *insulated* "body, or system of bodies, can continue to furnish *without* "*limitation* cannot possibly be a *material substance;* and it "appears to me to be extremely difficult, if not quite impos- "sible, to form any distinct idea of anything capable of being "excited and communicated in these experiments, except it "be MOTION." Ignorant of causes, Rumford explicitly stated that he was "very far from pretending to know how or by "what means or mechanical contrivances that particular kind "of motion in bodies which has been supposed to constitute "heat is exerted, continued, and propagated." With an intuition, however, seeming now almost prophetic, he further said, in relation to this subject: "But although the mechanism of "heat should, in part, be one of those mysteries of nature "which are beyond the reach of even human intelligence, "this ought by no means to discourage us, or even lessen our "ardour in our attempts to investigate the laws of its opera- "tions. How far can we advance in any of the paths which "science has opened to us, before we find ourselves enveloped "in those thick mists which on every side bound the horizon "of the human intellect?"

These words inaugurated a new era, and virtually struck the first blow at the materialism of modern science; but they have been overlooked and practically disregarded for half a century. Nevertheless, the illustrious Davy authoritatively announced again, in 1812, that "the immediate cause of the phenomena of heat is motion." It remained, however, for the modest physician of Heilbronn in Germany, Julius Robert Mayer, to publish, in 1842, a paper on the "Forces of Inorganic Nature," in order to breathe life into the discovery of Rumford, and stimulate the stubborn intelligence of our age to more just appreciations of the relations of physical force to mechanics, in contradistinction from inert moving matter, or *vis mortua*. But even the originality, acuteness, and value of Mayer's researches and views have been almost lost amid the rapid developments of the subject which now engross the attention of physicists. So unimportant was the burden of his simple but extraordinary paper at first considered by the mathematicians to whom it was successively submitted, that only the lofty genius and immoveable firmness of the illustrious philosopher Justus Von Liebig rescued it from obscurity; and, detecting its intrinsic value by a sort of intuition, he effected its publication upon his own responsibility, and contrary to the opinions of several distinguished critics. The communication of this fact to the writer by the great chemist himself, in an interesting conversation upon "the uncertain value of mathematical conclusions when analyses are founded upon inexact elements and theories," affords a pleasing recollection of his frequent encouragements of obscure merit, his original acuteness of thought, and his zeal for progress and exact research in all departments of science and philosophy; and I record it in this connexion, in order that that priority of discovery may be the more unreservedly accorded to Dr. Mayer to which he is justly entitled, and this through the prescience of Baron Von Liebig.

But the numerous experiments and persevering researches of Dr. J. P. Joule, of Manchester, conducted about the same period as those of Mayer, and published soon afterwards, entitle him also to the acknowledgments of the scientific world, inasmuch as the correlation, *equivalents,* and general conservation of force involved in the terms "motion," "heat," and "mechanical work," and *vice versâ,* first published by Mayer, were independently elaborated and perfected by him, and because the entire range of physical knowledge has been triumphantly extended by his searching experiments and the exactness of his computations and conclusions.

It remains for science to fill that space in the chain of physical discovery which allies motion and heat with gravitation, and to connect the phenomena developed in molecular and experimental physics with the general forces and dynamics of the cosmos.

21.

The discoveries of Mayer and Joule have profoundly influenced the scientific views of the age. While receiving, however, legitimate and solid development at the hands of several distinguished scholars, the wildest conjectures upon cosmical subjects have sprung from them also; and professors and teachers of commanding authority have even invoked the mathematics to transmute these conjectures into fact and stable doctrine. Indeed, to such an extent is this sort of speculation and empiricism carried, that solar heat and light are believed by many physicists to be absolutely generated by the impact only of solid matter, which, flying at random in space, incessantly falls upon the sun by virtue of gravitation; and the quantity and weight of this matter even have been calculated and announced with as much apparent confidence as if such phenomena were proved to exist by exact observation, and were beyond possible doubt. In like manner, the Zodiacal Light and the rings of Saturn are now declared to be composed of meteorites, and their different degrees of luminous intensity are ascribed to the alternate packing, or to the collision and diffusion, of these bodies, as they rush in larger or smaller masses in wild confusion against each other, while moving in established paths through space. In speaking of the Zodiacal Light, Nichol says: "No hypo-"thesis, indeed, appears *adequate to satisfy known dynamic* "*conditions, but the one at present generally received.* It is now "recognised, that the larger planets are not the sole consti-"tuents of our solar system, but that in all probability "masses of planetoids or *meteorites*—too small to be discerned "individually—circulate around our luminary *in streams.* . . . "We might refer also, as an analogous instance, *to Saturn's* "*rings. The periodic showers of meteors receive in such an* "*hypothesis their best explanation; and in the conception of* "*such a ring of meteors nearer the sun we find a key to all* "*established phenomena connected with the Zodiacal Light.*"

The last treatise on Astronomy published in the English language, so far as I know, is that by Lockyer, already referred to as a reliable authority. Speaking of Saturn's rings, the author says: "Of what, then, are these rings com-" posed? There is great reason for believing that they are " neither solid or liquid; and *the idea now generally accepted* " *is* that they are composed of *myriads of satellites*, or *little* " *bodies*, moving independently, each in its own orbit, round " the planet; giving rise to the appearance of *a bright ring* " *when they are closely packed together*, and a very dim one " when they are scattered. In this way we may account " for the varying brightness of the different parts, and for " the haziness on both sides of the ring near the planet, which " is *supposed* to be due to the bodies being drawn out of the " ring by the attraction of the planet."

Such descriptions and opinions, based mainly upon conjectures, are perhaps unavoidable in the present state of learning; but they expose the vacuum which really exists in physical science, while they also indicate the direction in which speculative inquiry tends in consequence of Mayer's discoveries, suggestions, and calculations. Other forms of explanation may, however, be legitimately deduced from the positive knowledge of cosmical facts of a chemical and mechanical character; and these explanations may be considered reliable and of permanent value, when free from hypothetical elements, and when based upon facts which no theory can ignore. Such must be the case, when the existence and periodical intensities of terrestrial phenomena are found, by every method of observation and inquiry, to depend upon special distances of our own planet from other cosmical bodies; and when the phenomena of motion, light, and actinism observable in bodies studding the more distant realms of space not only resemble, but also are identical with, those of a terrestrial and experimental character.

It is in view of the present state of scientific inquiry, and of the facts already accumulated in various departments of physics, grasping firmly such facts, and steadily rejecting all hypotheses, that I now propose to apply principles developed in the preceding discussions to a new series of investigations; and, proceeding from the known, endeavour to reach the unknown, and, if possible, fill the chasm now so patent in cosmical physics.

22.

Correlations and Equivalency of the various forms of Force.

The sum of exact knowledge attained in every branch of physical research is, when considered in the aggregate, vast and bewildering; but the *forces* which illustrate every class of phenomena constituting this sum are so simple as to be still more wonderful. The endless variety of geometrical and morphological development everywhere observable results only from expenditures of different measures of these forces upon matter in its *molecular states,* and from the submission of all consolidated forms to the aggregate energies of these forces. These forces, originally units of attraction and repulsion, were subsequently divided, and subdivided, and measured out to these forms, in order to square, cube, and limit their respective volumes, define their individual actinic capacities, specify their basic and combining functions, and adjust their relative proportions and positions to each other as elementary quantities; and, although separately observable and computable in each basic form and combination, as *molecular* agencies, they still act in the *aggregate* as potential energetics to bind all nature into one cosmical system, a single mechanical unit, however varied its planetary and chemical details. All this, conceived and executed by the Supreme Being in the unknown periods of the past, we now discover in virtue of a cognate spirit of understanding in ourselves, the nature of which appears to be not only kindred to but assimilative with absolute Deity itself. This is demonstrable, inasmuch as, by patient searching after truth, the profound secrets of space, time, and thought are successively revealed unto and made clear to us.

We may now revert with the greater boldness to purely physical particulars, and consider the extraordinary phenomena developed between spongy platinum and oxygen and hydrogen, in such a manner as to detect their connexion, and that of kindred phenomena, with the central energies of the planet—the great fountains of mechanico-cosmic force, from which they undoubtedly spring. The mechanical conditions of combustion and heat being once clearly understood, and their elements—the "elementum ignis"—discovered to be simple vibrations of the *same molecular forces* which, aggregated in planets, suns, and comets, constitute their central forces, in systems cosmical motion, and in the stars universal action, reaction, and eternal activity, the obscurity which has heretofore surrounded the study of these subjects will vanish, and a fresh impetus will be given to general physical inquiry.

23.

When spongy platinum, saturated by attraction with its eight hundred volumes of pure oxygen, is exposed to a current of hydrogen, the latter instantly begins to penetrate and combine with the former gaseous substance; and so swiftly developed are their conjoint molecular action and reaction—that is to say, so rapid is the antithetic diffusion and conflict, the polar action, of attraction and repulsion among the particles of these gases as they combine in elective numbers and weights to form water—that the metal quickly becomes intensely hot. The water thus formed is, of course, most attenuated vapour, and only detected by condensation. The fact of the chemical combination is well known. When the flow of hydrogen is stopped, the metal will again instantly attract and surcharge the interstices with other eight hundred volumes of oxygen, which the hydrogen in turn will again agitate, inflame, and expel in the form of water; and thus this process may be endlessly repeated, always with like results, and *without the destruction, loss, or transformation of a particle of the metal, or either of the gases.* The entire mechanism of the phenomenon, complex and inexplicable as it may at first seem, is one of the simplest in nature. And inasmuch as it is dynamical in all its functions, it will lead us to a clearer understanding of the elements of mechanical force, and of their method of action, as creative and destructive agents, in their relations with matter, with geometric and organic forms, and with nature at large.

In this problem five elements are presented—platinum, oxygen, hydrogen, molecular attraction, molecular repulsion; three of matter, two of force. Since the three first remain for ever the same in *quality and quantity* (the platinum only holding, as a sort of plastic medium or recipient, such relations to the oscillating actions of the two gases as I formerly described molecules to hold to the two radical forces of attraction and repulsion), it follows that the *forces* alone with which the gases are saturated are fluctuating quantities and the invisible and intangible powers which determine the actions and results above delineated. It is certain that not an atom of either of the three material elements changes its nature, quantity, capacity, or weight;—in fine, not a principle or property of its specific elementary substance. The *condition* of the oxygen is only so far altered that at first it seems to have alienated its molecular repulsive forces and set them free into surrounding matter, and to have temporarily accumulated internal attractive force by which its particles have been condensed and apparently compressed between the particles of platinum, as we behold a comet contract as if compressed when it approaches the sun and is at perihelion. This liberation of its repulsion, involving its submission to new conditions of inter-molecular attraction, so different from its normal ones, is a phenomenon developed with the quickness of thought. The rapidity of the change

X

is so swift as to be inconceivable. The phenomenon is one exclusively affecting force—not matter—for the oxygen is the same in every respect, except condensed in bulk, and is not even ozonized. When hydrogen particles, *charged as they are so enormously with repulsion,* come in contact with the oxygen particles, the latter instantly acquire new energies, and assert claims to affiliate and unite with hydrogen rather than to associate longer with the metal; and in this struggle to combine and mix their respective inherent forces the molecules themselves are impelled into contact and repelled with such subtle, rapid, and intense vibratory *impact and reaction* that heat and combustion ensue in the same manner as friction and impact: that is to say, molecular pressure and reaction produce heat under all other conditions and circumstances; the *combustion* in this case being demonstratively only *intensity and subtlety of molecular motion* without the consumption or loss of anything. Translations alone of matter and force occur, ending in the formation of water by the final union of the oxygen and hydrogen molecules in their definite and natural volumes and measures. The intense heat—the *fire*—thus developed by the *chemical* combination of these gases appears to be due solely to the *mechanical functions* involved in the motion of their molecules among each other—to the action, but *especially* to the *reaction,* of determinate volumes of molecules and molecular forces one upon the other; that is to say, to the propagation of antagonistic forces from centres to centres of molecules, and to their decompositions and recompositions between these centres. In this conflict motion appears as the parent and equivalent of heat, and becomes equally convertible into light, electricity, and magnetism; and each of these secondary agencies is not only correlated to, but absolutely inoculated with, and vivified and actuated by, the principle of *repulsion,* which principle is the chief basic element of mechanical energy—indeed the very dynamical energy itself. The inseparable, co-ordinate, and co-efficient relations of attraction and repulsion as abstract molecular forces have been long established. My purpose is to follow these latter into their cosmical expansions, and connect them with all other forms of force, and with all phenomena in experimental, terrestrial, and cosmical physics.

24.

In the phenomenon just described the rush of hydrogen molecules charged with repulsive forces upon and among oxygen molecules, which had previously liberated and alienated their normal quantities of repulsion through the influences of the spongy platinum, not only tends to establish mutual union of their respective forces, but wherever the forces are infused the hydrogen also penetrates, atoms of each instantly grappling and struggling with atoms of the other; and thus collision, friction, impact, reaction, and the most subtle vibrations conceivable ensue, and are propagated throughout the metallic mass. Dynamical conditions similar to these always cause heat, or some other form of secondary force, to appear in every elementary substance, and in all their compounds. The development of heat, the striking phenomenon of combustion in the spongy platinum, is a strictly dynamical process. And while this conflict of molecular energies—this rushing forward and backward of subtlest molecules with infinite swiftness—has been progressing, an equilibrium of the forces normal to the original elements is at last restored; and, as a *chemical* result, the condensed water is found to weigh exactly as much as the two gases from which it has been formed. Intense as this heat and combustion have appeared to be, not a molecule in either element has been destroyed or lost. Not one has even been disturbed, transformed, or touched in its substance or ultimate nature. Thus platinum sponge, after its attraction and absorption of oxygen to saturation, after it has condensed and accumulated within its dimensions not only eight hundred times its own volume of this gaseous substance, but also the capacities for potential and repulsive energy which, normally acting between oxygen molecules, expands them into this enormous space—thus this metal, in virtue of its peculiar character, becomes the focus of an extraordinary mechanical, chemical, and cohesive phenomenon, that permits us to demonstrate the process by which heat is generated and accumulated, and to learn what the mechanism of combustion—the "elementum ignis"—really is. The mystery seems to consist in the *rapid transfer and propagation* of the radical forces of *repulsion* and *attraction*. The intense *celerity of these antithetic transfers* and diffusions of forces through molecules and through masses *constitutes undulatory motion;* and the subtle vibrations resulting from these propagations constitute local heat and light, and, according to varying conditions of matter and circumstance, magnetism and electricity also. All experimental researches show that the forces of nature, in performing their eternal cycles of action, creation, and reaction, converge to, meet in, and radiate from this fact and law. Inasmuch as no matter is consumed, annihilated, or disappears in the phenomena described, it is evident that heat, fire, and combustion (when considered as one and the same physical act and principle, differing only in intensity) are not necessarily, nor in any sense, destructive agents. They involve the final facts of the attractions, conductions, and repulsions of antithetic forces through molecules, thereby generating insensible vibrations, which amplify and expand so as finally to become sensible motion, effecting at last absolute translations through space of matter loaded to repletion with potential energy, which energy vanishes and appears in ceaseless mechanical and cosmical phenomena.

Through the insights into nature gained by this experiment we discover that heat, fire, and combustion, whatever their intensity or extent, in any element or in any globe, become reducible in definition to the simple fact and mechanism of the *propagation through molecules and masses of attraction and repulsion as abstract radical entities and antagonistic cosmical forces;* and that the mutual interferences and resistances of these forces, of each to the other, beginning with atomic vibrations of infinite subtlety, end in decompositions and recompositions of every grade: thus, on one hand, establishing celestial motion, and on the other chemical action and reaction: at one moment effecting cohesion; at another, solution: and for ever subordinating the juxtaposition of atoms and the mutual relations of spheres to the unequal distribution and oscillations of these forces.

In the platinum phenomenon the effects of molecular motion—that is to say, the propagations of attraction and repulsion, as pure abstract forces, resulting in vibrations and intense heat—are simplified and singularly free from chemical complications and disturbances of difficult explanation. The platinum itself remains always unchanged. It is, so to speak, simply a clear field of battle, where the inherent capacities and forces of oxygen and hydrogen encounter each other. The most that may be said of the possible interference of the inherent platinum forces in this phenomenon is, that the metal is *sensitive to, and a quick conductor of, attraction,* as shown by its natural cohesion and density. But these physical qualities being deranged by molecular disintegration, it may be considered, in a certain sense, a neutral field of action. In this view, the metal may be eliminated as an active element from the problem. And since the oxygen and hydrogen particles are not in the slightest manner altered in quantity or character by contact, only mixed the more intimately, and ready at any moment to fly apart when a voltaic current abstracts the cohesive property acquired in the experiment, it becomes clear that only absorption and expulsion—the *antithetic propagation* of forces from molecule to molecule, and the action and reaction of these forces upon the gases and metal, producing positive vibration, friction, impact, motion, and nothing else, and all of a molecular character—are the secret cause of heat in this case. Not a molecule has been added; not one abstracted. Forces only, attraction and repulsion, have darted in and out, and through and through, swifter than quickest thought; and sensible heat as positive mechanical working energy instantly appears, the correlative and equivalent of reaction: and reaction is only an indefinite word for the expressive radical term, *repulsion.* And since heat, light, magnetism, electricity, and actinism, are only inflections of each other, regarded as mechanical agents or effects, all originating in like manner from, and dependent upon, molecular disturbances, impacts, and reactions—that is, *upon propagations of antithetic forces through different elements under different circumstances*—it clearly follows that all not only possess similar functional qualities of polarity, and of attraction and repulsion, but are also equivalents of the latter forces, disclosing not only the correlations of the former to gravitation through repulsion, but also the absolute conservation of gravitative force, as insisted by Faraday; while it is, at the same time, constantly employed as an exhaustless element of potential energy throughout the universe. Thus forces abstractly considered, and appearing in this way as metaphysical entities, become awakened into positive physical existence; and, inseparable from matter, they become the eternal mechanical powers of nature.

The intense heat excited in the platinum phenomenon is now discovered to be the result alone of propagations of attractive and repulsive force generating vibrations among molecules. The *chemical change* in the conditions of the gases is purely COHESIVE in character, governed alone by hidden laws of quantity and volume. Thus we are finally brought face to face with the great fact that *chemical forces* and *affinities* are after all only quantitative conversions into actinism of the basic molecular forces of *attraction* and *repulsion;* the *same forces* which in the aggregation and unification of material powers constitute cosmical dynamic energy. Mysterious principles of selection and quantity alone specialize and distinguish chemical forces from other conditions of cosmical force; and these profound and wonderful differentiating powers seem like so many independent, substantial, directive ministers of nature, whose essences, secretly linked with and diffused throughout vast fountains of crude chaotic energy, express to us, and indeed constitute, the very infinity, power, mystery, and majesty of the Godhead itself. A conception of these occult principles and powers is conspicuous in the earliest culture of Asiatic thought; and from this conception appears to have sprung those polytheistical ideas which, spreading with the dispersion of the ancient peoples, have shaped in one form or another the beliefs of successive generations and the influences of which, indeed, have not yet wholly disappeared from the nations and institutions of our own times.

25.

Thus all the secondary and sensible forces of nature—heat, light, electricity, and magnetism—in all their modifications, are traceable to the penetration and propagation of the insensible and radical forces of attraction and repulsion through molecules and masses. Molecular vibrations are the immediate result; and motion is the measurable exponent of the sum of these vibrations represented in heat, light, magnetism, electricity, and chemical affinity; which in turn are everywhere seen to be transformable into material development as mechanical power and productive work. Physicists have already announced, with undoubted correctness as a general fact, the numerical relations and equivalency which all the secondary forces bear to each other and to motion. Rumford determined, nearly seventy years ago, that the *molecular motion* involved in the combination of one pound of hydrogen with eight pounds of oxygen, to form water, will develop sufficient heat to raise thirty-four hundred pounds of water 1° C. This heat, as a volume of *molecular repulsion* and an offspring of abstract force, the opposite of gravitation, instantly transmutable and translatable into all sorts of effective motion through any form of matter, either for lifting, as by the *lever* or pulley, or projecting other matter, is capable, by computation, of lifting 47,000,000 pounds one foot high.

Now these same nine pounds of molecules, eight pounds of oxygen and one pound of hydrogen, when condensed from gases into steam, from steam into water, and then consolidated into ice, employ a force of *molecular attraction*—the antithetic of molecular repulsion—which can be equally reduced by numbers to approximative mechanical values. When thus reduced, they represent the mechanical power involved in the *inclined plane,* which, expressed in numbers, is equal to the effect of gravitation upon a ton weight falling through three successive descents of 22,320 feet + 2,900 + 430 = 25,650 feet, or upon 51,350,000 pounds from the height of one foot.

It is not my purpose to discuss the exactness of these computations, but only to show that the two expressions of energy elaborated in this work are absolute antithetics, with distinct although correlative bases, both of which are capable of *numerical* development in opposite directions from molecular into cosmical quantities and functions, in like manner as I have *graphically* developed them in prior chapters. By applying both methods of treatment to this subject, we discover and prove the positive and mighty energies of the molecular forces of attraction and repulsion, when, in propagating themselves in opposite directions through matter, they act and react upon molecules as local offshoots and functions of their concentrated cosmical cognates. Springing from perennial fountains of power—from the condensing and reacting energies of a planet surcharged with antagonistic forces unceasingly propagated through all its molecular elements from centre to circumference, and *vice versâ,* in order not only to execute elaborate local designs, but also to sustain its dynamical relations to the sun and other cosmical bodies—we further discover how impossible it is for matter anywhere to be inactive or forceless; that where heat, light, magnetism, electricity, cohesion, and solution exist, there attraction and repulsion, as molecular and cosmical energetics, are sleepless in their penetrative and propagative functions; and that *forces* can be no more really static than *matter* and *masses* can be really inert.

All this eternal activity of molecular forces—dynamical even in their inceptive propagation and action in crystallization—we shall, further on, be convinced is inseparably associated with cosmical force and motion, and wholly dependent upon the ellipticity of the earth's orbit, and the eccentric relations of the sun to this orbit—a point difficult at present to comprehend. Meantime we will continue awhile our minute study and discussions upon the mysterious method of transmission of force through matter, in order the better to understand the marvellous subtlety of these agents of Divine power, and the dependence of every class of phenomena—mechanical, chemical, and vital—upon simple modifications in quantity and action of these identical agents.

In the platinum experiment we not only saw the marvellous force of attraction demonstrated so far as to expose the secret property of adhesion and definite laws of quantity, as the eight hundred volumes of oxygen affiliated themselves with the spongy metal and tenaciously clung to it, but we also learned the amazing velocity with which molecules themselves—solid matter—can be transported by pure attractive force, and pulled into the minutest interstices of an element for which this force has special affinity, and through whose molecules it incessantly and silently flows with extremest fluency, inconceivable subtlety, and remarkable exclusiveness. The discovery of the true mechanism of heat—that combustion itself is simply intensest vibration of molecules—molecular motion—resulting from the antithetic propagation and interference of repulsion and attraction, as dynamic radicals, through and between all forms of matter, by which the latter may be disintegrated, decomposed, and transformed;—this profound fact is an incidental consequence of our critical analysis, and is highly important, inasmuch as it clearly demonstrates that heat, light, and combustion, magnetism, electricity, and chemical affinity, *per se*, are elevated above all substantial conditions, and that, as elements of motion convertible into mechanical power, through wheels, levers, steam, waterfalls, &c., they are absolute forces in themselves COMPOUNDED of *attraction* and *repulsion,* and capable of direction by human ingenuity into special channels for the development of intellectual designs.

26.

But we wish to grapple with these forms of mechanical force in their most profound conditions of existence, and in their most silent and hidden methods of propagation, and to detect them, if possible, as separate and differentiated molecular energetics stealing through, and connected by, material channels of communication with those central forces of the planet from which they flow, and a part of which indeed they are when considered integrally as cosmical units of repulsion and attraction. In order to expose these important conditions, we will experiment once more upon the same simple elements of nature, but in a somewhat different manner.

It is well known that when oxygen and hydrogen in any proportions are simply mingled together, they will remain in contact as disunited gases for indefinite periods, undergoing no physical change whatever. When, however, slips of pure, clean platinum are suspended in the midst of this mixture, the molecular and massive repulsion existing in the two gaseous bodies, as expansive potential energy, immediately begins to vanish; and enormous quantities of attraction, as positive dynamic energy, more or less rapidly replace it, pulling every molecule into that state of juxtaposition demanded by their union to form water. So prodigious is this amount of attractive energy, viewed as a dynamical agent, found to be when reduced to numbers—if, for instance, one pound of hydrogen and eight pounds of oxygen alone be submitted to this experiment—that its equivalent is represented by the mechanical power that will raise 47,000,000 pounds one foot high!—and the whole amount of this attraction, as abstract and absolute force, has evidently been induced to enter silently and insensibly from without in a very brief period.

Now, I inquire, whence comes so stealthily all this mighty volume of energy—the veritable *vis viva* in rational mechanics —whose potency as a positive dynamical agent, subject to human direction as working power, has been determined with an exactness which no physicist questions? The platinum undergoes not the slightest change. The water generated by the presence of the metal weighs the same as the two gases before their condensation and union; and when separated again into its original elements, neither oxygen nor hydrogen has lost a molecule in substance or an iota in weight. Forces alone have glided in and out with the silence and celerity of thought, replacing each other and showing conclusively that the platinum is the mediate agent which induces and conducts the efflux and influx of this vast and effective sum of exterior and imponderable power. The energy which enters to combine, condense, and liquefy these gases is the antithetic of, and is as a mathematical quantity dynamically equal to, that which separates and expands them from water into their original elements, volumes, and conditions. Where and how are these forces stored? Whence come these voluminous imponderable energies so suddenly, swiftly, or incessantly, in order to execute with such intensity, periodic alternation, regularity, and apparent intellectuality and foresight, the physical mutations of elements, the oscillations of the pendulum, the lifting and falling of the trip-hammer, the movements of the beam and shuttle, the processes of formation and decay and of life and death upon the surface of the planet, extending even into the developments of homologies and *reversing* homotypes in vertebrate and axial organisms, and even differentiating the sexes, and determining the numerical equalization in all the varieties of the human species? Vast and amazing questions! But holding firmly the thread with which we began this bewildering journey, let us trust that the earth and heavens will at last answer and give up their secrets.

27.

Were our planet composed exclusively of the three elements, oxygen, hydrogen, and platinum, charged with the same cosmical forces of attraction and repulsion now stored in it, similar processes of condensation and expansion, varied or repeated over and over again, might periodically occur with each rotation and sidereal revolution. But, composed as it is of multiform elements, whose atomic functions consist in differentiating fixed numerical amounts of these two forces which they respectively appropriate and continuously assimilate to themselves—functions constituting their peculiar differences of chemical character, and endowing them with their relative affinities and repulsions for each other, and which indeed constitute their very secret power of catalysis also ;—our planet thus composed, we behold at a glance that these two cosmical forces of which I treat are not only subdivided and distributed between a thousand different forms and conditions of matter in order to execute mechanical acts and unfold creative designs through all, varying their relative proportions in each (creating density and traction in gold and platinum, for instance, expansion and tension in hydrogen and nitrogen, &c., and mixed conditions in all), but that they are capable withal of darting with inconceivable swiftness from point to point in order to establish equilibrium when local or chemical energies are disturbed, and of transforming themselves instantly from molecular and chemical into cosmical and mechanical expressions of potential power.

Having discovered that attraction as abstract, metaphysical, imponderable energy, an absolute dynamic agent *per se,* can glide through platinum with such startling facility, subtlety, and swiftness as to pull irresistibly together and condense a volume of oxygen molecules into eight hundred times less space than they normally occupy, and also to pull together and condense eight pounds of oxygen and one of hydrogen, when mingled together, so as to exhibit an act of mechanical power equivalent to the lifting of a ton weight through a perpendicular height of 25,220 feet, we will inquire, on the other hand, if its antithetic, repulsion, is likewise living abstract imponderable energy, an absolute potential dynamic agent *per se;* what its numerical value really is, in comparison with that of attraction; and whence it also really springs.

It might be sufficient for my purpose to allude only to the reverse action of platinum, whereby repulsive force, separating itself simultaneously and with equal dynamic energy from these condensing gases, is conducted outwards in order to reunite with and diffuse itself into the general volume of terrestrial repulsion, which act I may here be permitted to say is proved by the *intense heat* generated in the metal in consequence of the intense vibration of its molecules produced by the swift propagation of these antithetic forces in opposite directions—a fact in itself conclusive of the existence of the laws and properties appertaining to matter and force of which I treat, and the dual functions of which have not heretofore been clearly defined and understood. To allude only to this reverse action of platinum suspended in these gases might suffice to establish the theorem, without adverting to other proofs; but various substances execute the same catalytic functions in a very definite and palpable manner; and it will not be long before scientific men will recognise that these metastases of force, darting in opposite ways, at the same time and with the subtlety of thought, are incessant, universal, and more or less active through all matter, and that they constitute the radical bases, the very mechanism indeed, of all natural phenomena, and of physical and vital mutation.

Solution—destruction of cohesion and continuity—is a palpable result of this instantaneous metastasis and absorption of *repulsive force from without,* as expansive molecular energy. It springs to the inner parts of bodies with greater or less quickness, and exercises in them the same function which it exercised in other matter before entering the former. This phenomenon is so common, constant, and universal, as to attract little attention or philosophical consideration.

Nitre and Glauber's salt, for instance, thrown into water, instantly begin to disunite; and there are many acid and alkaline salts which dissolve with remarkable rapidity. While in the former cases of condensation and crystallization, by the inflowing or catalysis of attractive force, more or less heat is manifested, we observe in the latter, on the contrary, that not only does the fluid itself become instantly colder, but also that all surrounding substances are simultaneously reduced in temperature and condensed, as they give up their repulsive or expansive forces in order to produce this phenomenon of solution or destruction of cohesion in crystals and solids. This fact is clearly established by the condensation of atmospheric vapour upon vessels into which repulsion has been induced to flow and concentrate during the processes of solution going on within them.

The volatilization of all the ethers is a phenomenon of the same character.

The instantaneous expansion of liquid carbonic acid is another example. And so, by multiplying similar illustrations, the swiftness of this metastasis or translation of external repulsive force can be traced from slower to quicker degrees of activity, until we at last discover it rushing into those compound forms of matter called gunpowder, gun-cotton, nitro-glycerine, &c. with inconceivable velocity, producing instantaneous expansion, and developing *measurable* sums of mechanical energy, as we observe in the condensation of oxygen and hydrogen into water.

There is another experimental fact of a mixed character, and so remarkable in its bearing upon this point of our analysis, that I will present it in the clear language of my friend, Baron Von Liebig. In treating upon our ignorance of the nature of physical forces, that illustrious philosopher says: "We have recently become acquainted with a large number "of phenomena of which we hardly know which of all the "known forces or causes have a share in producing them. "In former ages men would have hastened to deduce from "this the existence of peculiar forces till then unknown.

" This we no longer do, because we are conscious of our " ignorance in reference to the peculiarities of the known " forces, especially of the so-called molecular forces, cohesion " and affinity.

" If we place in a common champagne glass a solution, " saturated at a high temperature, of Glauber's salt in water " (two parts of the salt to one of water), and allow it to cool, " the salt crystallizes, and the liquid congeals to a thick mass " of crystals, like ice. If the same glass be half filled with " the hot solution, its mouth covered with a plate of glass, a " watch-glass, or a card, and then allowed to cool, the liquid, " after ten hours or more, deposits no crystals, not even when " the covering is removed. If we now immerse in it a " common glass rod, the most beautiful needles and plates of " Glauber's salt are formed from the *surface of the rod*, and " in a few seconds the congelation is complete. The liquid " is in a glass vessel, but, although in contact with glass, " does not crystallize; another portion of glass, that had not " cooled with it, instantly causes crystallization. This ap- " pearance is sufficiently remarkable; but still more striking " is the fact, that if we heat one end of the same glass rod for " a few minutes in the flame of a spirit-lamp, and then allow " it to cool, the rod at that end is quite without action on the " crystallization of the salt. It may be immersed in the " liquid, and moved about in it, without causing the slightest " change. But if we turn the rod round, and touch the " liquid with the other end which has not been heated, " crystallization at once ensues. To a superficial observation, " the rod seems as if it now had poles, like a bar-magnet. At " one end it retains a property which it has lost by the " action of heat at the other. If left exposed to free air, it " gradually recovers the lost power; but if inclosed in a shut " vessel, it continues inactive on the solution for ten or " fourteen days. Even after being dipped in water, and " allowed to dry in the air, it has not recovered its lost " efficacy.

" We have a sufficient explanation of the effect of motion

"on crystallization; but the action of heat on the property "possessed by the glass rod of causing crystallization is, up to "this time, utterly unexplained and obscure."

In direct connexion with the preceding phenomenon Baron Liebig describes another, which being also *à-propos* to our subject, I will cite, in order the more clearly to delineate the translation of both attractive and repulsive force as pure entities, specialized and contra-distinguishable from each other by their effects, and acting as absolute mechanical energy translating molecules, and showing at the same time the definite correlation of mechanical and chemical force, and demonstrating the differences in the abstract nature of elementary matter for accepting and rejecting these forces as radical mechanical powers or energetics.

"If we place a copper-plate engraving on the top of a "shallow open pasteboard box, on the bottom of which lies a "little iodine, and thus expose it for a few minutes to the "vapours of iodine, such as rise at ordinary temperatures; and "if we then press it firmly on a sheet of paper which, like "machine-made paper, has been sized with starch jelly, and is "moistened with very diluted sulphuric acid; we obtain, on "this paper, a most exact impression of the engraving in the "most beautiful azure blue. If this blue impression be laid "on a polished copper-plate, the blue lines gradually dis-"appear, and the image now appears in perfect distinctness on "the copper. A copper-plate, a drawing, even an oil-paint-"ing, when exposed for a short time to the iodine vapours, are "reproduced on a plate of silver; and when this plate is now "exposed, as in the daguerreotype, to the vapour of mercury, "treated in the usual way, a beautiful daguerreotype picture is "obtained, but without the aid of light. It is here quite "obvious that the dark parts of the copper-plate, or of the "drawing, have attracted and condensed the vapour of the "iodine in a much higher degree than the white paper. A "moist surface, containing starch jelly, attracts the iodine from "the dark parts; on the paper appears the blue iodide of "starch, as a blue copy of the engraving. A plate of copper, "again, attracts the iodine from the blue compound, and on "the copper appears a copy in iodide of copper.

"It is evident that the white paper, the black colour or "ink, the starch, and the copper have very unequal attractions "for the iodine; and that the cause of the condensation of the "iodine is identical with that which effects the condensation "of gases in general on the surface of bodies. The ink of the "engraving attracts the iodine, but no true chemical com-"pound has been formed; for the properties of the black "colour remain unchanged, and the iodine has only had its "volatility destroyed or diminished; it acts on starch like free "iodine.

"By these phenomena," continues Liebig, "we are in-"voluntarily reminded of one of the most remarkable occur-"rences in the human body—namely, the part played in the "respiratory process by the solid parts or globules of the "blood."

These examples will be sufficient to present the fact, and convey my idea, of the distinct and positive capacities of matter to receive and to be acted upon by translations of attractive force, whereby condensation and cohesion are effected, and to receive and be acted upon by translations of repulsive force in a contrary manner, whereby solution, expansion, explosion, projection, and radiation are effected; and to direct attention to the more profound truths, that all displays of chemical and mechanical energy upon the planet's surface are not only identical forms of action in the abstract, but resultants also of the same principles which spring from and return into those general volumes of binary and chaotic dynamic energies which constitute the central forces of the planet—forces which, on one hand, compel all its molecules to cohere into a ball, endowing it with local and chemical as well as with centripetal, centrifugal, and rotatory functions, and which, on the other hand, act and react with instantaneous mechanical effect in harmony with cognate solar forces adjusting its heliocentric movements, and determining the eccentricity of its orbit.

Call the local, experimental, and limited phenomena to which I allude "simple chemical processes," "modes of motion," &c., apply to them any name whatever, constructed out of any tongue, or indicated by any symbols, the point which I hold and aim to establish is that the *force* which effects solution, decomposition, explosion, and radiation anywhere and everywhere, far from being what physicists theoretically pronounce "*latent heat*" *in the body itself*, is pure abstract terrestrial repulsion, which, existing outside of it, suddenly darts into, and is transfused among the molecules of, the body in question; that this force springs directly, and with inconceivable swiftness, from, and is a part of, the same general stock of repulsive force, stored in the globe from its centre to its outmost boundary, which was apportioned to the planet in the beginning as a given quantum convertible into all sorts of uses for definite ends in the economy of the cosmos, and which, uniting and disuniting with its co-ordinate and coefficient of attraction, constitutes the veritable *vis viva* —the mechanical energy—so computable by numbers in its relations to matter wherever studied throughout its entire range of chemical and physical aspects and possibilities.

28.

Mutual Convertibility of various Forms of Force.

It is at this point that all the phenomena of electrolysis unite with our generalizations, and that molecular physics as a series of phenomenal principles are clearly seen to radiate from and converge into cosmical physics. From this point of view all voltaic, magnetic, and electric batteries appear like so many instruments arrayed to demonstrate the truth we propound. Here, moreover, the entire store of facts garnered by experimental and theoretical chemistry legitimately falls in to confirm and establish our deductions. For inasmuch as we have proved by plain uncomplicated platinum experiments that the forces which execute the composition and decomposition of the physical elements are simple attraction and repulsion, it follows that all the conditions and mutations of these elements and of their compounds are effected and determined by transfers alone of greater or lesser quantities of these forces, from one point to another, with greater or less degrees of rapidity. If heat be generated in virtue of these metastases, we are now in a position, in consequence of preceding researches, *to know that it is not heat* PER SE *which effects molecular, chemical, and mechanical changes.* For heat and cold are proved by experiment to be *only sensible results* of motion, or *secondary properties and affections* of matter, arising from instantaneous translations and mutual interferences of our two radical forces of attraction and repulsion. These forces, propagating themselves from molecule to molecule and from mass to mass in antithetical directions, thus generate vibrations which in turn combine and amplify themselves into sensible motion, and develop by their intensity, regularity, or confusion, the entire series of thermic, luminous, magnetic, electric, and actinic phenomena constituting the alpha and omega of chemical and mechanical science. All these multifarious states of matter have been shown by Faraday, and other reliable experimenters, to be more or less cognate and convertible one into the other, and into motion and mechanical power. And Mayer and Joule, with singular sagacity and untiring patience and zeal, have enlightened the world by extending their investigations so far as to determine, not only that each of these special functions of matter is convertible into the other, but also that they all are the exact numerical equivalents of one another as agents of mechanical power; and that as all spring from motion, all are convertible back again into motion; thus establishing the fact of conservation of force in one form or another as the element of mechanical power, and proving that, like matter, it is only a given quantity in the economy of nature.

29.

Conservation of Energy.

From our present commanding point of view we may see further, and discover even how motion can be ultimately resolved into its binary molecular elements of attraction and repulsion, and how these latter forces expand into their cosmical cognates, and so into mechanico-cosmic motion and cosmical phenomena of every description. And from the manner in which the entire subject of "force" has been developed in the preceding pages, we may behold the prophetic and grand idea of Faraday consummated in the discovery even of the conservation of gravity itself, and of the correlation of all forms of mechanical energy with gravitation, through its co-ordinate of cosmical repulsion. For we shall soon be convinced that all displays of thermic, magnetic, and electric force in the planet, and consequently every impulse of mechanical energy and every mechanical act upon its surface, are allied inseparably with, and dependent for their ceaseless activities and periodic intensities upon, the ellipticity of the planet's orbit, and the eccentric relations of the sun to this orbit.

To recur for a moment to what has immediately preceded, we learn that there is no such condition or entity in nature, strictly speaking, as "*latent* heat;" that heat, and its equivalents of magnetism, &c., are generated *at the instant* by the propagation of antithetic forces, and during the processes of their metastases or translations, thus exciting molecules and masses to action and reaction, developing and propagating mechanical energy of one form or another, as these forces dart from point to point, and from centres to centres, to unite with their cognates, in order to restore the disturbed equilibrium of cosmical, telluric, and local powers; that thermic, magnetic, and electric tension, and intensities even, are, after all, as has been long suspected, only temporary and local accumulations of microscopic wave-movements, or vibrations of molecules dependent upon translations of the pure unsubstantial forces of repulsion and attraction from one place to another, and upon the interferences of these forces one with the other in the process of their passage through, combination with, and separation from molecules; that such vibrations, subtilized, quickened, and intensified or otherwise by the specific character and capacities of the different elements of nature—as of platinum and oxygen, on one hand, for instance, and of bismuth and carbon, on the other—and by the infinite energies and capacities of attraction and repulsion, constitute physical and sensible phenomena corresponding proportionally in intensity to the sum of the causes employed in generating them; and, finally, that mechanical motion, as a numerical equivalent of heat, and so forth, is the immediate product and the exact sum of the actions and reactions of attraction and repulsion in molecules, generating at first insensible vibrations, which develop subsequently into sensible conditions and creative results. Heat *per se,* therefore, is never "latent," nor is it independent mechanical and productive energy. Neither can electricity nor magnetism be so. And since heat can only be generated in the fall of bodies by actual impact, and by the resistances, interferences, and reaction of molecules and of the molecular forces incident to the moment and matter in question, and never by the incomplete act of merely falling and approximating without collision, it must be apparent that repulsion is practically an essential element in the development of all thermic phenomena. Similar physical conditions co-exist in friction, compression, condensation, expansion, and every other form of material violence, whether it be applied upon resin, india-rubber, bismuth, glass, iron, copper, or platinum. Reaction of molecules co-exists with compression, repulsion with attraction, everywhere and under all circumstances. Hence all these phenomena and various modifications of force and motion are universally and necessarily the numerical equivalents of repulsion and attraction, and are the simple methods by which gravitation itself is conserved and maintained as an everlasting fountain of regenerated and regenerating power. Such is mechanical force in its abstract conditions, in its transitions through material connexions into sensible phenomena and numerical functions, and in its effective results of motion, production, and material creation.

30.

It is not the design of this work to enter at large upon the discussion of heat and other secondary forms of force as "modes of motion," or as agents or varieties of mechanical power. Recent elaborate researches upon this subject by illustrious physicists have demonstrated its importance, without exhausting its possibilities, reaching its foundations, or lifting it above speculation. Neither is my aim so much to combat the leading scientific doctrines and mechanical theories at present inculcated, as to call attention to the *assumptions* at their foundations, which, as experimental truths lead us into wider fields of contemplation, seem insufficient to support the conclusions derived from daily observation of natural phenomena, or meet the demands of exact science. The recent advances made by Mayer and Joule in new paths of discovery are rendering all antecedent hypotheses unstable, while the very hypotheses founded upon these discoveries seem already so fragile as to tremble or fall at the slightest touch. That the imperfect observations and preconceptions of Thales, Anaximander, Pythagoras, and Empedocles upon space, infinitude, number, antagonism, and affinity hold definite and prophetic relations to the solid discoveries of Copernicus, Newton, Faraday, Mayer, Kirchhoff, and Liebig, all will admit; and that the *truths* contained in the various ancient and modern systems of philosophy will sooner or later be generalized as results of simple conditions and comprehensible laws of nature, few will doubt. My object is, if possible, to reach the thread by which all physical laws are united to one another.

If the principles of mechanical force—that is to say, of the mechanism of attraction and repulsion, as the binary elements of potential energy and dynamic action thus far unfolded—be applied to phenomena, the most striking or the most obscure, taken at random from any department of physical nature, the soundness of our deductions may be conclusively tested. Approach the wonderful field of electrical phenomena, in order to understand the mechanism of the thunderstorm, for instance. Knowing that all secondary forces are convertible into each other, and so into motion, and *vice versâ*, and that the radical elements—the *vires vivæ*—of motion are molecular repulsion and attraction expanding from zero progressively to amplitudes and possibilities commensurate with matter and the oscillating functions of the cosmos; recognising the absoluteness and universality of these laws, and remembering that evaporation, the radiation and expansion of vapour, are directly or inductively affected by cosmic repulsive force; we may well conclude, when condensations suddenly occur in the higher regions of the atmosphere, attraction thus replacing repulsion, that the latter must be instantly liberated, and become an independent volume of active energy, ready to burst on the instant into new relations with other volumes of force and matter, or rush back to its original connexions and sources. That it does not become latent is manifest by evidences too sensible, striking, and positive to be mistaken. Compelled by natural affinities to unite with its cosmical cognate, in order to restore an equilibrium which has been momentarily disturbed, it darts with inconceivable swiftness to other places in the atmosphere, or directly to the earth itself. Conducted along the shortest possible paths by the atmospheric gases and vapours, the mechanical effects of its propagation are not only made sensible by sudden developments of light and sound—intense local vibrations dispersed into wide undulations of molecules, simple modifications of subtlety and intensity in molecular action and reaction—but by the production of ozone also; which condition of oxygen we know is generated by its simple condensation—or, speaking accurately, by the liberation of portions of the repulsive forces with which this gas is naturally endowed. Thus demonstrable translations of repulsive force not only create motion, light, and sound, but *odour* also—another undulatory phenomenon, or secondary force of matter, fitted to impinge upon and supply another animal sense with its necessary aliment. Moreover, this sudden snatching of repulsion from oxygen is effected in the same manner as from aqueous vapour and solid forms of matter. And, in order to show conclusively that the mechanical element of electricity is the pure abstract principle of repulsive force—the molecular and cosmical agent which is taught throughout this work to be an independent principle, yet the co-efficient of attraction, incessantly oscillating with the latter to maintain everywhere that same unstable equilibrium established primordially in nature, in consequence of the unequal distribution of these two forces in different elements, different worlds, different systems, and in different regions of the universe;—in order to demonstrate this great truth, it is only necessary to refer to the sudden formation of hail and ice in the higher strata of the atmosphere during electrical storms. These congelations could never take place without the instantaneous reabsorptions of repulsion as dynamic and expansive force; for, when aqueous vapour is condensed into water at temperatures above 40° Fahrenheit, repulsive force is liberated, and below this temperature it enters again as an active element of mechanical power.

31.

The Coincidence of Ancient and Modern Observations, and their Astronomical Bearing upon Mechanical Theories.

It is in this connexion that I recall with veneration the many discussions and marvellously acute speculations of certain of the ancient Hellenic philosophers. For Heraclitus, five hundred years before our era, distinctly taught (although, so far as is known, without experimental or mechanical knowledge) that the principles of cosmic and perpetual motion were those which constitute heat and fire. Fifty years later, Empedocles and Anaxagoras insisted that, in regard to both force and matter, LIKE *perpetually strives to affiliate and coalesce with* LIKE, *in order to form unions and unities;* and that elements separate and decompose by antagonistic forces only to come together again, and thus generate and maintain perpetual change and unending life and mind. Not only was the birth of modern chemistry and electrolysis foreshadowed in these wonderfully profound ideas, but also every discovery in natural and experimental science tends to illustrate their general accuracy. For even in the formation of ozone by the electric spark the chemist has now ascertained that only a closer juxtaposition, a simple condensation of oxygen molecules, takes place. And thus it necessarily follows, that a portion of the repulsion normally inherent in oxygen as molecular force is liberated, alienated, and transferred to the cosmic current of its cognate, as this latter propagates itself in volume swiftly and violently from one locality to another through oxygen, in the natural process of restoring equilibria in telluric energies which had been previously disturbed.

The secret mechanism of this phenomenon—the abstraction and absorption of repulsive force by its cognate, from material forms—made so sensible to the olfactories in the ozonization of oxygen, becomes equally sensible to the eye, and comprehensible to the mind, when traced in its results upon solid bodies. The brain, spinal marrow, and nerves, heart, muscles, and even the blood corpuscles, of all living things instantly alienate and transfer their vibratory or moleculo-motive force—pure dynamic and working energy—to passing currents, thus yielding the only *physical* condition, the *mediate* principle, so to speak, upon which life, instinct, and mind hang in their connexion with matter; while trees, and other inanimate things, split and burst open in the act of discharging the molecular repulsion of their innermost parts into the cosmic volume of kindred forces, as the latter are propagated through or near these objects toward or from the earth's centre. The frequent occurrence of ignition during these translations of cosmic force through combustible bodies, as oxygen and carbon, proves the active tension and reactionary function of repulsive force, as pure mechanical energy, in every molecule, and in every line and locality where thermic conditions are manifested.

Thunderbolts, therefore, like all other forms of fire, are only intensest agitations of molecules induced by inconceivably swift and subtle translations of repulsive force through them from one point to another; which force, encountering resistances and interferences of antagonistic forces and of its own counter-currents, generates that undulatory action and reaction in matter known as vibrations, heat, light, electricity, and magnetism, every form and degree of which are functions of the same forces possessing equal numerical values, and alike convertible and decomposable into repulsion and attraction, the *vires vivæ* of mechanical power, and of impulsive and projectile motion.

Thus the germs, existing in the subtle intuitions, speculative culture, and general scientific conclusions of Grecian thought, dormant for a thousand years, in consequence of social disorders incident to the fall of Imperial and rise of Papal power at Rome, revived by the Arabians at Bagdad and Cordova in the birth of experimental physics and alchemy, developed in the successive labours of Stahl, Black, Franklin, Rumford, Mayer, Faraday, and Tyndall, at last blossom in the positive philosophy of the present day. Thus the source of all transitive, mechanical, and creative power may be demonstrated to exist in that same secret *elementum ignis* which was suspected by the ancients, sought after during the revival of learning in the "philosopher's stone," dimly seen in "phlogiston," approached in "caloric," almost reached in the experiments upon "heat as a mode of motion," and which has now probably been grasped at last in the simple antithetic play through molecules of the conflicting forces of attraction and repulsion, creating vibrations whose infinitely subtle actions and reactions are *the absolute fire itself.* Thus the conflicting "love" and "hate" existing in atoms and in all things, so distinctly taught by Heraclitus and Empedocles, as the universal element of "fire" (English words, without doubt, inaccurately expressing the peculiar idioms and the real ideas of the ancients), is represented in modern physics by the words "attraction" and "repulsion" and "chemical affinities," analogous principles capable of uniting "like to like," and of maintaining perpetual mutations among the elements and their compounds. And when it is considered that the secret principles of mechanical power have only lately been comprehended as correlative, measurable, and numerical functions of the radical forces of molecules, and that it is only after several centuries of incessant experimental inquiry, and after a rigorous analysis of facts derived from every department of nature, independently of all theories, that we are now able to elevate the entire subject of physics from the realms of empiricism, we may well venerate the wisdom of the Grecian sages, whose hypothetical knowledge of first principles was not short of that of our own times. Even in the highest departments of physical inquiry modern scholars may not be less surprised at the penetrating genius of the ancients. For it is indeed astonishing to hear Aristotle, 2,150 years ago, when summing up the scientific observations and general learning of antiquity, declare that earthquakes are the results of the moving of "spirit"—a sort of unsubstantial wind—within the planet, and that these phenomena are in some way influenced by the path of the sun; a remarkable expression in view of the deductions drawn from the immense stores of facts collected by Perrey, Mallet, and myself, upon which I long since announced the cosmic law, that the general intensity of dynamic reaction, from the planet's centre to its exterior, is inversely proportional to the length of the radius vector. Equally remarkable is the declaration of Aristotle, that earthquakes occur more frequently in the night than in the daytime; since numerical results from modern data, noted upon my records for the whole planet, by hours and minutes, confirm that statement in proportions of 9 to 7, when the twenty-four hours are divided equally, from 6 A.M. to

6 P.M.; and for the four hours of midnight and mid-day respectively, as 71 to 53, or almost as 3 to 2. The earthquake catalogues of the British Association for the Advancement of Science so conclusively confirm the results of my own tables, that there is no room to doubt that Aristotle must have possessed, not only ample data for his statement, but also prolonged observations carefully discussed, in order to have reached so profound and important a truth. His conclusions, based upon events occurring previous to 300 B.C., and, as we must naturally suppose, limited to comparatively narrow regions of Europe, Asia, and Africa, are now substantiated by the combined facts observed *throughout the planet,* since the fourth century of our era, prior to which no hourly dates have been discovered. In order to settle this point beyond doubt, and at the same time to applaud the acumen of that ancient philosopher, I will subjoin the hourly data, which I have tabulated from the catalogues heretofore published by the British Association for the Advancement of Science, the authority of which is endorsed by common scientific consent.

TABLE I.

From June 20, A.D. 431, *to July* 19, 1784.

Day.	A.M.					Noon.						P.M.
Hours	6-7	7-8	8-9	9-10	10-11	11-12	12-1	1-2	2-3	3-4	4-5	5-6
Totals	60	49	64	41	41	54	51	37	57	43	58	51

TABLE II.

From June 20, A.D. 431, *to July* 19, 1784.

Night.	P.M.					Midnight.						A.M
Hours	6-7	7-8	8-9	9-10	10-11	11-12	12-1	1-2	2-3	3-4	4-5	5-6
Totals	51	57	41	68	79	88	81	61	63	72	80	63

In these tables the proportions for day and night, of twelve hours each, are respectively as 67 to 89. For the eight-hour periods of mid-day and midnight, they stand as 87 to 124; and for the four-hour periods of mid-day and midnight, as 61 to 103.

TABLE III.

From August 27, A.D. 1784, *to December* 4, 1842.

Day.	A.M.					Noon.						P.M.
Hours	6-7	7-8	8-9	9-10	10-11	11-12	12-1	1-2	2-3	3-4	4-5	5-6
Totals	60	70	62	55	69	61	44	49	52	45	56	53

TABLE IV.

From August 27, A.D. 1784, *to December* 4, 1842.

Night.	P.M.					Midnight.						A.M.
Hours	6-7	7-8	8-9	9-10	10-11	11-12	12-1	1-2	2-3	3-4	4-5	5-6
Totals	47	61	82	80	93	98	93	90	125	114	91	80

As we come nearer to our own time, we find the dates of seismic occurrences more accurately and particularly recorded; and the relative numbers for day and night are the more striking and remarkable. The discussions of the last tables present, for the twelve hours of day and night respectively, the proportions of nearly 2 to 3—or, in exact figures, of 676 to 1,054; for the eight-hour periods of mid-day and midnight, as 36 to 67; and for the four-hour periods relatively, as 96 to 161. The entire catalogues of the British Association, published in 1858, approximately show the results to read: 1,282 for the hours between 6 A.M. and 6 P.M., and 1,859 for the corresponding twelve hours of the night.

Beside the pleasure enjoyed in rendering just homage to the sagacity, refined scholarship, and acute philosophical spirit of the Grecian sages, many of whose theories would no doubt be substantially confirmed by modern experimental inquiry, if peculiarities of Athenian thought and idiomatic turns of ancient speech, long lost, could be now clearly understood and interpreted, I feel especial satisfaction in being able to link the researches and deductions of modern science and philosophy with what may have been solid in the observations of antiquity. The opinions of Aristotle upon seismic phenomena in connexion with cosmic possibilities, considered as the conclusion of preceding observations up to his time, are now seen to be verified by numerical analyses of modern data.

Embracing all these earlier and later facts and inferences in a single category, and reducing them to their absolute weight and just astronomical value, we shall perceive among the higher deductions flowing therefrom, not only that the intensities of earthquake and volcanic phenomena are modified and governed by the ellipticity of the earth's orbit and the eccentric relations of the sun to that orbit, but also that *the rotation of the planet itself* is directly connected with the reactions of its interior upon its exterior, and, through these phenomena, with its diurnal relations to the central body of our system. Herein we behold the fact demonstrated that the principles of mechanical power, which originally determined the planet's trajectory and still sustain its projectile motion, which originated and continue its axial rotations, and which instituted gravitations and reactions from centre to surface, producing all vibrations, perturbations, displacements, and transpositions locally and generally of its individual molecules and elements, are not only analogous; but really identical,—not simply correlated, but absolutely the same and inseparably united,—dual parts of dual wholes,—constituting different expressions of the same cosmic energetics only because the conditions are different under which these energetics act: and, *vice versâ*, demonstrating that reactionary phenomena—that is to say, the forces which propagate themselves through molecules and masses within the globe from its centre to its surface, causing earthquakes and volcanic eruptions; in a word, generating all terrestrial phenomena, from the motion and decomposition of a drop of water to the liquefaction of basalt and platinum, or of the entire globe, including its disruption—are the same which effect its rotation upon its axis, and its translations around the sun. Thus these different forms and degrees of material motion result directly from the ceaseless struggle of attractive forces in terrestrial molecules to unite with their cognates in the solar molecules, and the ceaseless effort of repulsion inherent in molecules and in all bodies to separate them individually more and more from one another, thereby effecting their centrifugal movements and elongations from one another. Thus mutual attraction and mutual repulsion, in virtue of the unequal distribution of these forces in cosmic bodies of different elementary constitutions and magnitudes, compel these bodies respectively to revolve around each other at various distances asunder, and to assume trajectories with various degrees of ellipticity, not only determining thereby the positions of the foci around which they relatively move, but also the universality of their sidereal correlations. I desire here to express this induction clearly, strongly, and definitively, in order to unite the discussions of preceding chapters with the present development of our subject, and establish it as a connecting link with all that is to follow. In the light of a generalization, the induction is, indeed, a pregnant and striking fact. But, broad and important as this truth may appear, it is at the same time simple; since the principles heretofore developed tend to bring order out of chaos, and to subdue, so to speak, every phenomenon of material nature, however wild and intractable in view of existing theories, into obedience to definable conditions, and to exact mechanical and mathematical laws capable of every degree of reduction or development. The relation of the molecule to the cosmos, and *vice versâ*, the equal subjection of both to the same opposing forces, the chemical and mechanical changes effected by the antithetic propagation and unequal action of these forces through and upon matter, and the correlation and mutual dependence of matter and force of every type and proportion, rise now from obscurity, and begin to assume definable and comprehensible shapes in the understanding of philosophers.

32.

The Methods by which the Secondary Forces are generated.

Now since heat, as active energy, is known to be the exact equivalent of, and convertible into, electricity, magnetism, and mechanical motion; and since I have shown heat to be actually and only the sensible result of the compound action of cosmic attraction and cosmic repulsion propagating themselves through molecules in antagonism to each other; it follows that electricity and magnetism are no more fluids or subtle forms of matter than heat: and since the action of all is equally subject to the same numerical laws (that is to say, of intensities proportionate inversely to squares of distances, and directly as the sums of molecules acting upon each other) which govern attraction and repulsion as radical material forces, further proof is afforded that all originate in the manner indicated, and that molecular attraction and molecular repulsion are their ultimate springs and elements of power. The platinum experiment, whereby was demonstrated the swift and sensible translation of these physical forces through special forms of matter from without to their interior, and *vice versâ*, in order to effect or destroy cohesion, is a plain and uncomplicated exponent of every act accomplished within the range of chemical science, whether in the laboratory or throughout the masses of associated elements called comets, planets, and suns, so extensively scattered through space.

Every chemical transmutation is as much the result of pure *induction* of one or the other, or of both together (although in unequal quantities), of the planetary forces of attraction and repulsion, from without to within, changing forms and affecting weights, call it electrolysis, affinity, actinism, what we may, as is every perturbation of the magnetic needle an index of translations of electric force from one region to another of the surface of the globe, and an index of mutations in equilibrium in the central forces of the planet and the sun. When these functions of nature ascend to such magnitudes as to assume cosmical proportions, the mind of the philosopher must equally expand to embrace their elements, and his deductions must extend upon known principles in order to comprehend the mighty causes of such startling conditions. If our globe, for instance, be an igneous mass of rock and metal, whose central floods unceasingly react upon and burst open its crust, often shooting its incandescent elements above its general surface, we must approach the phenomenon with the clear conviction that the forces which vibrate through the glowing platinum, or the molten quartz and metal, in the refiner's furnace, are only cognates and offshoots of those which vibrate through and fuse the planets, generate light in comets, and gild the stars with their dazzling lustre.

33.

Before entering this cosmical field—always interesting and inviting, though heretofore so obscure—we will briefly consider a few more experimental points, in order to complete the foundations for subsequent studies, and prepare the mind for clearer comprehensions of the origin, nature, functions, connexions, and manifestations of the "mechanical force" —that *working* and *creative* agent of nature—which is now so skilfully utilized by man.

Chemical manipulations of every class plainly show what has been already illustrated by the platinum, glass-rod, and iodine experiments, viz. that molecular changes, vibrations, and motions—solutions and decompositions of masses, however dense or light—cohesions of matter and of compound substances, however large and varied—are universally effected by inconceivably swift influxes and effluxes of attraction and repulsion as translatable entities, and in quantities proportional to the work performed. In applying the word "quantity" to imponderable abstract forces which are only known by their effects, I would by no means express or imply ideas of absolute substance in attraction and repulsion,—only the existence of subtle energies capable of infusing positive power into matter, and of effecting results in like manner as animal instinct and will, for instance, act through brain, nerves, muscles, and bones to effect physical and creative results with the crude elements of nature. No physicist questions the metaphysical principle of the energy existent in these forces; nor will it be questioned that one of these forces is attraction, the agent which constitutes the very principle of gravitation itself; nor will it be questioned, moreover, that this force is reducible from cosmic conditions and volumes through indefinite divisions and subdivisions, absolute disruptions and segregations and segmentations, so to speak, into molecular and infinitesimal quantities. For *weight* is only another word for "gravity" or the "attraction of gravitation;" and the mutual influences playing between the elements which develop this property so act in all chemical processes and subdivisions of matter as to present invariably the same sum of power and function. The constitution of compound bodies is not only governed by exact and unchanging laws of atomicity, proportion, and number, but the diversified weights of their constituents, when separated by electrolysis, are invariably equal, in the aggregate, to the gravity of the original mass. The force, therefore, by which electrolysis is effected is the opposite of attraction—a force which at the moment and instantly enters and penetrates the mass from without, and is the veritable force of repulsion, as metaphysical in its abstract character as attraction; not computable by weight, because it is the antagonist of gravity, but by its effects in overcoming cohesion and neutralizing chemical affinities; and which, when reduced to its numerical elements, is found to be exactly equivalent in its mechanical effects to the falling of masses of various magnitudes from different heights. But, as abstract principles, neither attraction nor repulsion *alone* is an effective mechanical agency. It is only when they have entered molecules or masses together, and are propagated in opposite directions to each other, causing molecules not only to move and act, but also, and especially, to react, that these forces develop sensible phenomena, and transform themselves, in combination with and through the motion of matter, into mechanical power. Such is the mechanism or process by which these vast binary volumes of energy, apportioned to the planet for its general physical activities and phenomenal developments, are divided, and subdivided, and contributed to the numerous elements and their compounds constituting its mass, and by which the same elements and compounds in turn contribute to sustain the fluctuating translations and vibratory propagations of this energy. Thus, further, it is demonstrated how disturbances of equilibrium are universally generated; and not only how unstable equilibria of various kinds and magnitudes are sustained, but how absolutely necessary they are also in the economy of nature, in order to perpetuate the mechanical activities of the universe and the creative and progressive designs of a guiding Providence. For it is in consequence of ceaseless disturbances of equilibrium, everywhere incident to such processes, that the radical forces of attraction and repulsion are individually inducted, transformed, and exchanged on the instant by various elementary and compound bodies, thereby generating sensible heat or cold, electric and magnetic attractions and repulsions, chemical affinity and decomposition, absolute mechanical motion, and dynamic power. The reciprocal convertibility of these secondary forces, and their equivalency as mechanical agents and numerical quantities, having already been determined by Faraday, Mayer, and Joule, are facts upon which science can permanently rest. Thereupon we stand, and extend our analyses in order, if possible, to grasp the still more profound conditions in which these secondary principles receive their wondrous birth. And in this search we descry more clearly how all forms of force are correlated and conserved; and how attraction, governed by those exact laws of definite, multiple, equivalent, and combining proportions discovered by Richter, swells by gradual accumulations from the zero of their infinitesimal and molecular states into those cosmical amplitudes, the numerical elements and reciprocal relations of which were reduced by Newton to laws of quantity and adjustment equally exact, constant, and harmonious, and co-extensive even with the entire cosmos. The veil, indeed, rises yet higher, and permits us to obtain the first distinct outline of the positive material connexion, and of the mechanical and chemical unity and inseparability, existing between all celestial bodies. For, now knowing that light and heat are everywhere and always nothing more or less than subtle vibrations among the molecules of elementary bodies unequally excited by antithetic propagations of attraction and repulsion—light always differing in colour with the difference of the element excited by electric polarities, and colour, in turn, always modifying vibrations, and causing different degrees of electric intensity and heat;—knowing these facts, we perceive, when the solar image, or any of its rays, is dissolved in the prism, and their vibratory peculiarities are sufficiently magnified, that we can have no other sensible phenomena presented than we should obtain from the same chemical elements under intense excitement in our own planet. The sagacity of Frauenhofer in first observing the spectral lines, and the genius of Kirchhoff and Bunsen in detecting their chemical indications, are the more worthy of honour, inasmuch as these eminent scholars were wholly ignorant both of the physical constitution of the sun and of the elements and mechanical motion which lie equally at the base of chemical and cosmical phenomena. Since light and heat are subject to identical laws of quantity and action, and must be everywhere generated by the radical forces of attraction and repulsion, and in the manner heretofore described and illustrated, it follows that *solar* light and heat, manifested to our senses as aggregate quantities and entities, must be constituted of the multiform and combined vibrations which have been excited and differentiated by the respective molecular forces of the various chemical *elements* and *compounds* composing the solar mass, whatever the number of the latter, whether two, "sixty-four," "seven hundred," or "two thousand," as many as may be indicated by their characteristic lines in the spectral photograph. Comprehending the full force of this induction, we shall clearly understand how every ray of the sun combines and represents the vibrating forces of the chemical elements which exist in and constitute the active internal conditions of the entire solar mass. And for the first time it now becomes explicable how one of these rays, dissolved and disintegrated by the analytical power of the prism, reveals in the magnified spectrum that wonderful alphabet of Frauenhofer, whose stedfast characters spell out the physical constitution of the sun, determine with precision the chemical elements of stars and comets, and establish the identity of matter and material forces in the earth and throughout the heavens. Inasmuch as "like" combines everywhere with "like," and opposite polarities exist in every known form of force, the alternation of bright and dark lines, and the replacing of dark lines by bright ones of various hues, in experimental physics, may be legitimately explained by the partial or total absorption and refraction of those vibrations—that is to say, of the mechanical elements of solar light which find their analogies or antitheses, their positives or negatives, so to speak, in the physical condition of our own globe;—all these molecular and chemical expressions of force composing in the aggregate the planetary forces whose reciprocal action and reaction as cosmical units perpetuate the unstable equilibrium of the celestial motions.

34.

All that has been said of attractive force in connexion with the preceding subject must equally apply to repulsive force; since not the slightest vibration can exist or be generated among molecules, nor can motion be perpetuated in masses independent of the co-operation of the latter, as an oscillating function in mechanical power. In like manner as sparks of electricity or fire, which are simple *vibratory foci*, or centres of intense molecular action and reaction—a species of electrophorus for exciting inductions of the attractive and repulsive forces into the different elementary bodies—in like manner as sparks will cause a diffuse mixture of oxygen and hydrogen, for instance, to *condense instantly* into a small space, forming either a liquid or solid; so will the same sparks cause compounds of various elements, such as gunpowder, gun-cotton, nitro-glycerine, &c., to *expand instantly* into diffuse and gaseous conditions. If repulsion, viewed as a distinct and quantitative element of energy, be instantly specialized, differentiated, and rejected in the first case, reacting with attraction, and propagating impulses to surrounding objects, generating heat within, and sound without, the condensing gases; so, in the last case, an antithetic expression of the same phenomenon takes place with similar vibratory results. Both phenomena present positive conditions of dynamic energy, the developments of which are equally measurable; and when the conditions are inverted, the physical effects are found to be numerically equivalent to each other. The existence of both forces is necessary to the phenomenal action of either of them. Every chemical act, whether experimental or natural, is accomplished by similar binary transfers and interchanges of these forces, either partial or total, either nstantly and in volumes with violent, or slowly and in minor proportions with insensible, or gradual, mechanical effects; these transfers being modified by the character of the elements undergoing mutations. As the entire phenomena of chemistry are simple results of motion, it follows that repulsion and attraction are not only the basic forces of matter, but also the radical elements of mechanical power, everywhere and under all circumstances. Since no other secondary forces than heat, light, electricity, and magnetism, in one form or another, are known or found to act as dynamical agents, either in atoms or planets: and since all these are reducible to "heat;" heat to motion; motion to the oscillating functions of repulsion and attraction, as these forces (viewed in their cosmic connexions as universal agents, acting under cosmic tensions, pressures, and tractions) are propagated through the molecules of the various elementary substances and their compounds: it follows that all mechanical force is cosmical, and dealt out from cosmic fountains, as the special requirements and inductive necessities of molecules, masses, elements, localities, machines, cells, plants, and animals may demand, anywhere and everywhere throughout the interior or surface of this or other globes.

When electricity, embodying positive mechanical energy, capable of effecting measurable material results, can glide through solid matter with a velocity equal to 288,000 miles in a second of time—a fact determined by Wheatstone—an approximate idea may be formed of the subtle action of the *radical* principles whose propagations are necessary not only to produce this agent in the first place, but also to create simultaneously its known functions of polarity and duality. Here, again, we clearly distinguish the existence of dual forces more profound, penetrative, potent, and creative than electricity or magnetism, light or heat; and reach further toward the veritable sources of dynamic energy, the fountains whence are eternally supplied the *vires vivæ* of mechanical and productive power.

35.

Chemical affinities and electrolyses of every class and magnitude, therefore, resolve themselves into pure inductions of attraction and repulsion from exterior sources, the interpenetration and propagation of which among and through molecules, creating physical action and reaction, are clearly demonstrable by their effects. The antithetic propagation of these forces through the multifarious elements and their compounds generates not only the correlative varieties of secondary force denominated motion, heat, light, electricity, and magnetism, but also those still more subtle, intense, specific, and differentiated forms or degrees of vibration and undulatory motion, pregnant with instinct, and almost with intelligence, denominated actinism, and measurable in the Frauenhofer lines of the magnified solar spectrum; vibrations the specific wave elements of which are for ever unalterable and characteristic, since the rate, strength, and resistance of the energies propagated are determined by the atomic structure—that is to say, the chemical constitution (the capacities for saturation of atomicities)—of the body through which the attractive and repulsive forces are incessantly transmitted.

Now, since the intimate structure of the different elementary bodies depends not only upon the *degree of cohesion* among their molecules, but, as chemists and physicists prove and declare, upon the *geometrical shape* of their ultimate atoms also; since all geometrical forms and crystallographic *tendencies* and *laws* establish clearly the innate presence and necessity of definite and unalterable amounts of attraction and repulsion in the molecules or atoms of the elementary bodies; since the weights and densities of each are specific and unchangeable; and since the combining properties and energies of each and all are determined by numerical laws of definite, multiple, and equivalent quantities and weights; it irrefragably follows that specific sums of attraction and repulsion must have been measured to each of these elements in the beginning, and must be continuously supplied and maintained in exactly the same relative proportions. But the universal distribution and juxtaposition or combination of these elementary substances, so numerous and so diverse in their physical properties, conditions, and affinities, must necessarily modify the intensities and impulses of the cosmical currents of attraction and repulsion, which, in order to supply these elements with their continuous actinic and specific powers and functions, are incessantly propagated by convective processes from the poles of the globe to its centre, and from the centre throughout its circumference, and which under the names of "centripetal" and "centrifugal" force are theoretically admitted by all physicists, at the same time that astronomers and mathematicians alike ignore the living action of the latter, and devote their efforts to substantiate the existence of the former only; and that without the slightest knowledge or understanding of its essence or mode of action, or an endeavour to ascertain its relations to other forces, or to the matter upon which it acts. The entire field of experimental chemistry and physics, and the universal store of facts observed by geologists and mineralogists, establish, however, the important physical generalization which I have here presented. And thus, avoiding and rejecting all hypotheses, and building philosophy upon incontrovertible facts alone, we discover abundant causes, introduced in the beginning, and for ever existing in the economy of nature, for disturbances and instability of equilibrium; causes which, commencing with molecular and chemical phenomena, end in those grand developments of energy that not only have generated and perpetuated planetary rotation, but have also initiated, and still maintain, those oscillating movements of celestial masses around each other so clearly determined by the ellipticity of their orbits, and the perturbating influences mutually exerted upon each other according to known laws of inverse distance and direct quantity. Thus it is seen how closely the permanent existence and specific difference of every element and phenomenon in physical nature are inwoven with, dependent upon, and inseparable from the molecular forces; how these molecular forces are inwoven with, dependent upon, and inseparable from the cosmical forces of cognate characters, with which the planet is loaded to repletion; and how these latter forces (seemingly universal, incomprehensible, chaotic abstractions, yet proven to be individual, living, potential entities in virtue of their interplanetary influences) may communicate vibratory impulses to the molecules composing cosmic bodies, and thereby resolve themselves into sensible motion, heat, light, electricity, and magnetism; and, furthermore, become differentiated, subdivided, appropriated, and worked up by the specific capacities of the different elements into those actinic transmutations which, originally established by the Creator in order to initiate perpetual being and activity, for ever oscillate with self-sustaining energy, and act the part of a mighty pendulum within every globe, and throughout the machinery of the heavens.

36.

Friction.

It is in this connexion that the introduction of other experimental data is demanded, in order to extend the foundations and strengthen the arches whereupon every class of physical and organic phenomena is successively unfolded, and every element of positive and inductive philosophy may securely rest. The facts alluded to are few and simple; but being generic in character, so to speak, they will exemplify principles, powers, and functions of nature which are equally observable in their extended and cosmical as in their limited and experimental conditions. They will elucidate with singular clearness the origin of those obscure phenomena of polarity and duality so manifest and striking in all known expressions of force. They will finally establish the separate and independent reign of the TWO principles which I have hitherto discussed and held to be the co-ordinate and co-efficient sources, both radical and universal, of all mechanical and phenomenal energy; principles without the compound and conflicting action of which in molecules and throughout planets there could exist no such remarkable phenomena as those I now present.

When a rod of glass is rubbed with dry silk, a force possessing peculiar and well-known properties, denominated *vitreous* or *positive* electricity, is generated. This always flows to the north.

When a rod of sealing-wax is rubbed with dry flannel, another force, possessing equally peculiar and well-known properties, denominated *resinous* or *negative* electricity, is also generated. This always flows to the south.

These two electricities originate from the same cause, FRICTION, simple disturbances of equilibrium in the relations of molecules, in the forces which bind molecules together—that is, in attraction and repulsion, the most profound and subtle of known physical principles. These two electricities, thus identical in origin, are practically similar in capacity and character: but they are opposite and mutually antagonistic in all their functions and effects. Denominated separately as "currents" or "fluids" simply for the purpose of comprehending and defining their respective characters by their adverse functions and effects, they are positively the exact equivalents and co-efficients of each other as energetic principles, and can be insulated, accumulated, and differentiated the one from the other, and managed as if they were substantial and ponderable entities, while they really are not so. These two electricities, however, combine and harmoniously affiliate to establish an equilibrium between themselves, somewhat as different gases diffuse their particles respectively among each other; and they would for ever exist in this state in molecules and throughout matter, and, so far as we can infer, be non-existent as secondary force, if all molecules and every other form of force were at rest, and no cause of unrest existed in nature. In these facts the principle of *duality* existing in what is termed electric force is distinctly exposed; and the same principle of antithesis appears to pervade the entire physical universe. But whence spring these dualisms—these antithetic manifestations of similar elements of energy? We shall comprehend this hereafter. It is known that the glass may be rubbed indefinitely with the silk, and the sealing-wax with the flannel, without the slightest loss of substance in either of them ; and still that the forces generated and proceeding out of them are persistent and inexhaustible, and capable of being measured, accumulated, and employed by propagation as creative and productive mechanical power, as long as friction continues. Observations in these respects correspond with those attending the generation of heat. Our platinum experiments only need to be recalled in proof of this. But Count Rumford established these facts in 1798 in a different way ; and furthermore came to the conclusion, through his famous experiments at Munich, instituted in order to measure the heat obtained by friction in boring cannon, that "motion" constitutes heat by being transformed into it in some manner which he did "not pretend to know," or presume to conjecture. I need not again refer in detail to the important steps achieved of late, by which the correlation and numerical equivalency of heat, electricity, magnetism, and mechanical motion have been definitively established. These are truths upon which all philosophical analysis may be as firmly based as upon the laws of gravitation : and proceeding upon these foundations, we may safely conclude, when we reach the origin and causation of one of these secondary forces and their attendant phenomena, *that we have grasped the root of all.*

If MOTION be transformed into *heat*, as Rumford conjectured, and as physicists in general now teach, then repulsion and attraction must be the *roots* of heat ; because there can be no motion without the antecedent existence and action of *these two* forces. This we have heretofore established. Faraday has proved that gravitation is not electricity nor magnetism ; and that the two latter are not congeners of, or convertible into, the former. And yet that there is some hidden link connecting their respective phenomena he could not doubt ; and this is conjectured also by all who employ themselves with the special study of these subjects, and particularly with that relating to their cosmical characteristics and periodical variations of intensity. The link, however, has not hitherto been detected.

37.

Having shown clearly the *duality* of electrical force—the glass and silk and the sealing-wax and flannel respectively differentiating and insulating the binary and antithetic elements of this force; the silk and sealing-wax specializing, separating, and temporarily holding the one called *negative*, the glass and flannel the one called *positive* electricity—we will now exhibit their *union and antithetic action* in other substances, and so present this union and antithesis as to elucidate the principle of *polarity* which is so striking and conspicuous in molecular and experimental physics, and so observable throughout the surface and crust of the globe as a universally pervading and cosmical condition of matter.

When a rod of iron or hardened steel held in a north-south direction, or in the line of the magnetic dip, is struck with a hammer, a force is generated possessing peculiar and well-known properties, denominated *magnetism*. Strange to say, this force thus generated in the iron rod combines the identical elements, distinctly affiliated, which were so clearly generated, differentiated, and insulated in the rods of glass and sealing-wax. These may moreover be separated and insulated also, and so specialized and manipulated as to act in all respects like the positive and negative electricities; thus proving not only their correlation, but indeed their *identity*. Instead, however, of displaying their antithetic properties with equal definiteness and energy over all parts of the iron rod, these opposite elements manifest a disposition to segregate, and to separate their respective potencies as far as possible from each other, in like manner as in the glass and sealing-wax experiments. In fulfilling this law, their greatest intensities concentrate at opposite extremities of the rod; and these intensities diminish toward the centre of gravity, where they become neutral and vanish. One of these magnetic elements, thus specializing itself, is called "*positive*," and flows from the south to the north, as if it were water or a subtle fluid; the other, called "*negative*" (names resulting from ignorance of their being the true elements in segmentation and creation), flows from the north to the south. Thus behaving, these two electricities establish what are called "poles;" and from this phenomenon thus displayed—that is to say, the distinct concentration and manifestation of opposite elements of apparently the same force at opposite ends of a bar of iron or steel—the idea and definition of *polarity* are derived and determined. If the iron bar be continually hammered, the magnetic force will be as persistently generated and poured out from it, ever acting as positive mechanical power, capable of moving machinery and effecting productive and creative work; and still neither bar nor hammer will lose a particle of their substance nor an iota of their weight. Thus we see that heat, electricity, and magnetism are generated in similar ways, viz. by friction and impact—*molecular agitation alone;* and that not a molecule is added to, or abstracted from, the masses which develop them. It matters not how large the masses are. The results are in proportion to their magnitude and their intermolecular action and reaction. Now, friction and impact of every sort and amplitude, examined from both physical and metaphysical points of study, are simply the compression of molecules into smaller spaces, and their reaction—a species of compulsory attraction between molecules, so to speak (as may be observed in comets approaching the sun, for instance), and the excitement thereby of proportional repulsion. Impact plainly communicates the force existing in a falling body to the body struck. This force is evidently the gravitative attraction of the falling body, which, propagating itself from molecule to molecule, is counteracted in turn by the repulsive forces of the same molecules. Thus the vibrations—that is to say, molecular motions—ensue which are, from the first, equivalents of heat and magnetism; and thus the transfusion of force into matter, traceable directly to the *crude radical* elements of molecular attraction and molecular repulsion transforming themselves into vibrations, the basis of all physical phenomena, mechanical motion, and productive creation, becomes manifest and demonstrable. Every form of analysis and treatment of this subject, however its elements may be handled, ends in the same result, viz. that the conflicting propagations of the crude antithetic forces of molecules—absolute metaphysical and incomprehensible essences of themselves—create motion. The correlated subjects of motion, heat, &c. have been ably and more or less thoroughly unfolded, in every mechanical and mathematical direction, so far as their laws apply to the useful arts and social economy. The fundamental origin of these forces, and the explanation of their connected action and interaction, are what we seek, in order to enlarge the ranges of human knowledge, and to perfect the theories and improve the applications of mechanical energy—the grand point upon which modern art and practical inventions are founded, and through which civilization is developed and refined.

38.

Friction and Impact universally similar in their Physical Effects.

Friction, however applied, immediately produces disturbances of molecular equilibrium analogous to those generated by impact, though more limited and superficial. Indeed, the forces communicated by the former excite vibrations the sensible and phenomenal developments of which are not only similar, positive, and productive, but also the equivalents, in every degree, of those generated by the latter. The repetition of facts so simple and well known as these, in their correlative and numerical results, might appear too elementary for a work so general as ours; but it is necessary to possess precise knowledge of the origin, causes, and fundamental mechanism of motion, heat, and electricity, as they are observed within limited and experimental ranges, and to have, moreover, clear comprehensions of the metaphysical subtlety of the forces by the antithetic propagations of which through molecules vibrations, heat, magnetism, electricity, and light are generated, so as to be the better enabled to reach safely into wider fields, and to occupy such a position as to determine with mathematical certainty the operation and mechanism of the same forces and causes, when we discuss cognate phenomena in their cosmical aspects. These facts, simple and insignificant as they appear, are the foundations and exponents of principles and laws, the action of which is universal, and the application of which to kindred conditions in the globe or most distant stars will unravel mysteries which have been heretofore studied from hypothetical points of view, and considered impenetrable.

Having traced heat, electricity, and magnetism backwards and downwards through the "MOTION" of Rumford into the *molecular forces of attraction and repulsion*—principles recognised alike by ancient and modern philosophers—which forces, propagating themselves oppositely to one another, infuse the power of vibration into matter; we discover plainly how motion was first generated, and how it has been perpetuated. The augmentation of the slightest vibrations into the grandest developments of motion by successive increments of force and undulatory impulses is a subject not only clearly determined by experiment within measurable ranges, but already reduced by many physicists, in one way or another, to exact mathematical law. The same laws, modified by circumstances and conditions, apply to every form of matter and motion, and may be indefinitely extended so as to comprehend the phenomena involved in the mechanics of the heavens.

Now it would be superfluous to array here the endless data recorded at the meteorological, magnetic, and electrical observatories which have been established in various parts of the globe, in order to ascertain the action of, and the laws governing, terrestrial forces; or even to discuss the cosmical value of their periodic and numerical variability. It is sufficient to state that all classes of facts long ago convinced physicists that the magnetic and electric forces observed everywhere upon, above, and below the earth's surface *are cosmical*, appertaining to the *entire planet*, and constantly influenced in their fluctuating intensities by the variable angular distances of the sun and moon. Not only, indeed, is this great truth established, but equally determined is the fact, that the forces which generate all electrical and magnetic phenomena in experimental physics are in some way connected with terrestrial forces of cognate characters, and *absolutely induced therefrom*. Of late, to be sure, some philosophers have imagined that *motion alone*—the mere mechanical act of rotation in electrical machines, for instance—is sufficient force of itself, and is converted into electricity independent of induction from without, and wholly disconnected from cosmical sources and alliances. This hypothesis cannot be long maintained in view of advancing acquirements. For when it is considered that the inoculating touch of an electrophorus is requisite to initiate electric development after rotation has begun; and that the speculative dogma of Empedocles, that *affinities invariably seek their affinities*, is, after all, when applied to forces, *an absolute law* (as proved antecedently in our discussions upon the nature and behaviour of gravitation); we are compelled to look deeper into the mysteries and mechanism of motion, and to consider always that while no electricity can be generated independently of molecular motion, so no molecular motion can exist or be generated independently of pre-existent antithetic currents of repulsion and attraction—distinct, abstract, metaphysical, living forces—correlated by their antagonisms—absolute centrifugal and centripetal essences. All vibratory molecular motion implies the necessity of pre-existent cosmical forces, acting as antagonistic principles; and these forces so unequally distributed through the chemical elements, and through the planetary and stellar masses, as never to be in equilibrium, but constantly in search, so to speak, of this state. Thus it will be seen *why* the forces of which we speak are constantly in process of translation from point to point, and from the centre to the surface of planets and comets, and *vice versâ*; and why these influences are mutually and delicately responsive from the centre to the boundaries of planetary systems, and *vice versâ.* Amplify this induction as widely as matter is distributed in space, and it will be found to apply with equal strength in support of the truths already reached by experimental study, and to prove, by the universal energy of gravitation, the absolute and necessary existence of the universal energy of repulsion also. Why electricity and magnetism are, also, forces of universal distribution, and correlated everywhere with gravitation through repulsion, and why they therefore exert mutual interplanetary influences, will be discovered further on. The labours of the immortal Gauss alone almost established this point, while all subsequent observation and discussion have confirmed its truth. Indeed, the incontestible evidence afforded by the magnetic needle alone shows that the very polarities of electric and magnetic force are absolutely cosmical principles affecting individual atoms as well as the entire planet; and doing this in consequence of atoms being a part of the planet, and subject to the universal influences of attraction and repulsion, which, acting from the centre to the circumference of the planet as radical motive forces, generate the thermo-electric and electro-magnetic forces upon cosmical scales by the incessant friction and motion—actual molecular pulsating action—existing between all molecules entering into the constitution of the planet, from its centre to its circumference. The objections which will immediately arise in the minds of physicists to this general statement are, that the interior of the globe is incandescent, and that magnetic force diminishes in intensity and vanishes in direct proportion to augmentations of thermal intensity. But these objections are met by our present exact knowledge that the same causes which generate heat, generate electricity and magnetism also; that the three forms of force are equivalent and convertible one into the other; and that if heat appears in the centre, magnetism in the cooling and crystallizing crust, and electricity in the surface and among the gases, they are only three different states of molecular pulsating action—only differences of mode and of intensity in the propagations of force among molecules, the mutual relations and affinities of which are various, and upon which the radical cosmic forces of repulsion and attraction act and react within the limits of definite necessities and laws. Whence these fluctuating intensities spring, and why they are periodical, will be hereafter understood when we discuss these conditions in relation to the ellipticity of the planetary orbits, and the eccentric position of the sun in these orbits, thereby tracing mechanical force directly to its fountain-head. All forms, properties, conditions, and periodic manifestations of these secondary forces depend upon these ellipses, and upon the mutual gravitation and repulsion (considered as internal reactionary forces) of the planetary masses; and if electricity, magnetism, and heat spring from molecular pulsations, and in turn be convertible into mechanical motion and effective working creative energy, in the ways determined by experimental physics, and observable in organic morphology, they must necessarily depend also, in all degrees and modifications, upon the varying distances of the earth from the sun—*not* upon the light and heat *radiated* from the solar orb, but upon the length of the radius vector; in other words, upon gravitation, considered as a *concentral* force, and acting as such upon every individual molecule in the planet, and upon repulsion, its eternally reacting cosmic associate, co-ordinate, and co-efficient, traced also into equally molecular infinitesimal distributions from the centre to the boundary of the planet.

39.

The binary nature, the dual characteristics, of electricity and magnetism—or, as they are at present denominated, positive and negative elements—are equally and definitely traceable to cosmic origin. The capacities existing in these antagonistic entities for mutual segregation and isolation, and their absolutely separate and individual inductibility and veritable inductions from planetary sources and congeners, in various physical experiments, and under all circumstances, and in all magnetic observations, are beyond question. These important truths are demonstrable in a thousand ways: since the *positive* element accumulating, and absolutely insulating itself, in *one end* of a steel bar, and identical with the force that may be differentiated and insulated by *vitreous* bodies, invariably propagates itself, and directs the steel bar, when delicately suspended and free to move, toward *northern* points of the planet; while the *negative* element, accumulating in the other end, and identical with the force generated and insulated by *resinous* bodies, propagates itself, and directs the steel bar toward *southern* points of the planet. These two antithetic expressions of force are so definite and fixed; the innate differentiating spirit and energy by which these binary and co-ordinate elements separate themselves, like oil and water, from each other, in so intuitive and intelligent a manner, are so replete with design, conveying to the mind a sense and consciousness of imperious necessity and commanding will, implanted in the very springs of their being and origin; and these entities and dualities are so positive in their action, inflexible in their character, and universal in their dominion; that needles subject to their influences are for ever restless when in the slightest degree deflected from the lines of their terrestrial currents.

Duality and polarity are, therefore, one and the same entity and power. They originate and define the principles of segregation and segmentation, so important in creation, so visible and wonderful in organic developments—principles first traceable in force, absolutely cognisable by their physical action and reaction in a crude and chaotic way among molecules, and known by the old forcible words attraction and repulsion; principles first seen to assume directive agencies and organic functions in electricity and magnetism after the above-named forces have been propagated, mingled, and agitated, so to speak, among molecules, thereby engendering new and compound forms of energy, and when they attempt to separate themselves again, in order if possible to regain a primordial state of isolation and mutual repose. However profoundly we are permitted to penetrate into this department of nature, we are more and more overwhelmed with the discovery that these forces are not only infinitely subtle, but absolutely metaphysical; that they are the nerve powers of a mighty mysterious agent; the ventricle and auricle, so to speak, of an infinite heart, without which no energy could circulate, no molecules pulsate, no masses roll and oscillate.

40.

Galvanism and its capacities.

Now, if at this point we permit ourselves to pass for a moment to the consideration of the forces generated by *galvanic* processes, embracing, as they do, the development of every form and shade of secondary force; running from the simplest electric excitement through the ranges and affiliations of electric, magnetic, thermic, and actinic phenomena; gathering these multiform expressions and results of molecular action and reaction into a single analysis, and applying the truth elicited therefrom to the same end; we shall be equally interested to find how delicately and precisely the experimental facts of chemistry and physics harmonize with mineralogical, astronomical, and mathematical data, and how all converge in and substantiate a final generalization, the principles of which are founded upon nature and primal law, in opposition to hypothetical assumptions and the hazardous conjectures of philosophers, whose speculations are based upon and limited to a *single* abstract idea. It is not my design, however, to trace the electrical elements through all their wonderful avenues of mechanical action, development, and law. The variety and modifications of their mechanical functions depend upon the number, quantity, and capacities of the chemical substances, the assemblage, association, and mutations of which complete the constitution and determine the sum and power of the central forces of the globe. My object is rather to fortify, by experimental proofs, the certain correlation of electricity with molecular repulsion and attraction; moreover, to identify magnetism absolutely with electricity, and to establish not only the numerical and mechanical equivalency, but also the identity in fact, derivation, and interior mechanism, of every form and expression of secondary force. But, above all, my aim is to simplify philosophical study by demonstrating how, in virtue of the dilatation of molecular forces into cosmical amplitudes, the actinic, thermic, electric, and magnetic forms of force expand also into cosmical conditions, transform their respective mechanical functions and antithetic characteristics into each other, and assume and exert reciprocal influences of an interplanetary character. By patient and persistent labour we shall at last reach the veil of the temple where aforetime was kindled the celestial fire, whose unquenchable sparks excite and sustain the motion and life of the world. Let us not fear for the character of our final attainments; but, recognising everywhere the same intelligent and omnipotent Creative Will, take courage from the past, and hesitate at no problem with which *experiment, exact observation, and the mathematics* can equally and unitedly grapple.

When the forms of force generated by *galvanic* experiments are carefully studied, they are found not only to possess similar dual, polar, and antithetic elements with a north and south, right and left, upper and lower, directive action and reaction—the very same segregative and mutually attractive and repulsive principles inherent and observable in the vitreous and resinous electricities, and in the magnetic forces—but also to exercise the most intimate and unlimited influences upon actinic compositions and decompositions, upon the subtlest molecular arrangements of geometrical being, assuming functional correlations even with germinal sensibility, the primal quickening of segmentive cell-life and embryonic motion; in a word, over the entire range of physiological conditions embraced in organic morphology, and upon which depend vegetable and animal generation, form, development, motion, existence, and continuance. They are found, moreover, not only to possess capacities for conversion into every variety and shade of thermic, electric, magnetic, actinic, and vital power, but also to be drawn or induced swiftly, instantly, and absolutely from the earth itself, to be really a part of the abstract potential energy appropriated to the globe as a cosmic mass attached to the sun; so that, indeed, the slightest galvanic action, the merest touch of a galvanic apparatus, however minute or vast its structure, is but an electrophoric inoculation giving instantaneous birth to influent and effluent currents which are continuous and inexhaustible while the slightest molecular motion exists. And, as in all organic, magnetic, electric, and thermic phenomena, not an atom of matter is consumed or lost, only translated in its relations to other atoms, during the swift and subtle influxes and effluxes of these antithetic elements of galvanic energy. In all these secondary forms and varieties of force are displayed the same definite principles of attraction and repulsion which exist in the ultimate molecules of crude matter, which impart to them their simple, fundamental, chaotic capacities of mutual action and reaction, and which follow them into masses, coalescing into centralizing volumes of energy, as formerly described; thus establishing the means of spherical agglomeration and cohesion, of cosmical motion, and of mechanical action and reaction throughout space. The convertibility of all the secondary forces into each other, the equivalency of their mechanical effects, the circumstances of their direct derivation from and resolution into motion—together with the profounder fact that the hitherto occult mechanism of motion itself is substantially the antithetic pulsative propagation of attraction and repulsion (considered in the light of compound volumes of cosmic energy) through the various chemical elements, generating at first molecular vibrations, the infinitesimal action and reaction of which gradually expand into sensible motion and measurable mechanical power;—all these demonstrable connexions and sequences of facts and phenomena experimentally, legitimately, and logically explain why the same laws of duality, polarity, equal and opposite action and reaction, variations of intensity with inverse distances, &c., so universally applicable to the phenomena of gravitation, especially to those of rational mechanics and molecular physics, and so well determined by the calculus, prevail through the entire series; and why all spring from and return into the same fountains of primal, subtle, and indestructible power.

Details of the pregnant capacities and varied action of galvanic force, its transformations into chemical affinity, thermo-electricity, &c., and *vice versâ,* may be obtained by reference to the observations, experiments, and discussions of innumerable labourers in these departments of physical science. Our special object is to condense and generalize all these labours, in order to deduce therefrom final results, and be enabled thereby to grasp the occult laws which govern potential energy as abstract dynamical and phenomenal power in its distributions to, and in connexion with, cosmical and organic nature. Comprehending the existence of the radical forces of attraction and repulsion, as infinite volumes of binary power in nature, at first undivided; then, secondly, segregative and subdivided in definite but unequal quantities between cosmical masses, and acting like abstract metaphysical agents, with affinities for molecules, and with the subtlest capacities for penetrating and gliding through matter; then, in consequence of these penetrative capacities and diffusive energies, that molecular vibrations and pulsating motions ensue—metaphysical principles thus passing into inseparable connexions with matter, and assuming functions of an absolutely physical character: comprehending these fundamental truths, we the more readily understand how and why the secondary forces are endowed with binary and polar functions; how and why they become segregative, and absolutely repulsive and attractive; how and why the mechanical element in each is and must be equivalent to that of the other, no more and no less; why and how they are correlated with, and convertible into, each other: and we may now clearly understand why, without the intercurrence of repulsion as a cosmical element, there can be no theoretical or mathematical recognition of the correlation of these secondary forces with gravitation; and why, through the existence and action of repulsion as a cosmical force and quantity, magnetism, electricity, and heat may amplify themselves into cosmical proportions with infinite capacities and with interplanetary and interstellar functions; and, finally, why both the primary and secondary forces of the universe are not only inextinguishable, but also capable of incessant mutation and eternal conservation.

41.

Dynamical Laws analogous in their action under various circumstances.

Conversant with the general laws of mechanics, and recognising the application of these laws to matter even under its molecular aspects, we may now intelligibly extend and apply the theory of the sagacious Ampère in such a manner as to embrace all known electrical phenomena, and even explain the mechanism of those wondrous *electro-magnetic* repulsions and attractions (first observed a half-century since by Œrsted), the practical development of which has ended in remarkable inventions and the most useful results to modern society.

When balls of crude *plastic* matter—dough, putty, mud, and the like, for instance—are projected against each other in mid air, or against any solid obstacle (that is to say, if they encounter any resistance whatever), they instantly *lose sphericity*, and *flatten out equatorially*. So, when the same bodies fall upon the earth, they flatten or fly asunder laterally in special proportions to the velocity acquired and the repulsive forces encountering each other. The energy which produces these lateral or equatorial movements of molecules was held, in the first case, at the first moment, or acquired, in the second case, at the last; and this energy of motion (Thomson's energy of position) virtually acts in the mass as so much central force, never latent and inactive, but always active in some effluent or influent way, until the ball strikes, when, encountering antithetic energy, the radial movement takes place from the centre, to which movement the force of every molecule either directly or indirectly contributes. This is a concise description of the instantaneous absorption and propagation of the simple molecular forces—distinct metaphysical and dynamical entities—as they are held or received from without (somewhat as platinum absorbs and conducts the positive electrical element, and steel both the positive and negative), and subsequently decomposed or transformed within the ball into distinct centrifugal and mechanical power. Now, it is not matter of itself alone which moves other matter, whether it be isolated molecules or molecules in mass. It is the antithesis, the conflict, of *forces* under tension and traction in matter, the action and reaction of attraction and repulsion propagated through the molecules of plastic bodies, which so acts within their boundaries as to change their shapes. Moreover, it is not the simple blow—the mere act of striking—of one body against another, considered in gross and producing a crude material effect; but the phenomenon is one where the entire force or motion (*motion being composite force*), held, communicated, or acquired, is instantly dissolved and divided up into molecular infinitesimals, and diffused concentrically through the entire mass, creating absolute tension with even a radiant atmosphere of force, so to speak, all around the body. This fact is plainly proven by cannon-balls in motion, which give out, at each successive instant, the repulsive force that was communicated to them, and with which they were overcharged, by exploding gunpowder. Not only do they gradually fall toward the earth, thus demonstrating this radial loss of acquired force on the whole; but the action of their "wind" in destroying animal life proves also the wonderful phenomenon of pulsating influxes and effluxes of force *during projection.* For men in battle who have been impressed by "wind-balls" and not irrecoverably injured, or who have lived long enough to describe their sensations at the instant the passing missile affected them, feel as if their whole frame was about to burst with a sudden fulness, accompanied by intense pricking throughout the surface from head to feet. These effects are frequently followed with partial or entire paralysis of one or both sides, or with such a sudden or gradual prostration of organic power that collapse and death become inevitable. Death from wind-balls is often instantaneous, as if by thunderbolts; and in neither instance has *matter* touched, or left the slightest lesion upon, the body. The sudden influx or efflux of force alone executes the terrible work. And this force we now know is pure, abstract, mechanical energy—dual or single—whatever may be its ultimate nature. But among the most remarkable facts connected with this phenomenon we observe the following. The part of the body nearest to the passing ball is acted upon in such a manner that the subjects for an instant feel themselves struck, or pulled, *in the interior of the body, equatorially to the line of projection;* and unconscious of the flying missile, they sometimes suppose themselves assaulted or insulted by some one, and turn to resent the injury or inquire the reason for it. Such events often occurred during the late civil conflict in the United States. For many interesting facts upon this subject I am indebted to Major Hall of Jamestown, New York, a scholar and careful observer, actively occupied in the loyal cause during the rebellion. Among others, he reports the case of a coloured soldier in his regiment, who, while lying on the ground, had simply the *heel of his shoe* struck in such a way as to knock the shoe from his foot, exciting only sudden surprise, giving no pain, and leaving not the slightest bruise. The man, however, quickly perceived himself paralysed upon the same side; and notwithstanding he conversed and laughed about the accident, supposing himself all the while not seriously injured, he was so profoundly affected by this subtle and sudden disturbance of his vital forces—that is to say, by the instantaneous propagation of dynamic energy out of his body into the passing ball, or, *vice versâ*, into his body from the ball, either exhausting or overcharging the nervous and conducting tissues—that he soon began to sink, and in a few hours died.

By such facts and observations we learn that the *forces* acquired by projectiles, whether they be balls of solid or plastic matter, and however set in motion, even if by the human arm, as when we whirl a sling or throw a stone, are so much positive energy, *living mechanical power*, infused into them from centre to circumference—veritable imponderable entities, capable of instantaneous absorption, propagation, and radiation from molecule to molecule, charging the ball to repletion, saturating it even to disruption in some cases; forces and antagonistic principles which, when traced sufficiently far, will be found also, like magnetism and electricity, to spring from and return into the very bowels of the planet, as a part of its conservable and convertible stores of attraction and repulsion, crude, mysterious, intangible agencies, but the demonstrable bases of all motion, the radical causes of all phenomena.

Considered in this light, we discover that bodies in motion, whatever their magnitude, from the celestial masses to the smallest projectiles, move and have their motions governed and measured not only by the impulses impressed simply upon their external surfaces, but also by the quantity of force virtually received into their interior substance as so much central force. Experiments, moreover, show that differences in chemical composition, and in the mechanical form of bodies, also influence and modify their receptive and radiating capacities, impulse and other circumstances being the same.

42.

Thus (to follow out my illustration and reach its application), when the plastic masses spoken of, moving in opposite directions, strike each other, the molecular forces respectively existing in each coalesce, and reactionary currents instantly assume mechanical functions, translating molecules into equatorial positions. It is matter and force, both under active phenomenal conditions—not matter at rest and force inert, but both matter and force in motion and in antithetic conflict with other matter and other force—which here establish a reflex system of currents, the energies of which combine and generate entirely new ones, whose mechanical functions act at right angles to, and radiantly from, the primitive lines of impulse. In this theorem lies the secret essence of *uniformity* so observable in the laws of incidence, reflection, refraction, polarization, &c.;—laws which apply equally to phenomena developed by all the secondary forces; and which, modified by circumstances, penetrate and govern the entire range of rational mechanics. Extending even beyond physical conditions, this mechanical principle, for ever supreme throughout nature, and linked with the very centre of the earth and solar system, follows molecules everywhere, and penetrates even the final germ of organic life and development, suggesting and absolutely demonstrating to the physiologist and pathologist the origin of those wonderful intersections of vegetable and animal fibre which not only generate crucial types and forms of structure, but also establish those decussations of cerebral and spinal tissue upon which homologies and homotypes, and all bilateral symmetry and perfection of vertebrate beings, depend. In proof of this we possess facts derived not only from general anatomical developments, and occasional inversions of internal organs kindred to the inverse action of crystallizing forces, but also from those constant reverse results of cerebral lesion which paralyse both muscular motion and the sources of molecular power by which cell-life and nutritive action and growth are sustained.

To confine ourselves, however, closely to the physical points: we observe in proportion to the resistance of the materials under experiment, or, rather, as their chemical and cohesive constitutions differ, that the mechanical effects generated by them vary; and that our senses are excited in different ways and degrees by the decompositions and transformations of the mechanical impulses propagated through them. Thus under different chemical conditions (and here, again, we see why chemical nomenclature should be based upon mathematical quantities of "force," and not on shapes, or numbers, or volumes of atoms or molecules) the same ultimate forces of molecules—simple attraction and repulsion—as propagative and reactionary elements of energy, resolve themselves into different states or modes of motion and action, and assume distinct polarities and different component functions replete with absolute and convertible dynamic power. By this reasoning upon, and treatment of, known facts and laws, we reach that point where final forces, molecular and cosmical energetics, transform themselves into, and contribute to the birth of, the electrical, magnetic, and thermic conditions of matter; conditions which, as all experiments show, never appear independently of vibrating molecular motion: and this species of motion is impossible without the antecedent penetration and mutual action and reaction of those opposing forces whose primal functions are exerted to bend every molecule to the centre of cosmical bodies in order to determine spherical form, while they simultaneously initiate centrifugal evolutions, perpetuate internal convective and eruptive processes, and establish external relations with other bodies of a universal and reciprocal character. In this manner it is that the laws of molecular physics, which impress uniformity of action and similarity of development and of mathematical quantity upon all local phenomena, expand into general ones; and that magnetism, electricity, galvanism, actinism, and heat become universal conditions, and depend for their planetary magnitudes upon the inalienable capacities of molecules to absorb, assimilate, reject, and propagate the volumes of antagonistic energy in which they are enveloped, with which they are saturated, and from which they can never escape; volumes which, as individual cosmic aggregates, exist in chaotic or unconverted states, so to speak, until the chemical elements (all of which, and all the compounds of which, require different combining proportions of these agents) change their relations and juxtapositions to each other. Approximated and alienated, whirled, pushed, and pulled, incessantly active, never static, never inert, vibrating between attraction and repulsion, molecules become not only the solid bases of all existing things, but also the agents, both in their individual and massive conditions, through which these forces convert themselves into absolute mechanical powers and into sensible phenomena with identical numerical values. And inasmuch as these *three terms* in the problem of nature are both *fundamental* and *universal*, and mutually influential upon each other by simplest mathematical laws of quantity, it follows that the phenomena unfolded by them, although various, must be similar in all infinitesimal or infinite conditions and degrees; and that all the secondary forces springing from them must partake of their elements, be governed by their laws, and be convertible into each other, and so back again into the occult principles whence they originate. These inductions are final, since molecules never lose their individualities nor the chemical elements their identities, and since we now know so much of the action, effects, and laws of force as to be certain that there are only fixed quantities of radical motive principles in any planet, any system, or in the cosmos; that these quantities are conservable in each body and each system as specific sums of working energy exclusively appropriated to them at their origin, in some way not yet ascertained; that no more are necessary to execute the mechanical designs of nature than what exist; and that, having sprung in the beginning from supernatural and metaphysical omnipotency, they were issued, propagated, developed, and expanded in such intelligent ways as to generate motion and segregation, and to be for ever conservable by the very processes which they originally instituted.

43.

Under the preceding treatment we discover how the simplest forces in nature yet known (not electricity and magnetism, but the radical molecular and cosmical forces, as abstract elements of energy combining with molecules as bases for the expression of sensible motion and other phenomena) assume projectile functions and dynamic power; how, in thus demonstrating themselves as living entities, they initiate compact rotund volumes and measurable currents of force, establishing simultaneously tendencies to move and straight lines of motion, in consequence of inherent capacities for union with each other; and how the sudden union of different sums of matter and force (the commingling and reaction of various or unlike increments with each other generating vibration, a sort of friction among molecules in mass, resulting from impact and contact) decomposes these right lines and antithetic currents, resolves them into reflex and radial ones, and initiates the fundamental principles and laws of motion. All this is illustrated clearly in experiments upon plastic matter, while the endless variety of decomposing forces taught in theoretical mechanics and observable in the action and reaction of elastic bodies collaterally supports our inductions. Indeed, this finality rises completely above metaphysical abstraction, and becomes simple and clear physical truth when we consider that nature does really abhor vacuum, that matter and force rush and cling and oscillate together as the opposite of vacuum, that something and nothing are definite antagonistic conditions, that entity and nonentity are the mutual repellants and ultimate correlates of one another. Thus we arrive at the root and causation of all mechanical law. The simple example of two plastic bodies rushing together under any conditions whatever represents in gross dual antithetic principles, two counteracting currents and volumes of projectile force, propagating themselves through space, encountering each other, and resolving themselves into equatorial currents of mechanical motion and conservable power—material effects of the irresistible tensions of subtle agencies which embody the principles, the energies, and the laws at the base of every dynamical phenomenon within the range of scientific observation and research.

44.

Further on we shall extend the principles heretofore elaborated into their broadest applications to cosmical phenomena and general causations. But it is important to our purpose at this juncture to point out the following facts—viz. that positive and negative electric and magnetic forms of force are wholly *telluric*, and that, however generated experimentally, they are but fragmentary portions of their cosmical congeners, in virtue of the circumstance that the molecules and molecular forces which generate them are themselves but fragments of planetary entities. And in this connexion we must allude for a moment, also, to certain phenomena upon which the discussions of Ampère are founded, in order to conjoin, as intimately as is consistent with brevity, the most pregnant facts of molecular physics with the final drift of our generalizations. While far from adopting the theory of Ampère, or any other theory, it is quite certain that the mathematical conclusions of that illustrious physicist are not devoid of foundation. But the point to be reached in every scientific path is *exact* and *final truth*, and the *expression of this truth* in terms so clear and definite, that all who run may read, and that no fact in nature shall remain without *legitimate* and *graphical* explanation.

We declare, then, that the earth, as a planetary ball, is charged with heat, magnetism, and electricity, in one form or another, from centre to circumference; considering it already proved by every class and connexion of experimental and observational research. Latest observations and experiments in submarine and aërial telegraphy settle the point of the direct, local, general, and perpetually fluctuating play of both telluric and marine electro-magnetic currents, as positively as the eye determines the fluctuating action of the polar needle and the polar auroras in magnetic storms. Accepting these truths, we feel equally certain that their causation and mechanism may be satisfactorily explained; that the important astronomical and geological data which we shall soon bring to bear upon this subject will afford solid elements for its exact mathematical development, and for the determination of universal laws; and that, once elevated from the realms of hypothesis, the dynamical problems of nature will no longer resist complete solutions.

The mechanism of friction and impact, as we have already intimated, consists in the contact and intermixture, in the mutual action and reaction, of different volumes of force and molecules—in other words, in the decompositions and recompositions of the forces of those molecules which, under any circumstances or in any way, become alternately alienated and juxtaposed. No force is absolutely consumed; no matter is changed in its final chemical elements by friction or impact. Particles only change local *vibratory relations with each other;* and herein attraction and repulsion, considered as distinct entities with substantive connexions, undergo such modifications, during their subtle translations and conflicts with *elementary substances,* that, while all radically remain the same simple terms with unchanging quantities, their functions are altered in consequence of the *chemical bases and plastic essences* of the latter, whose inherent sensitiveness and instinctive consciousness, so to speak, to the presence and accumulations of the former are various. It is at this point that the *specialized* and subtly *specializing* characteristics of molecules begin to be distinguished, the elective affinities of which, discovered by chemists and physiologists, inspire us with the idea that matter is really replete with mind as well as with motive force. This is indeed the point whence the discussions of the pagan philosophers unfolded the doctrine revived by Leibnitz, relative to the "monad" and the "elemental mind" of atoms. This elective, geometric, and morphological essence in matter, so palpably a directive spirit, and so absolutely intellectual, potent, and universal, is that principle which, in the present state of knowledge, I hold to lie at the foundation of the existing state of things, to be the true pantheistic expression of omniscience and omnipotence, and to be that condition of nature where physics and metaphysics, the substance and the psycho-logos, affiliate and become inseparably knit together, and ever unitedly active to effect definite and intelligent ends. The physical forces, themselves immaterial and without substance, combine with the substantial and various elements of nature; and in the manner above described the conversion is effected of one or the other, or of both, of these radical forces into heat, electricity, &c. during their conflicting penetration, and mutual interferences with each other, among molecules and in the midst of the masses through which they glide and circulate, when mixed and excited into reaction by impact and friction. Consequently, impact and friction, when analysed and resolved into their molecular potencies, are found to be only forms or modifications of motion, which, as equivalents of heat and electricity, spring in like manner from, and return equally into, the same fundamental principles of molecular attraction and repulsion. These factors, amplified to any extent, generate proportional sums of all other equivalent and qualified expressions of dynamic energy. The specific gravity of the various forms of matter and of worlds as individual masses is, therefore, unalterable; since heat and cold, positive and negative electricities, &c., are immaterial, and only subtle and metaphysically abstract affections of the vibratory characteristics of molecules. However all these forms of force may change and replace each other, their *final* functions—the decompositions and oscillations of which in masses result in cohesion, radiation, expansion, rotation, &c.—remain the same, and, being unsubstantial and imponderable, add nothing to, and subtract nothing from, the elementary nature of molecules or globes. And inasmuch as only specific sums of energy have been affixed to any celestial body (to the earth, for instance, as is proved by its being a mass of chemical elements which combine their individual specific weights and forces), these sums must for ever remain the same; and, notwithstanding we know that bodies moving through space are subject to more or less active and constant mutual perturbations, we must now be convinced that these excitations are wholly individual and internal, concentral and reactionary, and that the absolute quantities of dynamic and working energy in each and all are eternally unchangeable. The conclusions of Laplace and Adams, based upon the records of the ancient Greek astronomers and modern observations, heretofore alluded to, may again be adduced to confirm this induction, and cement all our positions more strongly together. This great fact becomes perfectly clear at this point, as a corollary of all that precedes; for since heat and electricity are not matter, but only polar affections of molecules—antithetic forces, worked up and modified by different chemical forms of matter, for ever attached to and inseparable from matter, as active phenomenal entities—it follows, under every aspect of our discussions, that heat, light, magnetism, and electricity must be concentrated in material globes, bounded by their gaseous envelopes, and be impossibilities and nonentities where molecules do not exist.

Nevertheless physicists, ignoring many data, have persisted in discussions and mathematical calculations upon the amount of heat and light which have been radiated and lost in the interstellar spaces. Such a wild and wasteful state of things can have no foundation in the economy of nature. Thermic and electric force and polarities are *equally* and *wholly* dependent upon the chemical elements—that is to say, upon matter and atomicities—for their generation and action; and must therefore (if we confine remarks to this globe) be limited within our planet's atmosphere. If radiated from its interior, these secondary forces must be incessantly consumed throughout its circumference in order to execute the mechanical work required to clothe, adorn, and elaborate its surface, and to modify from cycle to cycle its geographical aspects. The necessities of evident terrestrial designs are too numerous to permit such radiant wastes of energy as existing theories inculcate. Not only are these forces employed in the crudest metamorphoses of the earth's surface, and in the continuous support and progressive development of all organic types and successions of races, but also for the higher purposes of incorporating consciousness and intellectuality; of consubstantiating principles of the loftiest and subtlest thought, which are constantly and palpably striving to exhibit themselves in some substantial manner through every molecule in the planet. The geometrical and mathematical elements everywhere occluded in every known substance and every known physical force, all tending to crystallization, to equatorial, polar, segregative, and decussating developments, demonstrate not only the attachment of all force to all matter, but also the incorporated presence of psychological and directive principle in the formless dust and most chaotic conditions of nature, struggling to burst into symmetrical being, to manifest sensibility and antithesis, persistent progression, and indestructible life and mind. Dynamical energy, therefore, whatever its state, thermic, electric, &c., far from being radiated from the globe, as we have all heretofore supposed, is strictly confined to and conserved in it; and is furthermore allied with, and directed by, some other force or principle more profound, abstract, and potent than itself. This is self-evidently an intellectual and geometrical principle—universal and omnipresent—a positively demonstrable, superintending, indwelling, guiding Power.

45.

Electricity and Magnetism the Dynamic Counterparts of each other.

Now when Ampère, in 1820, conjectured *dynamic* elements to exist in electro-magnetism, the equatorial phenomena of which were first distinctly presented to the scientific world by Œrsted a year previously, and when he reduced them within the ranges of numerical law, it will be perceived that the foundations of his studies were solid as well as broad and deep. In them he recognised a *repulsive*, a *true* dynamic element, without however tracing it to its origin. And notwithstanding he suspected that magnetic induction was derivable in some way from cosmical sources, conjecturing that *solar heat and light, per se,* produced this phenomenon by exciting the surface or atmosphere in a spiral manner, and thereby transforming them into a sort of superficial and natural helix, and so creating a planetary magnet; it will soon be seen how the imperfect development of modern mechanical philosophy has not only curbed and dwarfed the profoundest genius, but also riveted error upon the scientific world.

Not stopping here to discuss the causes of the cosmical developments of electricity and magnetism, but admitting their existence as terrestrial entities, we will for a moment refer to them as dynamic correlates and equivalents of each other. Attraction and repulsion viewed as creative and creating entities, when once instituted and mixed together in volumes and as spheres of force, and when unequally diffused among molecules (these latter viewed also as *other* entities with plastic and vibrating capacities), the conditions for the generation of thermic, electric, and magnetic phenomena become established. The existence of these secondary forces becomes not only an immediate consequence of the first, and the exact equivalents of their transformed energies, but also an absolute necessity for the elaboration and multiplication of natural forms, and for the harmonious unfolding and progression of all geometrical designs, as we behold these forms and trace these designs in everything around us. Whenever and wherever one polar principle receives its birth, the equilibrium of nature demands its antagonist. Therefore there can be no negative element or current without its positive antithetic, and *vice versâ*. Universal experiment and observation sustain this truth. The widest ranges of telegraphic electro-magnetism—really a cosmical branch of study—confirm, indeed, the permanent and general reign of this law. The forces, consequently, that resolve themselves into dynamic currents and become sensible and measurable phenomenal powers, are at base the molecular forces of *repulsion*, which, primarily associated with attraction as cosmical aggregates, are in reality the co-ordinates and counterparts of gravitation. The oscillations of these entities, acting together and reacting *in conjunction with molecules*, become absolute mechanical or material and *molar* power, and the subjects of scientific treatment under the name of "rational mechanics." Considered, however, as *ultimate* antagonizing *forces*, associating but specialized and self-segregative essences, possessed of capacities for differentiation with each other and with molecules, exerting independent propagative tendencies, and displaying their energy through inductive currents; thus considered, these currents become necessarily capable of transformation into equatorial and actually repulsive functions. Secondary forces being really composite and combined expressions of these currents—the same forces, indeed, modified in their vibratory measures and compositions by the atomic constitution of the elementary forms of matter and their compounds—the phenomenal functions of these secondary forces must also follow the same numerical laws, and resolve themselves under different circumstances into similar conditions of active mechanical power. Thus it may be understood why the electric and magnetic, and the thermic and luminous, forces possess and manifest dynamic and equatorial reactions, and special correlations with each other; why and how mechanical energy may be converted into these various expressions, and *vice versâ*; and why each and all are endowed with a radiant and repulsive function. The bilateral or dynamical principle displayed by these secondaries is generated in a similar manner, and by the same mechanism which we have seen at work in the plastic masses heretofore presented with the special object of elucidating this very point.

We cannot here object to this deduction on the ground that heat possesses and displays an element of mechanical energy of its own; for we know that, while heat is an agent capable of generating mechanical motion, it is yet itself a *result* of motion, thus demonstrating its existence to depend upon the action of more radical causes. Indeed, it becomes but too evident from this fact, that physical speculations require restraint upon this subject of heat considered as a "mode of motion." For, running too far in one direction and falling short in others, philosophers are at present disposed to ascribe every material action upon the earth and throughout the heavens to this force alone, as a *finality,* venturing to affirm even that it is the antagonist of gravitation, without knowing what it is, and without interrogating its origin or mechanism, or even the nature of gravitation.

We cannot object on the ground that *cold* is, *per se,* pure abstract mechanical energy; for we know it to exchange its functions with those of heat: and although the mathematical development of this point is too much neglected in physical study, we must infer from actual data that cold is the numerical equivalent of heat, and, like that, is an agent in the expression of dynamic energy.

We cannot object on the ground that *galvanism* is, *per se,* mechanical energy; for we know it to be the correlative of heat and cold, and to be dependent for its generation upon molecular vibrations similar to those which produce heat, although not wholly identical with them. For, omitting allusion to the chemical action of its currents, and only viewing the convertibility of these currents into their radical molecular principles, it may be here stated that Dr. Edlund of Sweden has recently determined beyond question that galvanism will rapidly *expand* metallic conductors *independently* of heat: thus showing that, while similar to heat in its mechanical effects, its special power is still a different sort of entity, and that its vibratory functions are measured by different expressions of quantity; but that, like heat, it exhibits the same dynamic energy, derivable in both secondaries from abstract repulsive force itself.

Neither can we object on the ground that *electricity* is, *per se*, mechanical energy; for we have no knowledge of its possible existence independent of the pre-existence of absolute *molecular* action: and we have every reason, in the present state of experimental science, to infer that electrical force is not a radical principle of matter. Nevertheless, knowing that it will not only dislodge and separate molecules from one another, but also translate them corporeally across broken circuits as if projecting them through space, and knowing even that purely inductive currents will effect the same phenomena, we detect in these displays of force and functional action the presence of the very same principle which exists in the radical from which it has sprung, which gives it distinctive energy, without which it cannot be made manifest, and into which it vanishes when converted into motion. It is scarcely necessary to array here the experiments of Daniel, Foucault, Ruhmkorff, and others in evidence of these truths; for the opinion of physicists is now settled upon the mathematical and immoveable axiom that "mechanical force" itself—an unknown something—that something called by Mr. Rankine "potential energy,"—is capable of being converted by instrumental experiments into electricity, and of being modified, and made to pass through every variety and degree of radiant and repulsive secondary force—heat, light, magnetism, actinism, sound, and even odour—and finally of being converted back again into the same invisible, intangible, insensible, vanishing principle from which it is so mysteriously induced.

Now *this final principle of energy* we have already intimated and shown to be *repulsion* (an abstract entity, to be sure, but no more so than *attraction, which is universally admitted to be a cosmic force*), springing from cosmical fountains of an inexhaustible and self-sustaining character, *which, reacting with gravitation* among molecules, penetrates and glides through all matter everywhere, as water through a sponge, actually *simulating* the properties of a fluid and the "diffusive" properties of gases, and which in its action we may compare with the movements of fluids, because molecules, as chemical liquids and gases, are mobile and plastic, and are moved in broken masses and currents by these immaterial and metaphysical agencies. It is here that metaphysical and material entities meet and mingle to constitute dynamics and the mathematics, lay the foundations of physical developments, and constitute positive science and intellectual philosophy.

Thus forces capable of penetrating matter everywhere are, on account of their excessive subtlety and opposite characters, propagated from molecule to molecule, delivering them from inertia, and *stamping them all with fixed or transient polarities, and with living dynamical powers.* So much can be experimentally, mathematically, and inductively grasped, and safely affirmed: indeed, much more. Matter being accumulated in masses, every molecular principle of a kindred character would obviously affiliate, and thus magnify themselves into unitary conditions, with unique expressions and capacities; and with functions, moreover, which, if convertible anywhere or ever into equivalents of others, would be so everywhere and always. And this remarkable state of things must, upon palpable internal evidence, actually have happened in the beginning, thereby presenting the broadest foundations for the action and birth of general physical laws, and for those very cosmical quantities and universal dynamic phenomena, a knowledge of which was long since acquired through the equations of Newton, La Grange, Ampère, and Gauss, while those illustrious men were entirely ignorant of their correlations and equivalencies, their internal essences and mechanism, or their remote causations and origin.

46.

Terrestrial Origin of local displays of Mechanical Energy.

Convinced by present acquirements in physical knowledge that *all electric and magnetic currents originate in, are inductive from, and radiate either directly or indirectly out of the* GLOBE *as the fountain of every form and constituency of mechanical force*, and that abstract immaterial mechanical energy, as we have thus far discussed and developed its dual principles, is absolutely convertible through molecular motion into every form and expression of *secondary force*, passing successively from heat through electricity, magnetism, &c., and *vice versâ*, it follows *that this same mechanical energy itself, as* HYPOSTATICAL MOTIVE POWER, *must proceed out of the globe also;* and whether manifested through experimentation, motive or locomotive inventions, or even in animal motion upon land or in the water, must involve the necessity of oscillating currents and antithetic conflicts of both *repulsive* and *attractive energetics* reacting from the earth's centre, and emanating and radiating from its surface incessantly and universally. And since we can now demonstrate that all thermic and electric currents and tensions are generated by the subtle and insensible transmissions of these radical forces through matter, initiating molecular vibrations in the course of their passage to combine with their cognates in distant places, in order thus to establish or repair disturbed equilibria; these transmitted forces being interrupted, decomposed, worked up, and converted into sensible dynamic and material force, palpable *mechanical* and *phenomenal power;* having reached this profundity of knowledge, we can begin to understand how the two electricities (simple radical cosmic repulsion, and attraction decomposed and materialized, so to speak)—in other words, how dual, electric, and polar currents—by agitating molecules in conductors, instituting reaction, and generating tension and mutual leverage among those molecules, may develop equatorial, radial, and inductive phenomena, direct electric elements transforming themselves into crucial and magnetic ones, attractions into repulsions; and *vice versâ*, under reversed circumstances.

47.

So extensive is the possible application of these principles, that we can even trace the same telluric mechanical energy converting itself, within organic forms, into every variety of secondary force. Not only does it so act as to separate actinically and translate dynamically the very mineral and metallic elements of the planet into vegetable and animal organisms, developing the bulk of their structure, pushing them irresistibly upward and downward and every way by additional increments of solid matter, and even instituting locomotion by detaching countless races from the planet's surface, through occult processes of life demanding incessant supplies of pure *repulsive energy*;—not only all this, but mechanical force also so acts in organic forms and processes of life as to exhibit conversions of its composite essence into thermic, electric, and magnetic functions, in like manner as such conversions are effected, known, and measurable in crystalline and all other purely physical conditions of nature; assuming crucial, decussating, and segmentary offices, and even assuming directive agencies in unfolding the symmetrical archetypes of locomotive being which have been established throughout the lands and waters of the globe. Everywhere, and through immeasurable time, the same dual principles have underlaid and overshadowed organic creation, manifesting their dynamical elements and ceaseless activities in the progressive developments of diversified types of axial and crucial structure, and transforming their subtle energies into the absolute powers and equivalents of life and health. Facts derived from anatomical, physiological, and pathological studies of plants and animals, especially of the human frame—the most elaborate of all organisms—test the truth of these inductions, in both an analytical and synthetical sense. I need barely allude to what we now know experimentally of the action of galvanism, heat, electricity, magnetism, and light upon every normal condition of vegetative, germinal, and locomotive being, in order to fix attention upon a subject which only requires to be thoroughly studied in connexion with *primal* telluric mechanical force to remove it equally from the realms of superstition and materialism, and to impart even to its metaphysical sequences a mathematical basis as exact and solid as the mechanics of the heavens. Lesions of the cerebral hemispheres and commissures, and of the various columns of the medulla oblongata and spinalis, and their offshoots, prove the material capacities of these organs to be adapted to, and employed in, not only the receptivity and propagation of forces from without, but also the subsequent distribution, modification, and conversion of these forces into internal motive and phenomenal power. That influent and effluent currents of cosmical energy are in constant action and reaction throughout the microscopic structure of every organ is proved by numerous experiments and various sorts of observation.

In hemiplegia the conducting circuit is broken; but where the hemispheric lesion is not so extensive as to wholly cut off the propagative action of these currents—currents of pure mechanical energy directly and inductively absorbed out of the planet, in like manner as the electricities are inductive and absorbable—corpuscular connexions are sooner or later restored, normal circulations of all kinds are re-established, and molecular vibrations again effect harmonious relations between the vital and polar attributes of the organism and the central and polar forces of the material globe. In this restorative process we plainly detect the execution of the same mechanical laws which is witnessed in the broken circuit of electric currents, viz. the absolute corporeal translation of molecules from one pole toward the other—except that, under the instinctive guidance of life, the molecules in the former case are progressively organized and vitalized, and at last united, filling and consolidating the void, and restoring abnormal conditions to lawful and healthful ones. The mechanical work accomplished in both cases is similar; the forces employed in the execution of it are the same : all spring from and return into the bosom of the planet. And if one expression of these forces be destructive, and the other constructive, it is because life is a positive principle of energy, as superior in itself to physical forces as physical forces are more commanding and potent in their action than the molecules which obey their influences.

In paraplegia wider phases of the same phenomena are revealed. And here we see a full manifestation of the dual principles lending their energies in the completest manner to assist the plans, developments, progressions, and perpetuations of life.

Now, since we know by manifold experiments that influent currents of electric force are incessantly traversing the frames of animals through the extremities of their locomotive members, and that these currents follow mainly the course of the nerves and the decussations of the spinal marrow to reach the sensorii of the opposite hemispheres of the brain, we may clearly infer that the terminal swellings of these electro-organic conductors are legitimate growths from, and the segmentative formative results of, the inductive action of our two cosmical forces upon plastic and omnifarious molecules. In the course of their telluric reactions these radicals have generated thermic and electro-magnetic currents, the polarities of which, associating themselves with life, decussate as in metallic conductors, and, accumulating in terminal tensions, consentiently attract and transport matter; thereby building up ganglionic tissue at opposite poles, and establishing bilateral and equal hemispheres, the very duality and co-ordination of which are essential to mental, vital, and physical harmony and efficiency, and the reactionary and reflex energies of which are also equatorial, radial, motive, dynamic, and creative, as witnessed in vertebral developments and in every department of nature. These forces, moreover, in virtue of the indisputable capacities for life and mind existent in all matter, have thus self-evidently laid foundations for the origin and development of organic beings, and for the progressive mutations of generic types and locomotive races, the spontaneous appearance and countless successions of which, from cycle to cycle, throughout the lands and waters of an unstable planet, have been among the darkest problems in science.

The abnormal action of the ganglio-medullary organs, and of their offshoots and forces, whether ending in, or appearing as, exaltations or depressions of function—as inspirations or mania, neuralgia or torpor, palsy or convulsions; inflammation, heat, and hypertrophy, or anæmia, coldness, and atrophy—all alike reveal the subtle inductive and differentiating relations of the cosmo-dynamic forces to the material and sensitive structure of organic tissues. Studied in their broadest and profoundest references to locality, climate, altitude, sea, land, light, darkness, and occupation, the vital energies are found to be absolutely and constantly dependent upon special conditions of telluric force—*potential planetary energies*—which are often excited and modified periodically and palpably by distant planetary influences. These facts are too well known to be disputed. It is, indeed, through disease that we are led to a clear appreciation of the character and mechanism of the forces which maintain muscular motion, sustentation, health, and life. As *in all other actinic changes*, we detect the flitting passage and constant agency of the same insensible antithetic forces—abstract motive energetics—leaving marks behind them, and making manifest material and sensible phenomena. It is, moreover, through studies of the internal action and external perturbation of these same forces that the medical philosopher must seek to perfect his nosological systems, and the naturalist his classifications of organic developments. But it is especially through computations upon experimental data, derived from the health and activity of the human body, that this problem has been of late reduced within the proximate limits of physical law, and that animal beings have been virtually and palpably brought into their natural connexions with the central and excentral forces of the planet from which they immediately spring, and by which they are multiplied, modified, and perpetuated. The numerical results alluded to being seasonable to our present object, they have only to be plainly introduced in order to expose the real difference between abstract mechanical force (force *per se*) and mechanical power—between "potential energy" and "actual energy," as Mr. Rankine denominates these entities—and the exact relations which the former sustain to the latter, as motive, practical, and creative agencies in nature and art. Their consideration will also bring us legitimately and definitively to the conclusion of this branch of our study, and, completing the circle of terrestrial physics, take us back to the point from which we started at the opening of this pregnant chapter.

48.

My illustrious and venerated friend, Baron Justus Von Liebig, far in advance of all other philosophers, so far as I know, long ago proclaimed, and his numerous disciples have in turn taught the doctrine, that the force exerted by an organic being in performing the work within its own body, and externally also through muscular and mechanical labour, is altogether derivable from, and developed in similar proportions to, the food and oxygen consumed; and that this force is generated within the organism in like manner as fire in a furnace, and, like this also, is capable of evolving mechanical motion and effective work in fixed ratios to the fuel consumed.

Now, important a step as was this announcement, beyond the speculations and utter ignorance prevailing prior to his immortal discoveries in organic chemistry, it must be admitted that later experiments of different kinds, instituted by various physicists in order to ascertain by measurement the exact sums of mechanical power expended and effective work performed, internally and externally, by animal organisms, in comparison with the sums absolutely developed, and capable of development, by the igneous combustion and chemical decomposition of given quantities of different sorts of food consumed by the same organisms, exhibit results suggestive and conclusive—*that the sources of mechanical force,* per se, *are still more profound than the simple chemical changes* which transpire momentarily in the molecular structure of muscular fibre, and in the decarbonization of the blood, or which can be derived from the food consumed and assimilated with the circulating vessels. The most reliable observations and calculations thus far obtained upon this subject, analysed and summed up in a very able manner by Professor Frankland, in his researches "On the Sources of Muscular Power," published in the Proceedings of the Royal Institution of Great Britain, June 6, 1866, afford solid bases for philosophical induction, and establish this point beyond question. The plain mathematical truth reached by his investigations, setting aside all speculative conclusions to which he may incline, is that the abstract force required, and absolutely consumed, in the various kinds of animal work performed in ascending mountains, moving treadmills, lifting weights, and executing hard daily labour, is from one-third to one-half greater than can be derived from food consumed, or the same solid matter converted by combustion into heat and mechanical motion. "Thus," he says, "even under the extremely unfavourable conditions of "these determinations, the actual work performed exceeded "that which could possibly be produced through the oxida- "tion of the nitrogenous constituents of the daily food by "more than twenty per cent."

The final conclusion from all these facts is, therefore, simply this, that food introduced into animal organisms acts like the slip of platinum inserted into the jar of oxygen and hydrogen, and like the rod of glass into the crystallizable solution of Glauber's salt. It is not more the pabulum of material repair and support, than the medium and excito-motor through which crude, terrestrial, and metaphysical energy is induced insensibly to enter the body from without, in order to react in turn and be converted into sensible, physical, and phenomenal results. The primal presence of lifeless masses of carbon, nitrogen, oxygen, and hydrogen in the midst of life excites the entire compound of elements constituting an organism to begin instantly and actively to absorb additional increments of energy from the external world, to be employed within the structure itself, and be discharged again in the forms of heat, electricity, light, muscular power, and productive work. As an electro-magnetic apparatus, the molecules of which, once touched and jarred by the electrophore, and thus inoculated with magnetic vibrations and attuned into electric measures, is afterwards linked with the entire volume of electro-dynamic energy existing in the globe and cosmos, and thereby becomes capable of converting cosmical repulsion and attraction into electro-dynamic currents, and of receiving the former from telluric sources as mechanical force, and again distributing it as modified, phenomenal, and productive energy; so it is with living organisms. Mechanical energy—the compound mixture, so to speak, of attraction and repulsion—having once excited living structures into corpuscular and systo-diastolic motion, the same crude potential energetics existing in the planet as cosmic forces, and binding all parts into one varied whole, are induced, absorbed, subdivided, differentiated, specialized, and converted into the working power of the animal, brought to bear through the action of muscular fibre upon the shafts and cogs of the bony skeleton, and distributed in the performance of internal functions, and in the development of external phenomena, similar to those produced by the wheel and axle and the general leverage of mechanical machines.

The searching spirit of Liebig caught but prophetic glimpses of the sources of subtle motive agencies, when he said, "Plants . . . are accumulators of force. . . . And while "these tissues in their turn are resolved in the vital process "into formless or inorganic compounds, the force stored up "in them is manifested in the most various effects, like "a galvanic battery, the peculiar properties of which are "determined by a certain arrangement of its elements, and "which consumes itself in giving rise to new manifestations, "magnetic, electrical, or chemical." Now, the battery does not consume itself by its own stores of internal force. But it is consumed, if consumed at all, by forces the sources of which are external to it, and which, proceeding out of the globe and back into the globe, with a swiftness more subtle and delicate than lightning itself, generate electricity and muscular power in a similar manner, acting and reacting among atoms, rejected everywhere as quickly as received, alike in the battery and in the organisms of plant or animal, and connecting locomotive beings with the elements and centre of the planet, and with the central and excentral forces which at one and the same instant control all molecular motion and the universal motions of the heavens. An animal organism, considered also in the light of a heat-producing apparatus, derives its powers necessarily from the same sources. And while the food is a plastic medium for chemical changes, a basis for the generation of the mechanical power and the secondary forces exhibited in organic life, their primaries, the radical principles from which they spring, must be the cosmico-mechanical forces of gravitation and repulsion. "Respiration," says Liebig, "is the "falling weight, the bent spring, which keeps the clock in "motion; the inspirations and expirations are the strokes of "the pendulum which regulate it. Whenever it is vouch-"safed to the feeble senses of man to cast a glance into the "depths of creation, he is compelled to acknowledge the "greatness and wisdom of the Creator of the world. The "greatest miracle which he is capable of comprehending is "that of the infinite simplicity of the means by the co-"operation of which order is preserved in the universe as "well as in the organism, and the life and continued existence of organized beings are secured."

In the palpable action of repulsion as a *cosmical* agent, in conjunction, and in antagonistic conflict, with attraction, among the molecules of the various chemical elements constituting cosmic masses, we discover the subtlety of the dynamic secret so remarkably conspicuous in experimentation upon the secondary forces, so reducible to quantity by the calculus, and so accordant with the general laws of dynamics which pervade and govern the broadest domains of the physical universe. Upon these last points of law and numbers the immortal labours of Newton, Ampère, Liebig, and Mayer converge. And nothing is now lacking to remove exact science and positive philosophy wholly from the realms of speculation and empiricism, but the acceptance of cosmical repulsion as the co-ordinate and co-efficient of gravitation. The introduction of this principle into public education, as an element in celestial as well as in molecular physics, is, I hold, as indispensable to a clear understanding of the general economy of nature, as was its action in the beginning to unfold and perfect the plans and works of the Creator. We may indeed predict from this point of view the coming of a new era, as important to the progress of exact learning as that which superseded the scholasticism and dialectics of the Sorbonne.

49.

General Application of the preceding Principles.

Now, when we study the geological aspects of the planet—not merely mineralogical specimens and cabinets, but the granite rocks and palæozoic strata at large and *in situ*—in the light of all that precedes, arriving thereby at notions of time, periods, and cycles of ages, and displays of force which only the profoundest studies of geology and astronomy can impart to man, we shall begin to comprehend the grand logical sequence and fact, that the infinitudes and endless successions of organisms, the remains of which have been piled up thousands of feet in thickness throughout the surface of the planet, were so many temporary galvanic batteries, small or great, aquatic or terrane, employed in inductions and conversions of crude cosmic energy into mechanical and phenomenal, creative and productive, power. And now for the first time, in view of all this mighty and eternal work of creation and generation, we shall begin to doubt whether the heat, electricity, and magnetism generated in and radiating from the interior of the planet were, or are designed to be, diffused, dissipated, and wasted in void and desolate space, or to be appropriated, *just as they have been and are employed,* for the special local work and general purposes of covering, smoothing, adorning, vivifying, and utilizing the surface. It must be very plain that no generation or display of muscular or mechanical energy could begin to appear upon the primal surface of our planet, unless that energy in one form or another proceeded from the internal forces of the planet; for all molecular motion implies the absolute necessity of the connexion of the molecular forces involved in the phenomenon with their cosmical congeners, and *vice versâ*. We therefore perceive that all force disengaged from, and proceeding out of, the planet's interior mass executes now, as at first, a multitudinous work upon its surface, under some imminent and appreciable, but invisible, prescient, guiding Providence; that as organic forms are undergoing continuous metamorphoses through propagations of force and translations of molecules, the agencies thus induced and employed are returned again into the great reservoir whence they issued; that if refrigerations of the earth have taken place, and consequent depressions and irregularities of its crust have ensued, the heat liberated has been immediately converted into economical purposes in its envelopes, and been persistently employed in mechanical and creative work upon its surface; and that thereby has been sustained that equilibrium of the forces at first communicated to the earth as a celestial body, and which Laplace, Delaunay, and others have shown to have been nearly invariable, or only slightly variable, at least for the last two thousand years. Here, then, we may conclude that all molecular pulsating forces, and consequently all cosmo-dynamic energy, are directly local in their phenomenal action and design, and confined to the various individual globes of matter which, isolated as suns, planets, &c., move around each other in paths determined by the chemical constitutions of their respective masses.

Now, if molecular force be dual—that is, composed of attraction and repulsion—then cosmical force must be dual also; and inasmuch as the law of probabilities indicates that no two celestial bodies are chemically identical, considered in regard both to quantity and quality, whatever their origin, or however created or segregated, it follows that the equilibrium of their respective internal forces must be more or less variable, and always unstable, and consequently that their densities—that is to say, their central and excentral action and reaction—must be fluctuating from moment to moment, and for ever inconstant. And inasmuch, also, as the masses composing our system are held to the sun by the sum of their molecular forces mutually acting and reacting from their respective centres, it equally follows that their orbits must always be elliptical and differently elongated, their heliocentric distances various, their velocities never at consecutive moments the same; and that the instability of equilibrium distinguishing each must be for ever maintained, in virtue of the fluctuating functions of the forces and matter embraced within their various boundaries. These general conditions, long ago determined as mathematical facts by the successive discoveries and labours of Copernicus, Galileo, Tycho Brahé, Kepler, and Newton, are just what should exist as combined results of chemical action and reaction—that is to say, of the play of moleculo-physical laws. The *points* requiring demonstration in order to reach a complete understanding and true theory of the cosmos are, firstly, that which connects motion with mechanical force; and, secondly, that which resolves mechanical force into its final elements as specific causative entities, and fixed combining and separating quantities. It is of no consequence to science and philosophy if these forces be metaphysical and transcendental in their nature, and only discoverable and measurable in their powers and quantities by their effects through matter. They are no more so than the mathematics. And if all indicate sources of still higher and more abstract knowledge and power yet beyond our reach, even these are nothing more than what the common subtle consciousness of man has accepted through intuitive and prophetic glimpses from the earliest times.

Thus the mysteries of celestial dynamics begin to vanish, and physical and organic laws in their widest ranges to be better understood, through rigorous studies of molecular phenomena and the application of experimental data to astronomical and geometrical problems.

50.

Resumé and Conclusion.

But our investigations are not yet finished. It remains now to examine as briefly as possible the wonderful phenomena of cosmical motion lately alluded to in view of, and in relation to, the facts and principles thus far discussed and deduced, in order not only to establish beyond question the finality of our generalizations, but also to solidify the foundations upon which they rest. The system of the world is so geometrical, exact, and correlative a structure in all its parts, and is so nicely adjusted by its weights and measures, that if one globe be accurately known, and the specific nature and mechanism of the forces which act throughout that globe be also known, the entire problem of physical and organic development will solve itself to the enlightened sense of studious scholars.

In the preceding chapters we attempted to unveil space; and arrived at the conclusion that it was infinite absolute vacuum. In this vacuum we traced the limited existence of material molecules agglomerated into exceedingly numerous isolated round balls more or less dense, solid, fluid, or gaseous. We descried these balls moving around each other in paths exhibiting constant, but various, eccentric and elliptical elements, and mutually influencing each other by certain peculiar inherent principles of energy, which manifest a double form of centripetal and centrifugal action and reaction, developing internal and external dynamic phenomena of a local or general character. We studied these dynamic phenomena generally and in detail, and found their variations of annual and total intensity to be periodical, and strictly coincident with known cosmo-mathematical laws of "inverse distances;" and, so far as accurate and indisputable data derived from cometary and terrestrial observations can be weighed and appreciated, we found the internal physical excitations and eruptive reactions of bodies revolving around the sun to bear close and constant inverse numerical relations to the length and sweep of their radii vectores—*to be positively greatest at perihelion and positively least at aphelion.* We discovered all these variations of dynamic intensity, whether repulsive and eruptive or barometric and gravitative, to be associated with thermic, magnetic, electrical, or luminous phenomena partaking of similar numerical elements and governed by similar geometrical laws; and, so far as the thermic phenomena, strictly speaking, could be studied through volcanic and seismic data, we were compelled to conclude that the interior of our planet is incandescent and more or less molten from its centre to its crust. In our discussions of *force*, abstractly and exclusively considered—indeed considered every way—we traced molecular attraction and molecular repulsion by ascending and inductive steps into planetary quantities; and following these antagonistic principles rigorously into their voluminous association with each other and with matter, we discovered the elements of cosmo-dynamic energy and celestial motion to be also dual, and that the generation of mechanical phenomena, whether small or great, whether mundane or stellar, spontaneously springs from the lively exertion of some occult animus, immanent and self-existent in every jot of both these forces, to separate, segregate, and alienate each from the other, in order to affiliate, if possible, with their respective congeners existing *in larger volumes elsewhere.* In this mutual but constantly abortive exertion to effect the union of "like force" with "like force" we detected the mechanism of other concomitant elements of nature, the subtle secrets of elective and mutual affinity and of polarity—principles as clearly cosmical as they are chemical; and we began to see how an idea advanced at first by Empedocles in a distinctly speculative sense, 500 years prior to our era, embraced afterwards with indefiniteness in Grecian and Arabian learning, and vitalizing the progressive developments of alchemy and chemistry, became at last the positive and mathematical entity whereupon all modern chemical theories are based.

In the preceding parts of the present chapter we have traced these subtle principles of attraction and repulsion—so cosmical and universal in their widest ranges, and so molecular, actinic, and subdivisible in their ultimate aspects—into their associations and dissociations with atoms, and among the elementary forms of matter. We have seen how these forces, in the course of their immensely swift, incessant, and unrestrainable translations in *opposite* directions through the various elements, affect molecules with vibratory action; and, thus generating undulatory material motion, become not only absorbed into these elements to supersaturation, as water into sponges, but also transformed, in consequence of these combinations and antithetic reactions, into appreciable mechanical energetics assuming quantitative conditions and pregnant with measurable capabilities and results. In analysing these conditions, capabilities, and results, we became able to distinguish plainly the manner and mechanism by which the immaterial entities of attraction and repulsion are linked with matter, and how in propagating themselves from one focus to another, in order to unite with cognates or fly from antithetics, they loaded molecules and masses with motive tendencies and transformed themselves into sensible mechanical energy. We traced this mechanical energy through its developments into a multitude of phenomena, all charged with, composed of, inseparable from, and governed by the *same radical forces,* converted by motion, however—that is to say, their own action upon molecules—into *secondary forces* of diversified forms. We thus discovered the very mechanism by which heat, electricity, magnetism, and light are generated; forces absolutely immaterial in themselves also, but so appreciable to the senses and mind of man, so much within the reach of experimental and mathematical inquiry, as to show that they are the veritable instruments by which all departments of creation have been excited into development, and upon which all depend for their maintenance and progression. We were able, furthermore, to trace the generation of heat to its final sources, and to settle the question of its origin for all time, proving the existence of "fire" even, however intense its phenomena, to be simply a result of *the antithetic propagations of attraction and repulsion through molecules,* fixed intensities of action producing fixed intensities in effects, and showing that not a particle of matter, nor a jot of force of any sort, is consumed, destroyed, or lost in processes of combustion. At this point we were brought into connexion with the great discoveries of Dr. J. R. Mayer, Dr. J. P. Joule, and Professor Michael Faraday, of the correlations and numerical equivalencies of motion, heat, thermo-electricity, electro-magnetism, and mechanical force, of the mutual convertibility directly or indirectly of all these entities into each other, and of the general laws of conservation which pervade and govern all the transformations of mechanical energy. Accepting these discoveries to be of equal importance in the developments of philosophy to those of Kepler and Newton, we were enabled further to trace from this point the molecular forces into their dynamical elements or functions when converted through vibrations of matter into electricity, magnetism, heat, and *fire*—that is to say, into the phenomena of combustion; thereby demonstrating fire to be only the intense conflict of antithetic currents of cosmic attraction and cosmic repulsion, the crude universal forces of matter, flowing in and out of the globe with immense velocities, through such elementary substances, or their compounds, as are specifically excited *by sparks* TO INDUCE local currents and tensions of these fundamental energetics. Herefrom followed the wonderful truth, that a glowing spark capable of igniting carbon, oxygen, hydrogen, or any substance, is actually a certain species of electrophorus—strictly speaking, a *thermophore*—which, by its specific character, will inoculate every form of matter under certain circumstances with its own powers, and thereby communicate the occult capacity of inducing intense influxes of repulsion from the general fountain of planetary energy; and this to such an extent and so swiftly, as to push the particles of many substances asunder with terrific violence, repel them into the widest diffusion, and throw them into such new conditions as to assume other forms or motions under the influences of attraction. By graphical discussions of absolute and authentic data, following nature with studious care, chaining the discoveries of Empedocles, Copernicus, Kepler, and Newton with those of Rumford, Faraday, Mayer, Kirchhoff, Liebig, and Frankland, we were thus able legitimately to link crude abstract cosmic energy with chemical and all other forms of secondary force, with dynamic power, and with universal motion. Examining first the origin of motion, and finally tracing motion back to its fundamental causes, we found it, by processes both of analysis and synthesis, to be a *compound* phenomenon; that is to say, we found matter to consist of a multitude of specially sensitive plastic substances distributed unequally through space, and subordinate to the distinct measurable influence of *two* subtle potential entities, which, manifesting positive antagonistic properties and compelling masses to roll, gravitate toward, and recede from each other, resolve themselves in connexion with matter into the physical phenomena of motion and dynamic power. In the reactionary nature of these compound, decomposing, and recombining elements of energy we discovered the real essence of mechanical force—an *apparently* unitary principle when acting upon matter in motion, the duality of which, however, becomes self-evident by this very motion, inasmuch as matter only moves in one direction or another in virtue of the influence of attraction or repulsion, each of which is incessantly seeking to separate itself from its antithetic in order to unite with its cognate elsewhere. This antagonism being mutual, indeed *strengthened* and *eternized* in its mechanical and phenomenal results, in virtue of the fact that matter manifests as active an animus to rid itself of one or the other of these forces, or of both, as the forces do to separate from each other and from matter;—these conditions existing, and the unequal distributions of the various elements of matter and of the original sums of attraction and repulsion, as acting and reacting triangular entities, thus to speak, having been so arranged in the beginning by the Creator of all things that stability of equilibrium nowhere exists and can be nowhere possible, motion necessarily becomes universal and eternal; and, of course, heat, light, electricity, and magnetism become equally active, universal, and eternal, and the inevitable correlatives of the energies from which they spring: and, drawing their power from these sources and no other, their mathematical elements and dynamic functions must be similar and convertible also. Hence, and thus, we discover the nature of the laws through which all forces become conservable and convertible, and the correlatives and equivalents of each other; extending the conclusions of Mayer, realizing the predictions of Faraday, linking all matter with all forces, and the beginning of things with the present; and proving, finally, by the severest inductions, that force and matter are not only compound entities consisting of various, specific, measurable, and fixed quantities, neither of which is self-creative or self-destructive, or destructible in any way, but that these entities bind up also within themselves, *everywhere and under all circumstances,* the presence of a Supreme, Intellectual, Guiding Spirit, without which their mechanical action would end in uncertain and indefinite results. To this point philosophy, based upon facts and exact science, legitimately tends. And it is when we trace matter and force into organic phenomena, and especially their relations to vital and cerebral developments, that distinct *inductions* of instinctive and intellectual energy begin sensibly to appear: at first exciting crudest acts of vegetative, muscular, and mechanical motion; subsequently appearing more distinctly in the subtlest workings of an inner consciousness; and finally revealing a universe of metaphysics coequal and coextensive, at least, with that of physics and of mechanical power.

As a conclusion, indeed, reaching toward the occult origin of things, we remark, in closing this branch of our subject, and in order to connect it with a subsequent chapter, that science and philosophy tend to the exposition of an unique and all-pervading Providence in the cosmos: not, of an agency whose consciousness and energy are remote and doubtful essences or simple equivalents of atheism; but to a vast and comprehensive *monodeism,* so to speak, whose individuality and penetrative presence, prescience, and activity, everywhere demonstrable, exclude all possible accident or indefiniteness from the mechanical or material and organic developments of nature. Few who cultivate the sciences in detail will question their capacities to exhibit both direct and inductive proofs that this directive principle is an individuality, as definite in its being and character as the matter and the forces with which it is associated, and upon which it acts; and that it is in the same sense an absolute integral unit capable also of manifesting segmentative and absorptive functions, so to speak, yet exalted above all other principles in subtlety and power, and thereby establishing the universality of its reign in nature. And that this supreme unitary energy subjects itself, scientifically speaking, to analogous laws in its relations to matter as do the forces which it controls and the mathematical elements, the combining laws of which we know, is clearly established by observations upon the instincts and faculties of animals, upon the human understanding, and upon nervous and mental development in general in connexion with health, disease, and death, and with experiments of various sorts, such as freezing, excitation, anæsthesis, &c., upon the cerebral, ganglionic, and medullary tissues of both vertebrate and invertebrate organisms. In this manner science and philosophy, exact and positive, are able not only to identify the existence of the Divine Energy, but also to limit the field and define the modes in which this Energy works, and to reduce abstract inquiry from unknown chaotic and speculative outlines into comparatively narrow and fixed boundaries.

Rigorously following the mandates of Bacon through the entire domain of physics and into organic study, we at last enter that field of transcendental inquiry heretofore so profoundly and prophetically explored by Locke, Kant, and Hegel. Here, asserting the supremacy of our own nature, and lifting thought from matter and force toward that which is grander still, science encounters pure Mind, and detects the existence of One Supreme and absolute Being everywhere, affirmatively present and active throughout the infinite and the infinitesimal, and stamping all things with such a unity of purpose as to inspire us with the impression that

> "All are but parts of one stupendous whole,
> Whose body nature is, and God the soul."

Such is an incomplete summary of, and corollary from, the points and principles hitherto discussed. And, so far, we have not departed from the narrow paths of fact and logic, however imperious or seductive have been a hundred theories to awe or win us away from truth, and weaken our steps with doubt or error. It is here, then, that we can begin boldly to apply our principles to distant and profound realms of inquiry, and attempt the solution of cosmical problems which have heretofore defied the penetration of speculative mathematics.

CHAPTER XI.

CENTRAL FORCES AND CELESTIAL PHYSICS.

1.

THUS far we have proceeded, from step to step, with unrelaxing severity, as indicated in the summary of the last chapter. Ascending from chaos and from the initiatives of creation through demonstrated and accepted conditions of force and nature, subjecting elements and final principles, together with their inseparable union and reactions, to such critical inquiries and analyses as authentic observations and experiments have suggested and permitted, we finally arrive at the demonstrations, that mechanical force is a compound agent, and that its constituents, attraction and repulsion, are the bases of every degree of activity, and of every act of phenomenal development, observable throughout the realms of both molecular and cosmical physics. It is at this point, then, that we find ourselves prepared to enter more distant, and heretofore uncertain and theoretical fields, and apply the principles already established, in order to elucidate phenomena of a celestial nature, which, being strictly and entirely cosmical, are also cognate with those of an experimental character, and which correspond, moreover, with those observable throughout the physical structure of our own planet.

2.

Mayer's Theory of Solar Heat.

The unpretending but immortal Julius Robert Mayer, progressively extending his physical discussions upon "*forces*" as "*causes*," the final equivalences of these causes and their effects, and the convertibility and indestructibility of forces, at last, in 1848, applied the principles he had previously discovered to celestial dynamics.

It has been long supposed that terrestrial light and heat are derivable from solar radiation *alone*. The causes, however, which have persistently generated such intense light and heat in the sun as would be necessary to supply the globes constituting the solar system with warmth and lustre, have always been only subjects of conjecture, and *up to this moment are absolutely undemonstrated*. The Newtonian (and heretofore unquestioned) theory of celestial motion and dynamics *formally* and *unreservedly excludes* from consideration the *diffusive* and *convective* play of internal forces of active *repulsion* and attraction, as excentral and concentral, centrifugal and centripetal, oscillating cosmic agents; and thus sets at naught the natural and potential effects of the action and reaction of cosmo-molecular forces in a mass of chemical substances 1,200,000 times the volume, and 314,000 times the weight of our own planet. Nevertheless, some exhaustless radiant energy has been constantly, though paradoxically, insisted upon as necessary to enable our central orb to diffuse his rays so far and potently as to heat and illuminate his numerous and various appendages. The heat and light wasted in space, under all theories, have been incalculable, inasmuch as these purely local moleculo-vibratory motions of solar matter have always been considered in some way as diffusive and substantive objects. An unwavering disciple of Newton, and a strict believer also in modern optical doctrines, Dr. Mayer sought for adequate "causes," in order to reconcile his own discoveries with accepted theories. This was to "put new wine into old bottles." Knowing that the height of fall and the velocity of falling bodies bear special numerical relations to the amounts of heat produced by impact, he prosecuted his researches from this point of view, and developed a "*meteoric theory*" to account for the generation and persistent radiation of solar heat and light. It is not my design to invite attention to mere theories, nor to devote time to their discussion; but, inasmuch as Dr. Mayer's conception is among the grandest of speculations, and, as a legitimate consequence of acknowledged principles, not wholly imaginative, since the fall of meteoric stones and metals upon our own planet does actually occur, an allusion to it is not only just and proper, but also important, in order to set the truth deducible from solid facts and positive causes in a stronger light by the comparison of these facts and causes with Dr. Mayer's assumptions, and *through an inversion of his reasoning.* This acute philosopher has founded his theory and calculations of the causes of solar light and heat upon two postulates: first, the *common assumption* that gravitation is the ONLY active potential force in nature; and, secondly, that the interplanetary spaces swarm with solid meteoric substances of various sizes, flying and revolving like comets in all directions. From these two hypotheses he derives his third—that of *incessant impact* of foreign masses, which, falling upon the solar surface, from infinite or variable distances, thus generate *heat,* the intensity of which is beyond all calculation, and of course deprive the sun of any possibility of ever becoming a habitable body, or of developing its own internal mechanical resources for any useful local purpose whatever. The entire "meteoric doctrine" is fairly embraced and presented in this syllogism. I cannot present this "mechanical theory" of *cosmical* heat in its various aspects more clearly, succinctly, and justly than by quoting a passage from Professor Tyndall's reference to it, in his twelfth lecture upon "Heat considered as a Mode of Motion:"—

"It is easy to calculate both the maximum and the mini-
"mum velocity imparted by the sun's attraction to an asteroid

" circulating round him: the maximum is generated when the " body approaches the sun from an infinite distance, the *entire* " *pull* of the sun being then expended upon it; the minimum " is that velocity which would barely enable the body to " revolve round the sun close to his surface. The final velocity " of the former, just before striking the sun, would be 390 " miles a second; that of the latter, 276 miles a second. The " asteroid, on striking the sun with the former velocity, would " develop more than 9,000 times the heat generated by the " combustion of an equal asteroid of solid coal; while the " shock in the latter case would generate heat equal to that " of the combustion of upwards of 4,000 such asteroids. It " matters not, therefore, whether the substances falling into " the sun be combustible or not; their being combustible " would not add sensibly to the tremendous heat produced by " their mechanical collision.

" Here, then, we have an agency competent to restore his " lost energy to the sun, and to maintain a temperature at his " surface which transcends all terrestrial combustion. The " very quality of the solar rays—their incomparable penetra- " tive power—enables us to infer that the temperature of their " origin must be enormous; but in the fall of asteroids we " find the means of producing such a temperature. It may " be contended that this showering down of matter must be " accompanied by the growth of the sun in size: it is so, but " the quantity necessary to produce the observed calorific " emission, even if accumulated for 4,000 years, would defeat " the scrutiny of our best instruments. If the earth struck " the sun, it would utterly vanish from perception; but the " heat developed by its shock would cover the expenditure of " the sun for a century.

" To the earth itself apply considerations similar to those " which we have applied to the sun. Newton's theory of " gravitation, which enables us, from the present form of the " earth, to deduce its original state of aggregation, reveals to " us, at the same time, a source of heat powerful enough to " bring about the fluid state—powerful enough to fuse even

" worlds. It teaches us to regard the molten condition of a " planet as resulting from the mechanical union of cosmical " masses, and thus reduces to the same homogeneous process " the heat stored up in the body of the earth and the heat " emitted by the sun."

Professor William Thomson and other physicists have enlarged upon Dr. Mayer's assumptions, until they and the entire scientific world have arrived at his conclusions. Such are Professor Thomson's views, in his own words:—" The " source of energy from which solar heat is derived is un- " doubtedly meteoric." *

Thus it is seen that the conclusions of Dr. Mayer, *declared* by his followers after immense mathematical labour to be UNQUESTIONABLE, result from three successive *assumptions*, the only positive element in the equations being the secret magic of numbers, the mechanism through which light springs out of darkness: first, *meteors* (and these now *further* assumed to be concentrated into an area less than the earth's orbit); second, the influence of gravitation (*assumed, also,* to be the *only* motive force in nature) upon *these* meteors; third, the *assumed impact* of these meteors, the continual "bombardment" of which upon the sun can be alone likened to the incessant battering of hailstones upon the earth.

Offering no criticism upon this theory, and acknowledging with respect the earnest desire for solid progress and the eminent ability of its advocates, I nevertheless intrust it, like all other mere speculations, to the searching consideration of a coming epoch. Cherishing, however, so tender a veneration for the great philosopher of Heilbronn as to make a journey to his native town for the special purpose of offering him my personal homage, I am sure of *his* pardon at least for applying his own principles to *positive data,* in order to solve such vast problems in a less hypothetical manner.

* I am informed, since this work has been in type, that the opinions of Professor Thomson upon this point are undergoing some modification. But the theory still prevails, and no other, to my knowledge, has been substituted for it.

3.

In our chapter upon "Cosmical Repulsion" I presented a mass of evidence drawn from every class of celestial bodies absolutely conclusive of their *internal* molecular and mechanical agitation, and proving that this agitation, the motion and friction, the action and reaction, of their collective particles upon each other, maintained definite relations to the ellipticity of their orbits—that is to say, to the distances of these bodies respectively from one another. A recapitulation of these voluminous facts is here unnecessary. But, in regard to comets, it will be remembered how striking were the internal agitations of these misty and translucent bodies as they approached, passed, and receded from their perihelia. In regard to the earth, it will be remembered how detailed were our volcanic and seismic data, as rigorously numerical as the analysis of so vast and various a body of facts would permit, and yet so constant in the periodic differences of their total intensities at perihelion and aphelion, as to demonstrate an absolute oscillation of dynamic elements, and to establish beyond question that these fluctuations of internal mechanical energy depend, as a general law, upon similar physical agitations within the sun, the mutual actions and reactions of the two bodies holding invariable inverse relations to the length and sweep of the radius vector. These facts, as a body of evidence bearing upon any physical problem, are indisputable and of incalculable value. While *we doubt the continual bombardment of any celestial body from* WITHOUT, and consider such a phenomenon incapable of proof, we present *such evidences* of INTERNAL "*bombardments*," general commotions, collisions of molecules and volumes of matter from centre to circumference of cosmic masses, even of the sun and stars, as are sufficient upon known principles to explain clearly and conclusively every problem of cosmic motion, cosmic heat, cosmic light, cosmic electricity, and cosmic magnetism, however palpable or delicate to our senses the form or manifestation of such phenomena may be.

4.

From all which has been thus far discussed it follows, that the central forces of the globe are never at rest. We know, generally, the chemical constitution of the planet; and we therefore know that the forces of attraction and repulsion, as dynamic entities and affirmative quantities, are actively darting from molecule to molecule, combining with and separating from them, and generating vivid pulsations thereby, which are convectively propagated and incessantly multiplied into voluminous currents and massive irresistible impulses, from centre to equator and from pole to pole. In order for the *numerical element of intensity* observable in volcanic and seismic phenomena to *vary periodically*, and for the *maxima* and *minima* of these intensities to be respectively constant at *perihelion* and *aphelion*, we know that the energies engaged in developing such differences must also vary in like proportional manner and degree. We know, moreover, since the total intensities of terrestrial magnetic force are in like manner periodical, following the same general laws which govern the expressions of seismic phenomena, both in relation to space and time; since magnetic storms are universal—that is to say, since abnormal agitations of the magnetic needle are simultaneous throughout the planet's surface, on plains, at sea, in deep mines, and upon lofty mountains alike—and instantaneous; since magnetism, like heat, depends for its existence upon antithetic propagations of repulsion and attraction, and upon the mutual collision and reaction of molecules in virtue of such propagations; since heat itself, by the impulses which its elements impart to cold or cooling substances, is convertible into magnetism;—by these thermo-magnetic facts, and many others which will occur to thoroughly informed physicists, we may be sure that there can be no inertia or inactivity in matter; that a ceaseless friction among molecules, an absolute molecular impact and reaction, must certainly be a central and all-pervading phenomenon in the earth, and therefore equally so in *every other cosmic mass moving in an elliptical path around larger bodies.* Here, then, we discover the necessary and unavoidable causes of central heat and cosmic fire; for I have already shown that absolute intensest heat is nothing else than the violent *conflict* of attraction and repulsion, as opposing forces, in their incessant struggle, *under cosmic and universal pressure,* to penetrate and supplant each other in all forms of matter, and to throw individual molecules into new and evanescent conditions, neither matter nor forces being consumed, lost, or *fundamentally* changed in quantity or quality.

Hence, in applying further observations and the calculus to this new branch of mechanical astronomy, we may be guided by the general laws,—that all internal cosmic agitations, and local or periodical intensities of physical phenomena, are, as expressions or functions of mechanical energy, correlated *inversely* with the lengths of the radii vectores of revolving bodies; and that the degrees of ellipticity and of perturbation in the orbits of cosmic bodies result from the actions and reactions of the forces operating dynamically and reciprocally within the masses of connected spheres, whether such internal agitations be expressed externally and visibly, or otherwise.

These are unquestionable deductions from what precedes. And, so far as comets and Saturn's rings permit definite ocular observation, the evidences presented by their internal and external phenomena confirm these laws. These remarkable truths are furthermore confirmed by *telescopic observations* upon the variable aspects of many other bodies.

Why comets and the earth condense with increasing energy and activity as they approach perihelion, attraction between their respective particles and centres becoming absolutely more intense, so as to effect notable centralizations of matter, and excite the most violent internal reactions, at perihelic periods; and why, contrariwise, they relax this increased gravitation simultaneously with their retrocessions from the sun; and why they thus incessantly execute these mighty oscillations, like so many cosmic pendulums swinging from the centre of the solar orb, are pregnant and instructive problems, the solution of which it is here irrelevant to attempt, since the facts alone are the points which I wish to present explicitly, and aside from discussion, as all-sufficient causes and conditions for the development of cosmo-dynamic phenomena of every grade and character.

5.

Friction and Impact Equivalent Agents.

Now, it is well established that *friction is equivalent to impact,* as a cause or agent in the generation of the secondary forces, heat, light, electricity, and magnetism.

Linking this axiomatic truth with the cosmic facts immediately before introduced, it follows, that the amount of heat steadily generated, and absolutely developed into momentary, dynamic, and phenomenal results, by the *friction*—the mutual gravitation and reaction; that is, the *attraction* and *repulsion*—playing between the vibrating molecules of the earth's mass during its fall from aphelion to perihelion, must be equivalent to that which would be generated in gross by impact, if the same mass of matter, divided into many parts, should fall from the same height and collide at perihelion. The wonderful interests involved in this problem are all more or less capable of mathematical determination. Dr. Mayer, in a searching essay upon "Celestial Dynamics," elaborated in support of his meteoric theory, has subjected to severe numerical analysis a problem somewhat similar to my illustration, and stated the thermic results. Professor Helmholtz and other physicists have arrived at similar conclusions to those of Mayer, upon one assumption or another, all tending to the same point, of cosmic developments of heat by *external impact of gravitating masses.* Modified in one way or another, as the elements of the question may be assumed and treated by different analysts, the general result may be stated in terms similar to those reached by Dr. Mayer. And from whatever point this mighty problem is viewed, the truth elicited can only enlarge our considerations of the value of Dr. Mayer's and Dr. Joule's labours upon the equivalency and mutual convertibility of motion, heat, and mechanical force, viewed in relation to the development of physical phenomena, both in the earth and throughout the heavens.

The amount of cosmic heat, regarded as mechanical working power, generated by the central friction, to which I invite special attention, will no doubt be hereafter ascertained with great accuracy. Meantime, in order to stimulate inquiry, and give those who have never examined this subject an approximate idea of its prodigious sum, I will cite the remarks of one of the most learned specialists of our time upon this point, as it has been heretofore analysed. Professor Tyndall, in the course of his second lecture upon "Heat considered as a Mode of Motion," says: "From these considerations, I think it is "manifest that if we know the velocity and weight of any "projectile, we can calculate with ease the amount of heat "developed by the destruction of its moving force. For ex-"ample, knowing, as we do, the weight of the earth, and the "velocity with which it moves through space, a simple calcu-"lation would enable us to determine the exact amount of heat "which would be developed, supposing the earth to be stopped "in her orbit. We could tell, for example, the number of "degrees which this amount of heat would impart to a globe "of water equal to the earth in size. Mayer and Helmholtz "have made this calculation, and found that the quantity "generated by this colossal shock would be quite sufficient, "not only to fuse the entire earth, but to reduce it, in great "part, to vapour. Thus, by the simple stoppage of the earth "in its orbit, the elements might be caused to 'melt with "fervent heat.' The amount of heat thus developed would "be equal to that derived from the combustion of fourteen "globes of coal, each equal to the earth in magnitude. And "if, after the stoppage of its motion, the earth should fall "into the sun, as it assuredly would, the amount of heat "generated by the blow would be equal to that developed "by the combustion of 5,600 worlds of solid carbon.

"Knowledge such as that," he continues, "which you now "possess has caused philosophers, in speculating on the mode "in which the sun is nourished, and his supply of light and "heat kept up, to suppose the heat and light to be caused "by the showering down of meteoric matter upon the "sun's surface. Some philosophers suppose the Zodiacal "Light to be a cloud of meteorites; and from it, it is imagined, "the showering meteoric matter may be derived. Now, "whatever be the value of this speculation, it is to be borne "in mind that the pouring down of meteoric matter in the "way indicated would be competent to produce the light and "heat of the sun. With regard to the probable truth or "fallacy of the theory, it is not necessary that I should offer "an opinion; I would only say that the theory deals with a "cause which, if in sufficient operation, would be competent "to produce the effects ascribed to it."

Now it matters not where, how, or in what degree, collision between molecules is accomplished. Thermic results are always in direct proportion to the *radical forces* operating upon these molecules. One thing, however, is perfectly self-evident: that the *central heat and fires of our planet* are neither produced, nor sustained, by *external impact of meteors.* There are, then, *no other than interior* conditions for the generation of telluric heat. We must, consequently, abandon conjecture; and, submitting to the facts and evidence which have been brought to bear upon this subject, accept the conclusion that there must be sufficient agencies at work within the globe to develop every form of phenomenon connected with its existence, motions, and internal and external exhibitions of mechanical power. And since we now know the laws which govern the universal relations of heat and motion; and have determined that the periodicity and variable intensities of terrestrial dynamic phenomena depend inversely upon the length and sweep of the radius vector; the sequences are, that central heat springs from central motion, and that the molecular friction effected throughout the planet's mass in consequence of its progressive condensation during the passage from aphelion to perihelion, and its oscillating expansion during its return to aphelion (in like manner as comets condense and expand), would generate sufficient heat to reduce the globe to vapour and annihilate its cohesion, were this heat not instantly converted into mechanical power and employed in axial rotation, orbital reaction, seismic agitations and eruptions, in crystallogenic, magnetic, and electric functions, in tidal and aërial displacements, and in the infinitude of morphological processes so continuously progressing upon the surface of the planet. Thus the heat and other secondary forces of the globe exhibit themselves as certainly dependent upon the varying length of the radius vector, as does the ellipticity of its orbit upon the mutual reactions of the central forces of both sun and planet. These new forms of energy, thus acquired, being absolute conversions of the crude chaotic entities of gravitation and repulsion into central dynamic, and external productive and material power, and thus holding reciprocal and specific numerical relations to kindred conditions of force and action in the sun (conditions of quantity, determined by the immortal Newton); these new forms of energy thus acquired and thus sustained, the planet actually and incessantly develops its own resources of motion, heat, and mechanical power, its forces and phenomena oscillating in strength and function from aphelion to perihelion, and so continuing to the end of time. If a quantum of these forces be employed upon the surface in creative, organic, and mechanical work, it falls back again into the terrestrial reservoir, only, however, to be replaced with fresh relays from the same convective sources. For force, possessing affinities for matter alone, and being in virtue of its active nature the very antagonist of space as a nonentity, necessarily "abhors vacuum," clings to matter, and moves matter freely where and when the influences of greater volumes of *both united* draw or repel the lesser volumes, in order to effect union, segmentation, or elongation, and generally to execute the designs of their Creator. Here we catch a glimpse of the foundation of that speculative idea of the ancient philosophers, expressed so tersely in the phrase, "Nature abhors vacuum;" an idea the truth of which has been questioned in the late developments of an exclusively material philosophy, but which is as positively correct as the fact that matter and vacuum are the opposites of each other. Indeed, were not space an eternally sterile and infinite *vacuum,* which force and matter equally "abhor," the prodigious velocity with which cosmic bodies of every class fly through it would not only strip them of their gaseous envelopes, but also raise the temperature of the latter to such a degree that comets as mere gaseous bodies would vanish from the cosmos. Thus it is (and in virtue of this very truth alone) that worlds of all sorts could be, and have been, created; and that they continue to exist and move; and that systems of worlds perpetuate their motions and complexities—all mutually influencing each other; kept apart, and yet bound together: so that, knowing, as we do, that force is neither wasted nor lost, it becomes absolutely certain that no violent throb takes place in the sun, which does not in some way sensibly affect the earth and other planets; and, moreover, that no violent commotion ever occurs within the boundaries of Sirius or any other star, which is not responded to in some mechanical manner by the central forces of the sun; in the same way, indeed, that a pebble's fall into a lake will communicate palpable impulses all around its shores, or as an earthquake at Japan or Hawaii will disturb the tidal wave not only along the coast of California, but throughout the world.

6.

Certain objections will be raised at this point, by those mathematical philosophers who, assuming the position of mere idealists, and undervaluing Newton's real scientific character and aspirations, insist, because he demonstrated the numerical laws of attraction, and did not therewith discuss the existence of any other force or element of dynamic power, that consequently no other exists. Exclusively adhering to this doctrine, they will urge that, in order to reduce the orbital element and periodical phenomena of the heavenly bodies within the range and explanation of the laws of "rational mechanics," these bodies must have received, individually and at first, some impact or impulse in bulk or piecemeal, as an adequate means of primitive momentum and motion, and that in consequence of this initiative momentum all interior action of their own is rendered unnecessary, impossible, and inadmissible.

Inasmuch as these doctrines are entirely speculative, and founded upon *à priori* assumptions, they are not valid objections to the results of any inductive inquiry, and must therefore, vanish with the developments of positive physical truth. In passing from this question, I may, nevertheless, be permitted to say, that any *imaginary* force, whatever its origin and action, communicating the momentum assumed by mechanical astronomers, must have been opposite in its nature and functions to that of attraction, and must have been consequently actual repulsion, or a mechanical power equivalent to it, and (as we have already shown) its co-ordinate and co-efficient everywhere. All similar objections may, therefore, be finally discarded, since they are based upon hypotheses the truth or error of which can alone be reached by ascending steps, such as we have pursued from molecules and molecular forces, through experiment and observation, and the severest inductive and exhaustive processes, into the grandest developments of cosmic form and entity, and cosmic action and reaction.

7.

But another objection, growing also out of the historic element of science, and equally based upon a grand and authoritative hypothesis, will be raised against our conclusions, notwithstanding their legitimate induction from indisputable facts. It will be urged that if space be an infinite vacuum—that is to say, destitute of "a resisting medium," "attenuated matter," "imponderable fluid," "pulsating ether," or some other nameless substance upon which present optical theory is built—there is no agency whereby the heavenly bodies can propagate their radiant power, interchange their light, and mutually influence each other.

I might simply reply that repulsion—an entity of whose internal essence we know fully as much as we do of the essence of gravitation—an entity as imponderable as attraction, and as positive in its activity and results—is that agency, and there leave the question. But this would be dogmatical and unsatisfactory in a serious inquiry; and although a truth as positive as it is abstract, it cannot be treated as an admitted point, in the present state of knowledge, nor spoken of as lightly as we speak of gravitation—a term in such common use that physicists, astronomers, and mathematicians employ it as mechanically and flippantly as gold-beaters handle gold, without questioning its final nature, origin, alloy, or end.

On the contrary, I will meet this objection by experimental facts and authentic observations, the positive and collateral bearing of which, if not sufficiently potent to be immediately destructive of "materialistic theories," must hold them in grave abeyance until further evidence and discussion shall leave them among the splendid errors of the past.

While the cosmic forces of attraction and repulsion, reduced into their manageable and measurable elements of electricity and magnetism, are not only capable of generating, but are moreover *incapable of acting otherwise than to generate, both heat and light in connexion with molecules and the most attenuated gases;* while they manifest dynamic action and reaction, and develop sensible mechanical phenomena, when propagated in currents through molecules, however diffuse, attenuate, and imponderable these may be rendered;—while such are facts, it is proved by the latest and severest experimentation that electro-magnetic currents not only do not produce, but are also *incapable of producing, the slightest sensible mechanical or pulsating action in a vacuum.* At the same time (as remarkable and suggestive as the point is true) this vacuum shows a capacity for transmitting the ELEMENTS of heat and light directly through or across it, without obstruction or absorption, or, so far as we yet know, permitting their conversion into these secondary forces. Hence, therefore, inasmuch as these secondaries are purely compound varieties of two well-known radicals—that is, of attraction and repulsion—and are generated only in connexion with molecules and atomicities, it follows that the latter forces (the radicals) or their influences may cross a vacuum, either binarily without mutual interference, or singly without interruption, absorption, loss, or destruction in any way; the influences and functions of these primal forces appearing to act in a creative and mechanical manner only when in direct conjunction with matter, and to develop dynamical and undulatory phenomena only where molecules demonstrably and *substantially* exist. This is what we might naturally suppose; but it appears to be shown *experimentally* in the fact, that a thermometer, suspended *in vacuo,* exhibits sensibility—that is to say, expansion and contraction—to varying external temperatures. We know by countless observations that attraction, as abstract force—that is to say, gravitation—extends its power through a vacuum. We now know that repulsion, viewed also as pure abstract force, extends its special form of power in like manner; for if molecular changes transpire in a thermometer suspended in a vacuum, the forces bound up with those molecules must be influenced by the movements of similar forces in matter outside of the vacuum (herein lies the secret of calorific developments), such forces acting insensibly through the latter by their own subtle penetrative and expansive energy—a principle or property too near the origin of things to be comprehended otherwise than in physical results.

Now, if we may draw from these facts any inference whatever applicable to cosmical physics and optical discovery, and if we may be permitted, in view of our preceding positions, to hold them up, like flambeaux, in both hands, and therewith boldly explore not only our own limited system, but also the remotest realms of space, and test, moreover, every physical theory in the light thereof, we may legitimately conclude that cosmic masses, charged to supersaturation (and even to self-destruction in some cases) with conflicting forces, both of which are actively and exhaustively employed in generating the several internal, local, and relative motions of these masses, maintaining their manifold internal and superficial economies, and in perpetuating a general plan of self-conservation;—now, if we may draw any inference whatever from all which precedes, we are compelled to conclude that the celestial masses are as vividly, strongly, and mutually sensitive to the central activities of each other, though widely separated, as if in actual contact. Through the perturbations and ceaseless approximations and elongations of the heavenly bodies, and through the periodical fluctuations of the magnetic and electric functions of our own planet, we become acquainted with the extreme sensitiveness of the entire system of the cosmos to the presence and reciprocal action of its individual members. By prolonged observations upon these wondrous masses, and by the severe calculations of modern mathematicians, it is established, that for more than two thousand years, although there may have been fractional changes in the forms or relations of the earth's or moon's orbits, *the earth itself has not lost an iota of force* of any kind, unless an equal quantity has been received, from day to day and hour to hour, in the form of actual working energy; and it consequently follows that no other body in our system—not even the sun itself—has suffered loss in the slightest degree.

Such being a grand, final fact, the conclusion is positive that no member of the solar system (whatever the internal agitations and local revolutions to which it is subjected in virtue of its connexion with all the rest) can ever radiate and throw away into space any of the mechanical energy with which it was endowed in the beginning, neither in the form of heat, light, magnetism, electricity, nor of pure elemental force. If, in the course of their mutual physical transactions, either of these bodies has radiated any of its mechanical power, the same quantum must not only have been instantly and directly absorbed by all the rest, but all the rest must also have contributed definite and proportional sums to restore as instantly and directly the first displacement. Should this sequence be viewed by any physicist, for a single moment, as an absurdity or impossibility, he must be reduced to view every *hypothesis* as a still greater one, which declares heat and light to be radiated into and extinguished in space, or diffused and wasted, as undulatory and mechanical energy, through an infinite field of attenuated ether. Consequently, it may be declared impossible, as an inductive truth, for force of any kind to be in any degree diffused so as to be unemployed or inefficient in the interstellar spaces; and this is rendered the more positive, since the general equilibrium of our own system is absolutely undisturbed from age to age, and from cycle to cycle. Even if evidences of countless cosmographic changes exist (as observed wherever we look in our planet, or seen by telescopic aid throughout the heavens—as in the Moon, Venus, Mercury, Mars, the asteroids, Jupiter, Saturn, the Sun, in the blazing star of Corona Borealis, and in distant constellations), we have reason to conclude, from positive knowledge of the functions and laws of quantity and action of the mechanical forces in and upon our own planet, that every cosmical change and cataclysmal revolution elsewhere has been connected with some progressive plan and ascending order of special and local morphological development, mapped out in the beginning by the Creator of all things.

Moreover, finally, we know that heat and light are only the

local jarring of molecules upon the surface or throughout the interior of any mass of matter; that they are varying series of vibrations, effected by the incessant absorption and rejection, or momentary action and reaction, of attraction and repulsion, as these forces dart swiftly and by convective processes from the circumference to the centre of such bodies, and conversely; and that they are phenomena appearing anywhere and everywhere in direct response to, and reciprocity with, molecular and cosmic laws of *quantity direct* and *distances inverse.*

Considering these facts established, and our preceding steps beyond contradiction, we may safely conclude that globes of molecules, viewed comprehensively as vast heterogeneous masses of chemical elements, metals, gases, and compounds, so constituted and combined as to require, absorb, and inalienably hold the entire sum of potential and phenomenal energy which has been created; viewed as masses absolutely replete with *repulsive,* as well as with attractive force, mutually sensitive to each other's being at all distances under Newton's laws, flying through an infinite vacuum in definite, yet multiform, orbits, the elements of which are determined by the condensative and expansive energies inherent in their respective constitutions, and eternally oscillating, both internally and externally, as centripetal and centrifugal functions; —thus viewed, we may safely conclude that cosmic masses will and must move, act, and react in their dynamic relations, while sweeping through space and rolling around each other, as if in actual collision, or at least in elastic connexion and intercommunication; and that, in virtue of these reciprocal reactions of their central forces, they will and must develop internally, superficially, remotely, and universally, both the *molecular vibrations,* and *every form of* SECONDARY FORCE consequent thereupon, which are observable and measurable in any member of the cosmos. That these very internal and excentral undulatory phenomena do, in some occult way, extend and perpetuate their influences reciprocally upon each other, without coincident *loss of working power* in any body or any system of bodies by radiation, is very certain; and if the light and heat observable upon the surface, or traceable to the interior, of any of them, *seem* to be connected with radiations from foreign bodies, such phenomena must nevertheless depend upon the local action and reaction of molecules, at the points where the light and heat appear, and cannot be produced by direct impingements of *heat* and *light per se*, originally emitted from foreign bodies, and propagated as such by undulations, through a pulsating ether, to the object heated or illuminated. All this I know to be the rankest heresy. But in virtue of evidence derivable from the phenomena developed in comets by *their intercosmic* reactions (to say nothing of seismic data) we must finally conclude that *repulsion* is an active function in every form of dynamic energy, displaying itself in all those reactions which are observable not only in molecules and in experimental physics, but also throughout those globes of molecules called planets and suns. And thus it is, when excited by gravitation into antithetic waves and equatorial and lateral currents of local action, inducing molecular vibrations, that *repulsion*, the co-ordinate and ever-present antagonist of gravitation, transforms itself into palpable and measurable sums of luminous, thermic, electric, and magnetic force, and blossoms into the grand mechanical, phenomenal, and creative energy of the universe.

Even if the statement be wholly true that certain cosmic bodies, as the planets and their satellites, shine only by that form of light termed "reflected light," it by no means follows that their respective internal forces do not act as phenomenal agents in the manner indicated, and thus develop mechanical effects not visible in a luminous, but manifested to us in some other manner. For these same radical forces are convertible into motion, heat, magnetism, electricity, and chemical changes, as well as into local light. And if we consider the moon alone, the most intractable member of our system, there are legitimate and weighty reasons, derivable from known lunar influences upon the vegetable growths, the nervous life, and the magnetic, seismic, and tidal phenomena of this planet, to infer that the activity of the lunar forces must not only be positive internal quantities, but also very potent external ones. Moreover, data exist at present wherefrom to infer that the light of cosmic bodies depends upon the mechanical or chemical combinations of certain gaseous and metallic molecules in their envelopes, which conditions are essential (as in stellar photospheres and in comets) to the development of "direct light." The absence of these conditions would end in darkness, while magnetism, motion, and all other mechanical effects of the conflicts of attraction and repulsion would still exist. I suggest these facts and inferences in the present connexion because they are important both to meet objections which may arise in the minds of many learned men, and to excite more active inquiry upon points which have not yet been reduced to exact knowledge. I advance no theory. I state facts and draw inferences; and leave the development of the subject to future observation, experiment, and discussion.

However widely this doctrine may differ from accepted theories, however unacceptable it may be to physicists in general, there is sufficient weight in the facts we possess not only to cast the gravest doubts upon all "ethereal hypotheses," but to compel us into new channels of thought, experiment, and observation. There can be no further question upon the point, that mechanical force, as a compound product of attraction and repulsion, acts and reacts directly from the centres of all cosmic bodies upon the centres of all other bodies, and then again from the centres to the surfaces, and through the envelopes of these bodies individually, and *vice versâ*; the earth, for instance, responding to lunar and solar influences in movements of the magnetic and electric force, only because the entire mass of the planet is excited and agitated by the molecular movements occurring in the entire mass of the moon and sun. The observations and studies of Schwabe, Gauss, Lamont, Sabine, Carrington, and others have proved this to be a general truth. The philosophical development of the subject remains for that grand generalization of cosmical physics which is sooner or later to gild the chambers of science with the radiance of its imposing glory. Here, therefore, I leave the last and most serious objection which might be urged against the application of dynamic laws, discovered in our studies of the earth, to the developments of the mechanics and general physics of the heavens.

8.

Now, when we attempt to ascend from the known to the unknown, in order if possible to acquaint ourselves with the operation of forces inherent in foreign globes of every class, the elements, combinations, quantities, distances, and motions of which are more or less measurable or observable, we shall find the steps legitimate and the results certain. In view of present positive knowledge, I would not declare, simply because the earth's interior may be proved to be incandescent, that therefore the planet was originally formed by successive accretions and impact of falling "meteors," as the only way of generating its central heat; and, ascending from this *hypothesis,* announce dogmatically that the sun was formed, and that solar heat and light are "undoubtedly" still generated, in the same manner; and therefrom further compel the mathematics into the humiliating attitude of declaring, that in process of time these meteoric bombardments must cease for lack of ammunition, that the heat and light of the sun must be extinguished, the solar system become dark, terrestrial life perish, all nature congeal, one star after another vanish, the universe collapse, and the sublime conceptions of the Infinite and Eternal Consciousness utterly fail in final wisdom and beneficence. On the contrary, rejecting all hypotheses, I would stand upon facts alone: and now, comprehending the dynamical action of attraction and repulsion as abstract yet living entities, and the laws which govern their relations with molecules, and with the chemical elements everywhere; knowing their mutual action and reaction, and the phenomena generated thereby, in our own planet, as cosmic facts, we may " ascend up into heaven," as if with "the wings of the morning," sure of the "hand" which leads, and of the "right hand" which holds, and that "even night shall be light about" us. We shall, indeed, find the physical phenomena observable throughout the distant realms of space subject to the action of the same laws, generated by analogous causes, and coincident with similar facts and conditions existing and operating in our own planet.

9.

I have already alluded to the important observations of Lamont, Schwabe, Wolf, and Sabine, in regard to the physical and numerical connexion of the solar spots with the periodicities of telluric agitation; in other words, to the coincidence of the reactionary and dynamic movements of solar force and solar matter with the dynamic movements of terrestrial magnetic force and terrestrial matter. Admitting this direct and palpable physical relation of these two cosmic masses (a fact which no physicist will now venture to dispute), the diurnal and annual variations of magnetic intensity—even transient anomalies of electric and magnetic phenomena coincident with earthquakes and volcanic disturbances—all point directly to the fact, that vast convective movements are continually taking place in the chemical constituents, metals, gases, and compounds, of the central orb. The discussions of Mr. Carrington's observations upon the solar spots, recently published by the physicists of the Kew Observatory, Messrs. Delarue, Stewart, and Loewy, also show the vast mechanical movements of solar matter which occur in connexion with the fluctuating heliocentric longitudes and distances of Jupiter and Venus. Observations like these leave no longer any doubt of direct and mutual reactionary relations between the internal matter and central forces of the *sun* and the internal matter and central forces of the *planets* which revolve around him.

Scientific men are sufficiently familiar with the violent central agitations and reactionary phenomena generated in *comets* when approaching the sun from all directions, and moving through perihelia at diverse distances, and in every degree of heliocentric latitude and longitude. These cometary phenomena, viewed as direct responses to similar dynamic movements and molecular agitations, incessantly active within the solar mass, are lucid and cogent facts, the bearing of which upon the problem now undergoing solution must not be overlooked. While this reciprocity of connexion with reactions of the solar matter cannot be *directly* proved, it can be proved *indirectly* with a weight equivalent to that of positive demonstration. For I have already shown the identity of phenomena generated *constantly* within the circumference of all comets and *annually* within the circumference of the earth, as these bodies move through their respective perihelia : and since we can prove reactionary connexions between the *interior of the earth* and the interior of the *sun*, it irrefragably follows, that the agency which causes central reactions in our planet is identical with that which excites the interior motions of cometary bodies. And inasmuch as all observations and computations show that solar forces are the agents which control the molecular forces of every cosmic mass moving around the central orb, the conclusion is irresistible that the motions, both internal and general, of all comets are correlated with and dependent upon translations of force, and consequently translations or vibrations of matter, *within the solar mass.* Of the degree of this internal solar action and reaction we are at present ignorant : but we now know, as a grand induction from Rumford's and Mayer's discoveries, that molecular friction can nowhere take place without the generation of specific quantities of heat ; and that, therefore, any *motion* of matter *within* the solar circumference must be equivalent in thermic effects to the impact of a limited bulk of that meteoric hail *imagined* by Dr. Mayer and insisted by Sir William Thomson to be the "undoubted" cause of the sun's heat and light. Thus a true and legitimate *internal* cause is established, and substituted for an imaginary and improbable *external* one.

In addition to the general inductive bearing of this class of facts upon the subject in question, there is a series of *visible* phenomena, the value of which is of the utmost importance in every inquiry upon the conversion of solar mechanical energy into solar heat and light. Sir William Herschel, Schwabe, Wolf, and other distinguished telescopists, and even I, a casual observer from several different points upon the earth's surface, have noticed that not only the spots, but also the *faculæ* and the general *cloud-like involutions*, termed by various observers "shallows," and the "wavy," "mottled," and "willow-leaf" appearances of the photosphere, are undergoing frequent and more or less constant changes, and fluctuations of light and shadow, all indicating alternate periods of intense agitation and marked comparative quiescence. When the vast area of many solar spots and groups of spots is considered; when the rapid manner in which the photosphere breaks up, rolls over, divides, and coalesces, like clouds of vapour, snow, and smoke, in order to form, change, and obliterate these ragged chasms, is further estimated; and when the mind of the philosopher is fixed exclusively and comprehensively upon this class of phenomena, uninfluenced by theoretical prepossessions, the fact becomes palpable that all these physical disturbances are the result of the convective action and reaction of attractive and repulsive forces upon the solar elements—a mere counterpart, indeed, of the action of analogous forces within our own globe and in its envelopes.

All these solar mutations are but so many phenomena of *molecular collision* and *friction*, of absolute *impact* of vast volumes of matter, the mechanical equivalent and positive product of which are light, heat, &c., whether the clashing elements be iron, magnesium, sodium, nickel, zinc, or any other substance discovered in the sun by Professor Kirchhoff, and whether they be in a solid or gaseous state.

Superadded to this conspicuous and convincing evidence of dynamic reactions *within* the solar globe, I will now present those interesting series of telluric facts collated and classified during many years, independently of each other and for different purposes, by Professor Perrey, Messrs. R. and J. W. Mallet, and myself. As they have been discussed at length in a preceding chapter, in order there to establish the existence of repulsion, as a cosmic entity pregnant with powers and consequences co-ordinate with those of gravitation, they need only to be recalled in this connexion to give a wider importance than they have heretofore signified.

The connexion of earthquakes and volcanic phenomena with magnetic disturbances, announced some years since by Dr. Kreil, of Austria, has been clearly established by many subsequent observations in various parts of the globe. The further connexion of these two classes of dynamic phenomena with the solar spots has been more recently asserted, and partially proved upon discussions of numerous data, by Dr. Kluge, of Saxony. The laborious observations and discussions of Lamont upon the variations of intensity in the magnetic and electric elements ; those of Wolf and Schwabe on the numbers, intensities, and variations of the solar spots, during the same long period ; and the coincidences of periodic variations of intensity between these terrestrial and solar phenomena, established by the discussions of Sabine, are not only well known and acknowledged, but they are confirmed by continued observations to be connected with some hidden intercosmo-dynamic law.

Now, the constant concurrence in space and in time of the *maxima* and *minima* quantities of the annual variations of magnetic intensity, of barometric phenomena, and of telluric reactions, as exhibited in earthquakes and volcanic eruptions —that is to say the constant dependence of total intensities, greatest and least, of these several classes of physical phenomena upon the planet's perihelic and aphelic positions (concurrent and periodic variations adjusted to the earth's eccentric circumsolar motions), each class following a law of quantity *inversely* proportional to the length of the radius vector;—now, this variety of association of dynamic constants proves *more* than the simple local fact, that oscillations of mechanical force—*i. e.* gravitation and repulsion—exist between the centre of the earth and its own atmosphere. Inasmuch as the numerical proportions of quantity and intensity observable in these phenomena hold fixed relations in all their aspects and variations, diurnal and mensual, indeed, as well as annual, to the eccentricity of the earth's orbit, their variety and association appear to be connected with some profound solar mechanical excitation. They point to a physical union, of a fluctuating character, between the centre of the earth (and, consequently, of all other revolving bodies) and the centre of the sun, and lead to the demonstration of a universal intercosmic law. They indicate, even, that some working and more creative power, some grander function, is actively immanent within their respective centres than mere quantitative mathematical sympathy—an unknown undefined something, termed gravitation—which only so influences them somehow in bulk as to leave them, as chemical masses, still inert and mutually unresponsive to each other's internal forces. They demonstrate, in fine, that a palpable and measurable connexion really exists between the sums of mechanical force inherent as working powers in these bodies respectively; and that the systems of attractive and repulsive currents which maintain their individual central activities, thereby developing solar and planetary heat, light, electricity, magnetism, and general dynamic and eruptive phenomena, are the same which, being exactly meted and adjusted to each in their sums and qualities of matter, establish and maintain their mutual external relations to one another, and general orbital distances asunder.

That this induction may be clearly comprehended, I may here remark that these intercosmic influences are not such as are taught by those professors of physical astronomy who liken the motions of the heavenly bodies to the projection and gravitation of cannon-balls, but such as are conspicuous in all great comets when they pass perihelion, and which were visible in Biela's comet immediately and for a long time after it divided, and whenever its segments have been subsequently near enough to the earth for definite observation.

The sun being so vast a body, so much larger than all the other members of our system combined ; and the latter differing in their masses, heliocentric distances, densities, absolute weights, elemental constitutions, and in the ellipticity and eccentricities of their respective orbits ; it follows that the internal solar forces must be constantly fluctuating, and necessarily transporting the molecules constituting the solar mass by multifarious convective impulses in every radial direction.

There exist laws which are fundamental, positive, unchangeable, and eternal—not simply because so proved by the calculus, but for the all-sufficient reason that they are universal, and embody the presence of an infinite, absolute, and transcendental principle, which addresses itself alone to the highest human consciousness : this is one of them. Not only the solar system, but the entire system of nature also, is built thereupon ; and it becomes self-evident, because facts of every class harmonize with it. Geometry and the algebras being, as I hold, the bases, the essences, and the mental expressions of this infinite principle, they are likewise its final abstract judges, and must, sooner or later, confirm that which is physically and metaphysically true.

10.

When, therefore, we consider the facts clearly within our knowledge and inductive reach, and apply the calculus to develop their elements of power and quantity, we shall find sufficient agencies at work *within* the solar sphere to generate its heat and light, without resorting to meteoric and speculative causes, the truth or error of whose existence it must be utterly impossible for human art to ascertain, and the probability of which, indeed, is so strongly refuted by the calculus itself as to reduce the assumption to absurdity. Rejecting all hypotheses, we now discover, in view of our preceding studies, incessant causes, which, mathematically developed upon the principles enunciated by Dr. Mayer himself (principles which have been elaborated and confirmed conclusively in manifold ways by Dr. Joule, and which are now accepted by every class of scholars as decisively as the laws of gravitation), will be sufficient to account for the origin and continuance of solar heat and light. When the reactionary impulses of the interior of the solar mass are regarded as results of momentaneous transmissions and transformations of mechanical energy—of swift and incessant antithetic movements and conversions of gravitation and repulsion into molecular motion—of forces acting and reacting from its centre to its outermost gaseous boundary in immense waves and eruptive agitations, as analogous forces act and react in our own globe, and in all comets, and in Saturn's rings;—when the action of the solar forces and the mechanism of solar phenomena are thus regarded, we may definitely comprehend the origin of that wondrous sum of heat and light, the unfailing sources of which cannot be explained upon the existence of gravitation alone, nor upon the prevailing theory of celestial dynamics.

When, furthermore, we recognise the fact that the variable heliocentric distances, not only of Venus, the Earth, and Jupiter, but also of all the other planets and their satellites, and of every body which approaches and moves around our central orb, are from instant to instant dependent upon the reactionary movements continually and complexly transpiring *within that orb*, as well as upon their individual reciprocities of molecular agitation, we shall discover therein still more extended causes for those mechanical excitations which end in the development of solar heat and light.

But it is when these principles are extended into their broadest cosmic applications and possibilities that we begin to comprehend the real magnitude of our inquiry, and the important consequences of our philosophy. For while insignificant storms in the solar photosphere, so visible in its spots, may indicate the approach and play the part of outposts to the flitting passage through space of a few clinging worlds, the flaming splendour of the entire orb reveals the formidable energies of its interior functions, and demonstrates the ceaseless activity of its molecular forces, as their pulsations, swelling into stellar strength, respond to the throbs of other suns, and thereby assist in sustaining those chemical mutations and that fluctuating instability of equilibrium upon which the dynamics of the heavens, the central heat of planets, the light of comets, and even life, death, generation, and organic and physical changes of every class upon the surface of our own globe, depend.

In this wondrous internal economy of the starry hosts, we now not only discover the mechanism whereby their external relations with each other have been established and continued, but also descry the inexhaustible sources of the light, heat, magnetism, and electricity of the cosmos, and arrive at the fountain-head of all mechanical energy. Philosophical inquiry of every kind may hereafter be conducted with more positive aims; and the physical and natural sciences, vivified by a new principle, may go forth in search of intellectual conquests as much greater than those we now possess as our present acquirements transcend those of the mediæval ages.

11.

Emissions of Heat and Light.

The manner in which, and the means whereby, the heat and light, so persistently generated in the solar orb by the mechanical processes above discussed, are communicated to the earth and other bodies, appear as obscure as those which promote or permit intercommunication and sympathy between their respective attractive and repulsive forces, and which bind all worlds to a common centre. When the final conditions of either of these phenomena shall have been discovered, those of the other without doubt will be also; because they are inseparably connected and dependent one upon the other. Great questions spring up at this point, not to be so easily or quickly decided under existing theories. But experiment and induction assisting, we are compelled to conclude, in the light of present facts and knowledge (however theories may conflict), that, space being vacuum, the action and effects of space are *nil;* that distance and motion, as components of time, lay the foundations of law upon this subject; that matter is essentially necessary to the generation and exhibition of *every* secondary force; and that no force of any sort whatever is wasted in, or can be radiated into, space in such a way as to be absorbed only at haphazard by bodies scattered in confusion throughout an immensity which is replete with intelligence and yet (an inconceivable paradox!) devoid of economy. We now know too much about the fixity of elementary bodies, the natures and quantities of the molecular forces, and the correlations, equivalencies, transformations, and conservation of all forms of energy, to admit longer any such fallacy.

All this has been heretofore passed sufficiently in review; and the requisite knowledge is yet lacking by which the question can be definitively settled. The entire subject is therefore open to doubts and discussions. The simple fact that time is necessary for the light of distant worlds to be recognised by the mind does not conflict with our deductions: since the numerical laws of light spring from those of gravitation and repulsion; and since the perturbations of the latter must affect the former. The animal senses are subordinated to both; and, so far as succession in the order of facts can strengthen argument, light, as a reactionary phenomenon, follows the dynamic conflicts of attraction and repulsion, viewed as radical forces propagated through molecules, and our senses become a tertiary condition, so to speak, in the order both of facts and of time.

12.

Cosmic Light and the Frauenhofer lines.

Another subject, however, to which special investigation may be directed from new points of view (a subject relating to the action of mechanical forces exclusively *within* the solar mass, and involving the generation of phenomena of great magnitude as successive and correlative sequences of one another, among which is the one just named), is the *manner* in which the *photosphere* is kindled into such dazzling splendour as to be conspicuous through immense breadths of space.

In opening the treatment of this branch of my subject, I am aware that I shall be immediately met by objections founded upon existing theories of celestial dynamics and of spectral philosophy. When I deny that the data we possess are sufficient to prove beyond doubt that the sun is an uninhabitable body, and that it is a *flaming* globe of iron, sodium, &c., the advocates of prevailing doctrines will declare that the sodium and iron flames observable in our experiments, taken together with Professor Kirchhoff's explanation of the Frauenhofer lines, are abundant and satisfactory evidences upon these points, and ought to set the entire subject at rest. Now this hypothesis is founded on bases similar to those upon which its parent, the meteoric theory of solar heat, is founded: and it may be safely asserted that, if one is inaccurate, the other is also in all its details, for the reason that the experimental study and the mathematical discussion of both series of data have determined that the laws governing the relations and phenomena of heat and light are identical, and as positively interwoven with each other as those which govern the relations and phenomena of electricity and magnetism. From this point of view it will be immediately seen that if only one half of the light generated in the solar orb is emitted, and the other half absorbed by the metallic vapours alleged to exist in its photosphere, so only one half of the heat generated by the sun can be emitted. Therefore all preceding calculations upon this point must be wrong, and the amount of heat actually generated in the sun must be twice as great as heretofore stated under any assumption.

My respect for the immortal philosophers who have been the discoverers and pioneers in these fields of investigation, from Newton and Descartes to Kirchhoff and Huggins, and whose very hypotheses exhibit not only their great learning but their penetrating and masterly genius; without the expression of which, indeed, their successors could not take a single step either in extending or elucidating this department of physics—my great respect for all this class of philosophers compels me to allude to the doctrines which now prevail. And in referring particularly to the theory in question, I do so not so much for the purpose of attacking it by a lengthy discussion, but rather in order to set in bold relief the foundations upon which it rests, to open new avenues of inquiry, into which the data already acquired may be more profitably directed, and to meet the objections to my own interpretations of the phenomena the intercosmic connexions of which are now so well observed and so fully admitted.

Nothing was positively known of the chemical constitution of the sun until the remarkable discovery of Kirchhoff and Bunsen: and it may be truly stated that the actual mechanism by which the illuminating and heating effects of solar action have been communicated to the earth has always been, and is still, involved in obscurity and doubt. But the elements upon which rests the prevailing theory of the physical structure of the sun and the emission of solar heat and light may be fairly stated in the following syllogistic terms: and I will premise that, while not questioning the facts, I dissent from the conjectures and reject the inferences involved in the deductions therefrom.

The foundations upon which the entire theory rests are, when reduced to their simplest terms, the two following propositions.

The vapours of the several metals, iron, sodium, &c., exhibit, when these metals are burnt in the earth's atmosphere, certain bright lines in the spectra of their light which correspond to certain dark ones in the solar spectrum.

" Kirchhoff found that a sodium flame which gives out on " its own account the double line D absorbs a ray of the " same refrangibility when it is given out by a body of a " *higher temperature* than the sodium flame, *thus producing* " a dark line D instead of a bright one; and he therefore " CONJECTURED *that the dark line D in the light of our* " *luminary was occasioned by the presence of the vapour of* " *sodium in the solar atmosphere, and at a* LOWER TEMPERA- " TURE *than the source of light.*"

Hence (all now say) the sun is " *a globe of the fiercest fire,* " *compared to which a mass of white hot iron is as cold as ice,*" a molten globe hot enough to hold even iron itself in a permanent state of extreme gaseous subdivision, and to lift and float its sublimated particles high in the solar atmosphere, and there maintain them as incandescent metallic vapour.

Such is the opinion of the most learned men at present living; and it is even declared that the probabilities are 300,000,000 in favour to 1 against the truth of this hypothesis.

Thus it is believed that our sun both heats and illuminates the earth (and even Neptune) by direct emissions of rays of heat and light *per se*, as absolute entities, in consequence of the intense *ignition* of the various metals and gases which constitute its mass and photosphere.

In stating the substance of the prevailing theory (as well the facts and conjectures as the conclusions), I have partly availed myself of the language of the most recent writers upon astronomy and physics, in order to show the influence which theories exercise upon the judgment of any, even the most enlightened, period, before new facts of an astonishing and exciting character have received their fullest developments. When several theories, thus equally unmatured, are linked together, and the mind is possessed by them, other doctrines when first presented to us seem absurd, and the attempts to interpret phenomena in a different way from that of their discoverers and of their disciples appear not only objectionable, but revolutionary and offensive. The truth, however, is what science demands above all preconceptions; and while aware again of the magnitude of the heresy involved in the rejection of a theory where the probabilities, mathematically considered, are so many millions in its favour to one against it, I am nevertheless assured that the subject is of so much importance that the form of treatment which I bestow upon it will receive a just consideration, even if the explanations presented do not obtain an immediate acceptance.

There are two general conditions of space, besides that of its vacuum, which enter into our discussions of the premises. These relate to the *temperature* and to the *luminosity* of space, abstractly considered.

First: I hold that, upon known principles, it cannot be assumed that the cold of space would ever be at a higher degree than the lowest ever observed upon the surface of our globe. "As to the proper temperature of the spaces at present " around us," says Nichol, "our determinations greatly differ. " Mr. Hopkins has recently adopted −38·5° C.; Fourier " estimated it at −50° C.; while, according to Pouillet, it is " as low as −142° C. It cannot be doubted that Captain Back " found the thermometer at Fort Reliance down at −56·7°, and " colds of −60°, −66°, and −70° have been reached in Siberia. " The accurate determination cannot be said to have yet been " attained; but the estimate of Mr. Hopkins seems the pre- " ferable one." This is the highest, that is to say the softest, degree of the *permanent* cold of space made out by physicists after taking every circumstance or theory of stellar radiation into their calculations. Thus it will be seen that the rays of solar heat are supposed to come in pulsating waves through the mean distance of 90,000,000 miles of this intense cold before they impinge upon our planet, and bestow on us what remains of their thermic power. The high degree of temperature prevailing at midsummer and in tropical latitudes we know: but upon these points I shall make no remarks in this connexion. I simply record the facts here. Their relative and absolute values will be more clearly perceived as our subject is unfolded.

Second: In regard to the *luminosity* of space, abstractly considered, little need be said; since observation and all experimental and inductive data not only indicate it to be a state of permanent and absolute darkness, but also establish the facts that no light exists or can be generated anywhere independently of the union and direct interaction of molecules and their respective forces.

These points are beyond question. The only *theoretical* point even, however remote its bearing, which could be insisted upon to conflict with these conclusions is that which has been conjectured to arise from the retardation of Encke's comet; but, inasmuch as this stands alone in the entire realm of cosmical nature, it may be safely set aside and be allowed to vanish, since I have already shown the existence in all bodies of a repulsive principle which may operate in such a way as to produce the effects observed in the motion of that body.

I have already shown that heat and electricity, as secondary forces, are derived from the interaction of molecules charged to saturation with the radical ones of attraction and repulsion—the former of these last elements, as a whole, constituting cosmic attraction, and the latter cosmic repulsion; and I have indicated the manner in which these, as the central forces of the heavenly bodies, act throughout the masses of these bodies to generate the general conditions of cosmic electricity and magnetism, and of cosmic heat, as *local* phenomena affecting these bodies respectively, and adhering to them individually as their fixed and unchangeable quanta of dynamic and productive energy.

Now we know that the laws affecting the phenomena of heat are the same in all mathematical and quantitative respects as those observable in the phenomena of light, even to those of polarization; and we have no data upon which to reason or conclude that light is generated by other forces or in other ways, than by those forces and in those ways through which heat is developed. Moreover, as far as induction from experimentation and analogies can be fairly adduced to prove any condition of things so distant as to be beyond direct personal examination, it may be affirmed, without a fact to indicate the contrary, that no electric currents, or any other form of force, can generate light in a vacuum, and independently of the positive presence of matter at that point where light appears.

Such are some of the conditions which surround the theories of the constitution of the sun and stars, of the origin and action of cosmic light, and of the Frauenhofer lines (the general theory of optics) which now prevail, and from which I find reasons to dissent.

Disregarding the *assumptions* of authors, and admitting all *facts*, I shall endeavour to interpret the latter in exact accordance with the laws and principles now well established by observation, and by experimental and mathematical researches.

The simple point or picture presented for analysis is this. In the midst of general darkness the sun appears to us as a globe of dazzling molecules. While this cannot be otherwise in virtue of the fact that space is *nil*, the question is, whether this solar light which acts dynamically upon the earth and eye, and thus generates terrestrial light, is actually emitted by the sun so as to be darted through space as a luminous impulse; or whether it acts directly upon the matter of our atmosphere, thus exciting lumiriferous reactions within the gaseous and metallic elements and compounds of our planet, independently of luminous impulses, as such, generated in and *apparently emitted by* the sun.

I hold to the latter view, in virtue of experimental data, and of a law similar to that which Kirchhoff discovered, relative to the equivalency of the force absorbed to that emitted from the sources of light.

Even in view of Mayer's and Joule's discoveries and

mathematical determinations, and of all which precedes, it may be shown, if molecular laws of a dynamical character can be traced into their cosmic cognates, and *vice versâ*, in respect to *one form* of secondary force, that all other forms of force which are both equivalent and correlated to this, in their origin, action, and effects, and which are incessantly convertible into one another and into this, will be traceable also into the same cosmical conditions, and *vice versâ*. When one series of cosmo-dynamic functions is found to be strictly planetary—that is, to originate only in virtue of local and internal reactionary agencies—then all must be and do so, as a necessary consequence of physical laws now well established as inflexible ones.

Even if I take as a basis for my discussions of experimental and optical data the conclusions at which physicists, mathematicians, astronomers, and geologists have arrived, and upon which all classes of scientific men appear to be united, relative to the conditions and exchanges of *solar and terrestrial* HEAT, I have solid grounds for my deductions. For (adopting the terms of high British authorities), "whatever may have "been the original condition of our globe, or whatever its "*present internal* constitution, all effect *on the temperature of* "*its surface as arising from that condition has long ceased.* "Taking even *the extreme supposition*—viz. that the earth was "*once liquid through fusion,* it is *indubitable* that, since the "consolidation of the outer strata, *the heat communicated by* "*solar action and that lost by radiation* BALANCE *each other.*" Such are the conclusions and the language of the most eminent scholars of our time, expressed clearly and definitely, in relation to solar and terestrial *heat;* and whether I assent or not to the hypotheses upon which these determinations have been reached, the statement and the facts are the points which bear upon the issue of our discussions relative to the origin and phenomena of solar and terrestrial *light.*

Now, since we know that light is generated in the same manner as heat is generated, and is subject in every respect to the same laws, it should follow from the preceding determinations that the same amount of light is radiated toward the sun as that which is communicated or induced by solar action. If so, then the absorption of the light which is indisputably indicated in the Frauenhofer bands is due to the local molecular action and atomic functions of the chemical elements of our planet; and not, as my illustrious friend Professor Kirchhoff conjectures, to the absorption of half the solar light by incandescent metallic vapours suspended in the flaming atmosphere of the sun, and which absorption, as he asserts, is effected because the flaming sublimates are cooler than the raging globe of molten metals from which they are incessantly projected.

I have already shown that our globe, viewed as a cosmic mass of elementary and compound chemical substances (and the same proofs apply to all other globes), is charged to extreme tension with repulsive as well as with attractive force; and Kekulé's discovery of the chemical fact of "atomic saturations" (if I may so express this important step in molecular physics) confirms the not less important one of cosmic saturations which I present. Indeed, one supplements the other: and when both series of facts are united into and viewed as a single group of interdependent and correlated phenomena, physical doctrines will become less unstable, and natural philosophy, as a system of positive truth, will the sooner possess a firm basis for its general developments, and its successive conclusions will be placed beyond controversy. While every electrolytic experiment tends to establish the former of these truths, the simplest dynamic experiment proves the correctness of the other; for no mass of matter can strike the globe without producing a series of *jars*, which are communicated to every molecule of the sphere: and it is out of these vibrations that proceed *light* and heat and electricity and magnetism, and all actinic changes. In order to prove that what is called the ray of *solar light* is a pure terrestrial phenomenon, resulting from the decomposition and intermixture, so to speak, of the impulses or influences of terrestrial and solar action and reaction, we may cite those early experiments of Faraday (confirmed by all subsequent ones, and by those of all other physicists), in which he proved the direct action of magnetism and electricity upon the ray of solar light through the agency of his analysing prism of boro-silicate of lead. These facts (so famous when first announced, and the most brilliant of Faraday's discoveries), establishing the influence of electric currents on polarized light, and the correlation of these secondary principles with one another, and proving their mutual convertibility, are now well known; but their significancy and the reach of their consequences upon physical theories are not yet fully developed. The important points in our subject elucidated by them are, that *if one is a terrestrial principle*, THE OTHER IS ALSO; and that the same chemical elements which generate electric impulses can also absorb the luminous ones, and *vice versâ*, in proportion as the equatorial, bilateral, or crucial expressions—the dynamic element of the ray (a point heretofore alluded to as among the most important of the fundamental conditions of matter and force, in all their forms and reactions)—are acted upon by the antithetic currents of the battery and the earth.

Now all these phenomena, expanded into their cosmical magnitudes, are, as I hold, legitimate local consequences of the fixed states of *saturation* of molecules and of globes with the attractive and repulsive forces; the facts about which are laid down by Professor Kekulé in regard to the former, and developed by myself throughout the preceding pages in regard to the latter. I have already alluded to the action and reaction of these globes, as chemical masses thus full to repletion of associated and opposing principles, as they roll through a vacuum, balanced against each other by the most delicate adjustments of their indwelling energies; and all experiments and observations show that these adjustments cannot be disturbed in the slightest degree without corresponding developments of light and heat, magnetism and electricity, of actinic changes, and even of the volcanic and seismic phenomena heretofore discussed. However various in form, expression, or magnitude these phenomena may be, they are all equally dynamic in their origin, character, and functions, correlated by the same laws of fluctuating quantity and intensity, and unitedly dependent for these variations upon the form of the orbits and the relative distances which these masses hold inversely to one another. So delicate, indeed, are these adjustments and the mutual interchanges of the influences between the sun and the planets (as is now established by the Kew observations, and as has recently been set forth in the interesting papers of Dr. Stewart and Mr. Lockyer), that the rolling of the heavenly bodies in relation to each other may be represented as perfectly analogous to the rolling in direct contact of the wheels of a mill or of a clock, the cogs and couplings of which are perpetually interlocking and unlocking, or to the motions of the screw, the correlative couplings of which are incessantly playing into each other and working out the mechanical designs of creation. For the fact of their relative connexions and motions *in a vacuum* must end in reciprocating local effects, which in every issue must not only resemble, but absolutely be, those of actual contact. That the influences and action of vacuum must be *nil* requires in the present state of knowledge no demonstration. Since, therefore, the great cosmical masses are, as we know, adjusted, balanced, correlated, and moving by the inflexible laws which govern their aggregate molecular energies, the phenomena generated by their mutual influences must be *local;* and the secondary forces and the various forms of mechanical power thus developed and made sensible during their temporary existence and conversions into each other cannot be subject to variation, or any form of absolute alienation whatever. Grinding together, so to speak, as these bodies perform their eternal circuits around each other and their revolutions around their own axes, and in a certain sense upon the surfaces of each other, their forces, more than their substance, mutually affect their relative conditions of density and molecular friction; and thermic and luminous, electric and magnetic polarizations, as dynamic or equatorial functions and phenomena, unfold themselves as an inevitable consequence of the chemical constitution of the cosmic masses, and the vibratory and electrolytic action which is incessantly taking place in their interior, and throughout their surfaces and gaseous elements. All observation, experiment, and analogy tend to prove these points; and I know of no fact to the contrary. The silent and picturesque images or revelations, so to speak, of these interchanges or oscillating functions of dynamic energy, established by the action and reaction of material forms charged with forces and playing the part of "couples" to each other (a fact which infuses a cosmical significancy and verification into the mechanical theory of Poinset)—the revelation and certainty of these mutual interactions of material forces we behold in the black and compensating bands of the prismatic spectra; and, thus viewed as photographic pictures, so to speak, of the absorptive and projective essences of matter, they become transformed into indisputable indices of the existence and presence of attraction and repulsion as the radical agents of the most subtle functions of intercosmic nature.

When we further consider the experimental researches of Faraday and of other observers in connexion with the effects of light upon metals in an extreme state of subdivision, and *vice versâ*, and the facts that certain forms of electric action upon gases will generate luminous bands complementary to the Frauenhofer lines without the development of sensible heat; and that even metals and alloys may be disintegrated, and their impalpable particles translated, and associated with other metals not in immediate contact with one another, and accompanied by brilliant electric radiation independently of destructive heat, or what is called combustion: when these and numerous other facts tending to the same point which are on record as having been manifested in our own atmosphere are considered, it may be declared at least doubtful if the prevailing theories are sound respecting the constitution of the sun founded upon Professor Kirchhoff's explanation of the causes and indications of the Frauenhofer lines.

Objections raised upon the grounds that the moon and planets do not yield spectra of their own can have no weight in view of the facts, principles, and deductions heretofore presented. For what is a polarized ray of solar light? Light is not original force *per se*, but an effect of force; the resultant of our two radical forces acting among the elements of matter, metals and gases, and generating specific sums of vibrations: and vibrations are only the equatorial local plays of these two antagonistic forces where and when they meet in atoms and resolve themselves, so to speak, into light, heat, electricity, and magnetism. The fact discovered by Œrsted of the dynamic element in these secondary forces, and which, as is well known, was developed mathematically by the immortal Ampère, runs as a principle through every form and manifestation of force in its relations to matter. Thus physical events occur in Jupiter and in the moon, as they do in the earth. As these bodies roll and grind, so to speak, relatively to the sun, their respective sums of matter and energy are measured against, and conflict with, each other. Attracting and repelling, pulling and pushing, in order to execute effectively the designs embodied in each of them, a succession of molecular impulses (that is to say, in the language of optics, an invisible ray) of polar force encounters another series of molecular impulses of planetary force, and, as I have heretofore illustrated, these impulses mutually resolve themselves into the equatorial phenomena discovered by Arago in the polarizing prism, a picture of dynamic energy complementary to the phenomena discovered in the solar spectrum by Wollaston and Frauenhofer. The planetary and the lunar rays, or molecular impulses, thus become truly secondary, subordinated, or reflected light. It is a solar impulse crushed, so to speak, broken up, and partly absorbed; and, mixed with other impulses, it is turned at right angles and projected right or left as radiant light to meet the eye of the beholder. Polarized light, like solar light, is demonstrably composed of the same elements of attraction and repulsion; for it is ready instantly, as discovered by Faraday, to mingle with a magnetic or electric current, and go back again into the body from which it at first proceeded. If Verdet and others have not been able to unravel this mystery, it is not because they have not laboured perseveringly and patiently, but because they have lacked the instrument with which to attack the problem.

Thus it is, I hold, that the stars, including our sun, composed of chemical elements more or less like those of our globe and other planetary bodies, and surrounded with gases illuminated by electrical action (as all spectroscopic observations indicate) —thus it is that the sun and stars glow with overpowering splendour. In virtue of their vast dimensions and of their mutual relations with each other and with their attendant spheres, their mighty stores of internal energy burst out upon their surfaces, and, agitating their gaseous envelopes, act at a distance, and are absorbed again into their own centres without the local loss of substance or of power. Thus it is that their appendages, smaller in dimensions but constituted of similar elements differently combined, and charged with the identical radical forces of attraction and repulsion (forces capable when excited by the larger masses of developing phenomena similar to theirs), shine by what is called secondary, reflected, or polarized light. The activities of their molecules, responding to the activities of similar ones in the sun, both give out and accept vibrations which are attuned in their capacities alike for absorbing and rejecting definite types and sums of energy, and which become sensible as light in consequence of their dynamic action being unfolded transiently to our eye as they dissolve and separate, and mingle with the kindred principles from which they sprang.

13.

Light and Cold.

The relations of heat and darkness, as correlatives of each other, are much better understood than those of light and cold. While the first two terms express the existence and conditions of agencies which can be measured numerically as radiative and absorptive functions, and which are the indisputable representatives of repulsion and attraction as abstract but effective principles in nature, their opposites, light and cold, as associative and correlative conditions tending also to peculiar and positive phenomenal developments, have not been fully investigated. While the genius and labour bestowed on the subject of light in all its forms of exhibition have been altogether marvellous, the subject of cold has been comparatively neglected, as a sort of negative entity having no direct connexion with dynamics; and even now its existence is doubted by almost all philosophers as anything but *a state of matter* indicative of the absence of heat. Whatever the final states of nature may be, they can be determined only by positive inquiry, and not by ignoring them because they seem unfathomable, or because preconceptions have existed for ages among men that they are so. Even in the present state of knowledge it cannot be denied that cold under certain circumstances magnifies itself into an active agent, and that it always appears to manifest relations to, and to hold direct associations with, light, influencing the action of the latter and facilitating the expressions of its energies. We have abundant evidence to show that these conditions extend beyond experiment, and that they present themselves in wide cosmical aspects. For while the passage of an opaque body between the sun and another body cuts off the mechanical action of the sun, or a ray of solar impulse, so to speak, at the point eclipsed, and thus suppresses that vibratory excitement of the gaseous and metallic molecules which generates direct light, and so produces what is called a dark shadow, the passage of comets between the sun and other stars gives evidences of the very opposite state of things. That the varying intensities of cometary light are due to the influences of the mechanical movements, both internal and external, of the two bodies in relation to each other, I have already elsewhere indicated; and that the tails of comets play the part of luminous shadows penetrating the profound depths of the interstellar darkness, through the subtle correlations existing between the mechanical affinities of light and cold (positive dynamic conditions of universal nature not yet fully understood), few who have studied these subjects in connexion with one another can seriously doubt. And it is from this point of view that I invite the attention of physicists and of spectroscopic astronomers, especially the eminent men whose achievements have adorned this generation with such peculiar glory, to new observations and a modification of their speculative ideas.

While my own interpretation of intercosmic action from the studies of the Frauenhofer lines tends to unveil the material links of connexion between all the members of the cosmos, binding telluric molecules to distant worlds, and the remotest stars and nebulæ which the aided eye can reach with the subtlest gases in our own atmosphere or in the darkest depths of our planet, there is nevertheless nothing in what I propose positively adverse to the *fundamental principle* of the theorem laid down by the illustrious Kirchhoff. This *principle*, expressed in my own terms, is that *the* SOURCE OF LIGHT *is a mass of matter charged with energies in a more intense state of vibratory excitement than those of the gases (and even metals in extreme molecular subdivision) upon which these energies act, and where reactions are ensuing.*

Under these aspects of the subject, distant conditions of nature are brought into harmony with experiments and with observations on the luminous phenomena, generated in connexion with cold, in the upper regions of our own atmosphere, where light, as a vibratory principle playing among attenuated gases, is more vividly expressed than at the surface of the planet: and thus we may be led to infer that the visible evolutions of the gases constituting comets are connected more positively with the mechanical conditions of light and cold, than with those of light and heat. Moreover, I am persuaded from a careful review of various data that, while the "opaque globes" may be proved to be similar in physical constitution to that of the solid globe—that is, to the dark body—of the sun, the comets will be found to be similar to the photosphere, to be cold, and to be excited into luminous conditions in like manner as is the case, as we have many reasons to infer, with the photosphere itself. That intense light may be generated without the direct presence of intense heat is a fact well known; and it is to this point that I shall now direct the special attention of the reader.

So much at least it has been necessary to set forth in order to meet the objections which will be immediately raised against my inferences: and if I do not extend both facts and arguments, it is not because the present state of science will not permit me to do so, but because I expressly wish not to complicate my discussions with attacks upon hypotheses, or delay the reader longer from the data which I have to present in connexion with this subject, and from the final conclusions of this chapter.

14.

Solar and Terrestrial Observations compared, and Inferences therefrom.

The interesting discoveries and deductions of Alexander Wilson, of Glasgow, published nearly a century ago, respecting the probable solidity of the dark surface of the sun and the elastic changeable character of its luminous envelope, have been continually confirmed by subsequent telescopists, and have in no serious manner been modified by any recent astronomer, notwithstanding all improvements in the means of observation. No spectrum observations, and especially no simple *speculation* upon the igneous state of the solar globe founded upon prismatic experiments, can overthrow his conclusions, inasmuch as it is apparent that the solar forces must act in bulk —that is to say, as a unit or as concentrated points—upon this planet, and thus represent in each isolated ray the *combined molecular activities* and aggregate chemico-mechanical atomicities, so to speak, which exist in, and are constantly acting and reacting throughout, the entire solar globe, in such a manner as to generate its refulgence. By this I mean to be understood that the chemico-molecular vibrations constantly active in the solar mass are there mingled so thoroughly, that when they act (as indicated in the preceding section) upon the earth's surface as points of repulsive force, and here encounter cognate forces and vibrations, their constituent elements fall asunder in the prism; and that the dynamic functions of terrestrial repulsive and attractive forces so act therein as to reject or absorb one series of vibrations or another, and in this manner demonstrate the identity of both matter and force in the two bodies.

In virtue of all that precedes, and of this deduction therefrom, it will be seen that spectrum phenomena, considered as indices of the *internal* conditions of stellar, planetary, or cometary bodies, afford no *positive* proof that the sun is an incandescent globe, or that its photosphere is an envelope of *flaming* metallic sublimates. Doubtful or revolutionary as these delineations may appear, they are nevertheless truthful; and experiment and observation tend to confirm their accuracy more and more. The necessity, therefore, of maintaining the irrational hypotheses, that the surface of the sun is a surging molten ocean in consequence of the incessant impact of meteors or of any causes; that the photosphere is constituted of metallic fumes, blasted into white heat or fed with meteorites and comets; that here and there black clouds of condensing smoke or vapour form in double strata, and float across its central latitudes, in constant coaptation, constituting the *umbra* and *penumbra* of the so-called solar spots, notwithstanding telescopic evidence to the contrary;—now the bases for such opinions must vanish in view of our positive knowledge, and from what we may infer from the action of terrestrial forces upon terrestrial matter—that is to say, our knowledge, of terrestrial physics, and the application of this knowledge to solar physics.

A momentary review of the action and effects of telluric forces will show that high degrees of luminosity may be generated in the *envelopes* of our own globe, *unaccompanied* by heat; and not only so, but also in the midst of the intense cold of polar latitudes, and during arctic nights; that intense light may at least be generated independently of glowing heat; that, contrariwise, great heat may exist in mineral and metallic matter without the display of dazzling splendour; and that all these states may be active independently of meteoric impact.

In order to give all forms of evidence that weight which they really possess when practically considered, and by way of illustrating the causes of solar light and heat, I will introduce at this point a few co-ordinate facts and inferences which, while they relate especially to the subject of heat, are collateral in their bearing upon that of light and the physical constitution of the heavenly bodies in general. And inasmuch as the laws governing the generation of both heat and light are identical, any fact relative to the sufficiency or insufficiency of this or that physical action as a productive cause of one of them will apply equally to the origin of the other.

Physicists and geologists who have made volcanic phenomena a subject of special observation *in situ* are aware that lava upon its eruption at white heat turns almost instantly cherry red; in another moment (in the daylight at least) is almost black; and in an incredibly short time, even while red, will sustain great weights, and might be traversed, were not its radiant heat so intense and suffocating. As facts abound upon this point, I will not here intrude in detail my personal observations and experiences at Hawaii, both in the burning lake of Kilauea, and upon the great lava stream of 1855 and 1856, which broke forth near the summit of Mauna Loa. It is well known with what rapidity incandescent lava bridges itself over with a brown or black crust, while yet a flowing river at white heat underneath. The radiation of heat under such circumstances is intense.

In view, therefore, of this experience alone, we may safely infer that it is not necessary for the solar globe to be so incalculably hot (as theories assert) as to maintain *iron* and other metals suspended in states of vapour in the higher regions of the solar atmosphere in order to radiate great sums of heat. Nor, indeed, would it appear to be necessary, in view of modern experiment, for the solar mass to be at white heat in its envelopes, nor even upon its surface, in order for its calorific power to be felt by our own globe. The experiments of Melloni, Tyndall, and others in diathermism conclusively demonstrate that intensest heat and fire can be generated by the dark calorific rays of the sun alone, when concentrated and transmitted through media of utter darkness.

The polarizing action of repulsion and attraction on matter, modified by atomic or chemical laws, appears to be the mechanism which develops such results—the action of the same forces, indeed, which under other conditions of matter, or the opposite expressions of their own energies, would produce and intensify light to such an extent as to manifest palpable mechanical and actinic power. Enlarging experimental data into terrestrial and solar magnitudes, and viewing these magnitudes as solid terms in the problem, we may now begin to comprehend a fruitful field for theory, and finally reach truth through facts, discussion, and induction—the only reliable means of positive knowledge.

The method by which heat is generated *within* the earth, and the fact of its mechanism being correlated with the periodic intensities of volcanic eruptions and of seismic phenomena, and with the relations which all these phenomena hold to incessant variations of heliocentric distance, have been already alluded to. That this heat, and the action of the mechanical forces generating it, maintain, moreover, constant numerical relations of intensity with similar phenomena existing in the sun, we have also sufficiently set forth and proved: and it now remains only for science to follow these facts to their final consequences.

By applying this brief array of experimental data—an array capable of indefinite expansion—to solar phenomena, we see no reason to question the remarkable discoveries and conclusions of Alexander Wilson upon the black and solid structure of the sun, as determined by the solar spots, and the possibilities of its habitable conditions indicated by those discoveries. Indeed, they are sustained by every species of observation and induction. For, however white and dazzling the molten interior of a cosmic mineral mass, and however violently agitated with convective motions this interior or nucleus, may be, the surface itself immediately loses this dazzling splendour when exposed to the cold of space, and its radiant heat appears to be then converted by gases into light. We have, furthermore, sufficient experimental means to show that the *thermic* power of the solar globe would not be extinguished in its peculiar influence upon the earth by a change of superficial aspect, involving a total loss of lustre, even were there no atmospheric envelopes to constitute a photosphere of transcendent brightness, and although perfect darkness might reign throughout the surface of the solar ball. That thermism, as excited upon the earth's surface by the sun, and discovered in the refrangibilities of the prism, is a different function of physical energy from *colorific* light, or colourism, and actinism, we know by daily experiment; and that all these material and immaterial powers are bound up together in the solar ray, and yet entirely separable, representing the manifold and separable elements of force and molecules and the atomicities of molecules existing in the central body, lenticular and spectral discoveries have positively determined.

Concluding these brief remarks upon solar heat, pregnant with amazing and undeveloped import, we will proceed to consider the manner in which its correlate, solar light, may be generated; treating the phenomenon from other points of view than those of the meteoric theory, and considering it as a result of dynamic conditions incident to the *internal* constitution of the solar mass, and the elastic character of its compound gaseous envelopes. Whatever the facts or inferences relative to the origin of solar light which may be reached by any process of investigation, they must apply with equal truth to the origin of cosmic light throughout the universe.

15.

The enormous magnitude of the sun is the first element in the problem. Its connexion with other stellar masses of equal or greater bulk is the second. Sir William Herschel was the first to demonstrate the relative angular motions of the sun and other so-called fixed stars. And while, as early as 1783, he approximately ascertained that our central orb, with all its worlds, is moving with prodigious velocity toward a point in the constellation Hercules, which he fixed at 257° right ascension, and 25° north declination, Otto Struve, of late, not only determined this point to be nearly the same—to wit, R. A. 250° 9½′ and N. D. 34° 36′—but that our system is flying at the speed of 150,000,000 of miles a year, or 17,123 miles per hour, the incredible rate of five miles per second, to that region of the heavens. It is in view of these discoveries that we now begin to perceive the wondrous potency of the mechanical forces at work between the different suns and systems in the cosmos. And inasmuch as we have already traced the laws which determine the internal commotions of the earth and of comets, as these bodies sweep from year to year and age to age in eccentric paths around the sun, having their individual, and even their local, agitations excited and sustained by their central responses to similar central forces ever operative in the solar mass, we shall be at no loss to detect in the solar globe itself sufficient mechanical causes to develop not only its internal heat, but also the intensely luminous phenomena manifested in its photosphere. And in this connexion it must be further remembered, that there can be no motion of bodily translation which is not directly dependent upon, and reciprocal with, the internal action and reaction of the attractive and repulsive energies which are incessantly oscillating from centre to circumference in these flying spheres; and that, as the sun rolls ponderously through this immeasurable circuit, both internal and external motions are all determined in obedience to and conformity with a unity and universality of law, which alike embrace molecule, sun, and cosmos in their simplicity and extension. The phenomena displayed so conspicuously in the new star in Corona Borealis, and the inferences drawn from them by Messrs. Huggins and Miller, tend to confirm this conclusion. Even the general phenomenon of scintillation or "twinkling," which Arago endeavoured so ingeniously to explain, only strengthens it. The various colours of the stars also might be adduced to show that the greatest differences of chemical combinations existing in the celestial bodies subordinate all their elemental and atomic laws to the absolute supremacy of this unique and universal principle.

Having stated these inferential elements of the problem, we will next array a few authentic facts and phenomena, in order to elucidate and simplify our analysis, and thus arrive by inductive and indisputable proofs at the immediate mechanism of the agencies engaged in generating solar light.

Since we may safely conclude, through the discoveries of Kirchhoff, Frauenhofer, Bunsen, and others, that the chemical constitutions of the sun and of the earth are similar, and in the main identical; and by all other physical observations and mathematical deductions, that the forces existing and operative in both bodies are also identical, differing only in measure, and this in proportion to the quantity of matter respectively constituting the two bodies; we may be equally sure that the conditions which generate light in the terrestrial envelopes are similar to those which generate it in the solar envelopes, and *vice versâ.*

That the solar photosphere is an atmosphere of elastic and moveable matter *similar in its physical capacities* to our own —capable of elevation and depression, of condensation and expansion, of bursting and closing and rolling, like clouds of vapour, smoke, and dust; and capable, moreover, like our own, of holding crystallizable substances in solution and suspension—is furthermore rendered certain by general telescopic and spectral observation, and by inferential forms of proof.

Now we know that the various terrestrial gases, and that

all metals in their most attenuated states of solution, are capable of high illumination, when excited by the action of electric currents ; and we must not forget throughout this discussion that electric force, magnetic force, and luminous force are neither more nor less than antipolar and antithetic translations through molecules of attraction and repulsion, viewed as mechanical principles, acting under pressure of cosmic and infinite quantities, causing various degrees of vibration, and thereby developing sensible phenomena. I introduce here no hypotheses, rejecting wholly the fancyings of the illustrious Biot, and admitting only as more reasonable the speculations of Kœmtz. I admit and use facts alone, and positive inductions therefrom. That which is achievable in experimentation, we find to be actually developed in nature upon a cosmic scale. Terrestrial auroras (which are in reality cosmic aureolæ) are often so bright as to illuminate the entire globe, and transform our atmosphere into a sea of pulsating and waving flames. In both polar regions their intensity is very great, and their light is more or less constant. Whalemen, navigators, and explorers of various nationalities have described their constancy, brilliancy, and wonderful coruscations. M. Lettin, of the French Scientific Commission, who spent the winter of 1858–59 at West Finmark, in 70° N. Lat., declares the refulgence of these shooting auroras to be at times so intense as to surpass that of stars of the first magnitude, and that the entire sky blazes with dazzling splendour, fringed with an horizon of jet, while an ocean of snow and ice below reflects and renders the light more resplendent still. That all this intensity of light is generated by the same dartings of attractive and repulsive forces through the gases and vapours of our atmosphere is not only rendered probable by our experiments, but positively proved by constantly accompanying fluctuations of the magnetic needle. And even when similar auroras appear in more temperate latitudes, either as circumscribed or general phenomena, we can now prove, by means of telegraphic connexions, that their origin depends upon internal mechanical agencies; thus establishing the great cosmic facts, that terraqueous electric conditions are universally associated with these aërial phenomena, and that both spring from the same profound telluric causes, sources, and forces. I have already shown in an exhaustive manner what these causes, sources, and forces are, and the nature and methods of their action and reaction, as effective instruments in all classes of physical developments. And when, in view of the premises, we consider that the earth, with its given volumes of forces, is less than 8,000 miles in diameter, and that the sun, in comparison, with its proportional stores, is 882,000 miles in diameter, and agitated throughout its prodigious mass by its interposition between all the other stars of heaven, to say nothing of its relations to its own circling system of worlds, we can no longer doubt that abundantly active and sufficient conditions exist *within* our central orb to generate that refulgency of its envelopes which, so remarkable and constant, is yet variable also, like our auroral telluric light, as shown in the observations of the Kew telescopists, Schwabe, and others. I might enlarge much upon these points; but physicists will comprehend their bearing: and that is sufficient for my argument.

Omitting specially to speak in this connexion of the violent *electric* discharges and *illuminations* which *attend sudden condensations of aërial vapours,* suspended at all altitudes (for I have noted them in the Andes, floating apparently higher above their summits than the Andes are above the sea), and omitting, also, to dwell upon aëronautic reports, all of which clearly establish the development of *luminous* phenomena to be coincident with, and dependent upon, the tranquil electrical changes, of a positive and negative character, which *invariably attend the oscillatory condensations and expansions of atmospheric vapours, at various altitudes,* I will call attention to another series of facts, and specially mention *the remarkable and constant illumination of the borders,* and apparently the remote surface also, of the *clouds* which form at certain seasons of the year along the coast of Peru, and float at low altitudes between Sachura and Tumbez. I name this limit because, during former travels, this particular region of sky was more or less open for many months to my continual observation. These nebulous phenomena always excited my curiosity and attracted my attention. The light, which was never less bright than that of a mild aurora, and was sometimes very strong, formed a border of considerable depth, entirely surrounding isolated clouds, and always adorning the upper edge of the endless banks of sea fog which collected at evening, and nightly rested along the coast. Were our little earth enveloped, like the sun, in continuous cumuli, or rolling masses of cloud, undergoing similar violent action, incessant changes of density, elevation, and depression, or of condensation and expansion, there is every reason to believe, from what we know of the reactionary influences of terrestrial forces upon gases, that this nebulous envelope would appear more or less luminous to foreign observers, and that our planet would even make itself conspicuous by its own light to remote regions of space. Facts, derived from observations of comets and other bodies in our system, render this conclusion almost certain. As the planet Venus has been sometimes seen at its inferior conjunction, and as it was especially observed under very favourable circumstances, in December 1866, by my friend Professor Chester S. Lyman, of New Haven, there exists every indica-cation, not only that our sister planet possesses an atmosphere, but also that this atmosphere is mildly illuminated by the action of its own internal cosmic forces, in like manner as is the earth's atmosphere in auroras, and as are the clouds of Sachura and Paita, by the activity of their electric conditions, and their connexions with telluric forces.

Heated aqueous vapour, suddenly condensed, is strongly luminous. This fact is daily demonstrated in the breathing and whistling of locomotives as they pass my window, filling my study with gushes of light and momentary cheerfulness during these dark days of a Heidelberg winter which I am at present experiencing. That the highly electric energy which steam is known to possess as it escapes from every steam valve is in these cases transformed into luminous phenomena through the influence of cold is more than I venture to declare; but the fact is constantly visible.

Beside these facts, there occurs a not unfrequent phenomenon of sudden and transient or long-continued illumination of the surface of both land and sea, often occurring in dark, and even during starless nights—a phenomenon observable everywhere upon the planet's surface, noticed by all physicists, which is wholly independent of the light of the stars. Although this transient appearance has never been explained, it is unquestionably generated by the action of the forces and agencies now under discussion upon the gases which envelope our earth.

The sea, no less than the atmosphere, affords abundant evidence to show that the radical forces, radiating from the endless stores of energy with which the interior of the planet is charged, can so act as to generate luminous phenomena in its saline and moveable envelopes.

The remarkable luminosity which occurs at various periods over great areas of the ocean, and in all latitudes and longitudes, sometimes increasing to a strong and even fiery effulgence, and frequently (but not always) preceding or attending violent aërial movements, a phenomenon which Humboldt and others have ascribed to phosphorescent infusoria, I have long observed to have an origin exclusively physical. Phosphorescent infusoria and small crustaceans intensely brilliant on their inferior surfaces abound at certain seasons in many regions; and, during the passage of vessels through waters thus inhabited, they sparkle, and often mark the wake with many little star-like points. But they do not produce that remarkable glow and general refulgence of the salt water, which at particular times, during dark nights, define the track of vessels for many hundred yards, and which resemble the tails of great comets streaming into the black depths of space. The doctrine that the phenomenon called marine phosphorescence is wholly dependent upon vital forms, or putrescent organic matter, is a great error, reposing upon high authority but imperfect observation. Its most remarkable exhibitions are connected with telluric causes, often attended with ascending currents of repulsive force and atmospheric molecules, and followed and accompanied with the most terrific cyclones. My friend Captain Shipley W. Crosby, who some years since conducted a whaling cruise among the Aleutian Islands, narrated to me a fact which, on account of its authenticity, scientific importance, and direct bearing upon our subject, is worthy of permanent record. The afternoon and sunset were without portent on a certain day, when he went to pass the evening with the master of a neighbouring cruiser. When he came upon deck toward midnight to return to his own ship, he was struck with the uncommon stillness of the air. The ocean was equally motionless. He observed the sky to be unusually dark. But when his sailors dipped their oars in the sea, it flamed so much more than he had ever previously witnessed, illuminating his boat and men so strongly as to make them visible in the general darkness, that, filled with fearful forebodings of some uncommon disturbance in nature, he instantly, on reaching his ship, commanded the sleeping watch on deck, and put his ship in the snuggest order for a storm. Both officers and men thought it unnecessary prudence; and the night wore quietly away. But the following day the most terrific hurricane he ever experienced fell upon that region. His ship was thus saved, while the one he visited was lost, and the entire whaling fleet suffered great damage. All winds are but the gravitation of air which falls more or less violently to replace that which repulsive force, emanating from telluric fountains, has lifted toward space; and we now know not only that "the wind bloweth where it listeth," but also *why* it listeth, and, moreover, "whence it cometh and whither it goeth."

Beside personally witnessing similar phenomena in a less marked degree, in various longitudes, and through latitudes varying from the equator to 56° both north and south, I have often, in dark nights, experimented upon the strength of this light by noting the time upon my watch at the paddle-boxes of ocean steamers, and have remarked that the degrees of phosphorescence are constantly changing; sometimes being very great and luminous, sometimes slight or altogether absent. Indeed I have now no doubt, as the result of prolonged observation, that this phenomenon should be considered as a marine aurora, connected in some way with the action of terrestrial forces, as they play from the planet's centre toward space, and *vice versâ*, generating light in the oceanic envelope in the same manner as they generate it in aërial auroras. When the ocean is in this excited condition, the tracks of fishes may be traced by lines of light. The wakes of porpoises are remarkably distinct, flashing like fire; and it so happened, during my voyage, many years since, around Van Diemen's Land, that on a tranquil night I traced the form of a whale in his monstrous outlines and in all his motions by a pale greenish light, as he slowly moved his fins, rising to the surface now and then to breathe, and falling again a short distance below it, and keeping time with the sluggish movement of the ship as the latter sailed before a light unruffling wind. All this order of phenomena depends unquestionably upon the subtle influence of terraqueous forces, connected in some occult way with evaporation, and with ascending atmospheric movements.

Those remarkable exhibitions called *corpora santa* clearly demonstrate that electric light can be not only generated in the sea, but also radiated from the mast-heads, yard-arms, and every angular projection of ships, as they sail through regions temporarily excited by radiating currents, or exhalations of repulsive force made visible by atmospheric vapour; for *corpora santa* are true aureolas of marine electric flame, and their light is often very intense.

Since the order of phenomena to which I here advert has not, to my knowledge, been observed in rivers or lakes, it may be legitimately inferred that the solution of calcium, magnesium, sodium, and other metallic salts constituting the aqueous envelope of the globe, contribute to the generation of these peculiar displays of luminous energy. The luminous energy of the sun indicates by the spectrum analysis the presence of similar substances in that body. Upon our earth their exhibitions of light are not associated with intense heat.

It has been long observed that electric light often ascends in fearful flames from the surface of the solid ground, illuminating vast fields of overhanging vapours, and transforming our sky into momentary photospheres. The same phenomenon has also been observed to take place from the surface of the ocean. A fact of this kind is particularly recorded by a Kingston (Jamaica) correspondent to the *Boston Journal*, who, in describing occurrences attending the fearful West Indian earthquake of the night of Nov. 11-12, 1867, makes the following statement as an eye-witness :—"About eight P.M. "on the 11th the sky looked excessively black to the south. ". . . About half-past eight a fearful thunderstorm was ob- "served to the S.S.W. The vividness of the lightning was "far beyond anything remembered to have been witnessed in "Jamaica. It even at the distance of several miles lighted up "the whole surrounding country, and in the darkness, by the "aid of one of these grand momentary illuminations, H.M.S. "*Phebe* was plainly distinguished in the distance making for "the harbour of Port Royal. *The lightning looked as if it "sprang upward from the sea; at one time it was fork "lightning, at another it appeared like large balls of fire rising "up from the water and bursting with great brilliancy in the "clouds.* There was no appearance of the storm approaching "toward Kingston. . . . At 11.21 came the tremendous "heaving of the earth."

As the air and ocean afford such abundant, diversified, and indisputable proofs of the generation of terrestrial light, as a pure cosmic phenomenon; so the land also contributes its quota in turn, pointing to internal mechanical causes for such external conditions.

Facts similar to that last noted upon the ocean, and only differing in intensity, have been everywhere and at various times observed upon the land. Among them is one resembling the phenomenon just described, which has been recorded by Reid in his work upon "The Laws of Storms." Omitting details of the fearful ravages of the elements upon this occasion (another West India hurricane of 10th and 11th August, 1831), I will only cite a single sentence *à-propos* to our subject: "A few minutes afterwards the deafening noise "of the wind sank to a solemn murmur, or rather to a distant "roar; and the lightning, which from midnight had flashed "and darted forkedly with few but momentary intermissions, "now (3 A.M. 11th) for nearly half a minute played frightfully "between the clouds and the earth with novel and surprising "action. The vast body of vapour appeared to touch the "houses (at Barbadoes), and *issued downward flaming blazes,* "*which were nimbly returned from the earth upward.*"

Another fact, observed during the present year, and reported by M. Decharme to the Academy of Sciences, is still more remarkable; and it shows how actively and how intensely cosmic light may be generated independently of incandescent conditions, and without combustion or destructive action upon the surface of a planet or star. I will use the description of the phenomenon as it has been condensed by the scientific reporter to *Galignani's Messenger* for 28th August, 1868:—

"The Academy of Sciences has just received a letter from "M. Decharme on a phenomenon which occurred during the "great storm which visited a large portion of France on "July 25th last, and particularly Angers. The day had "been exceedingly hot, the thermometer marking 37° Cent. "(98° Fahr.). About eight P.M. distant flashes of lightning "appeared in the west, amid heavy clouds coursing the "horizon. About nine o'clock a curious spectacle attracted "attention: besides the usual flashes, which followed each "other in quick succession, ninety-two of them being counted "in the course of a minute, another kind of luminous phe-"nomena was remarked at intervals of about ten minutes. "It seemed to rise from the earth like a bouquet of fire-

" works, or like the flames of a volcano issuing from a crater, " or at other times like a fiery vapour, spreading in all direc- " tions, and successively invading the various strata of clouds, " over which the luminous effluvia seemed slowly to glide, " the eye being perfectly able to follow the progress. From " nine to a quarter-past eleven the phenomenon continued " visible in its fullest intensity. A few seconds before the " storm broke over the town, almost every flash of lightning " was accompanied by white vapour, slightly tinged with " violet, illuminating the whole space, and apparently pene- " trating into the apartments the windows of which were " open. Still no smell was perceptible, no commotion was " felt, no indisposition or peculiar sensation experienced. " This continued till the storm rose to its height, at a quarter- " past eleven, from which time the whole sky appeared " literally on fire. The hurricane itself lasted till half an " hour after midnight, and then intermittently till an hour " later."

All general readers are familiar with the observations recorded by English tourists through the High Alps, of the electric light flashing from crags and icicles, accompanying even rains and mists, as I have already described similar phenomena to be generated in the sea. Both, and all these conditions, are more or less local and temporary; but they are secondary results of the fluctuating translations of telluric forces from interior regions to exterior ones, and *vice versâ*, charging and discharging the envelopes of the planet, and displaying luminous phenomena by the subtle activity of such translations.

The luminous marsh-gases (known as *ignes fatui*) afford another species of evidence, however insignificant their weight, upon possible causes of cosmic light. But such or similar will-o'-the-wisps exist not alone in marshes, and are not the only examples of spontaneous aërial combustion or luminosity that have been seen. So remarkable a phenomenon of this class was witnessed by myself in South America, in May 1865, that I am induced to give it permanent record in this connexion, inasmuch as so important a fact is not only a contribution to knowledge, but may also excite more active inquiry into the nature and causes of such spectral lights. At the time mentioned, I was engaged in a geological reconnaissance along the coast region of the province of Piura, Peru, accompanied by two American companions, and an Indian attendant. We halted one day at dark, about half way between the river Piura, or town of Sachura, and Point Aguja, within sound of the waves of Sachura Bay. The leaden sky and damp bleak wind common to that locality between sunset and sunrise were chilling us long before we made our beds on the drifting sand of that Sahara-like region. Sleeping little, I observed for much of the night that the cloudy sky only broke sufficiently to permit occasional glimpses of the stars. Toward morning it was more densely overcast, and bleaker than ever. Tired of discomfort, I summoned my companions before daylight in order to get breakfast and prepare for an early start. We had barely risen, when one of them, an old resident of Paita, exclaimed, "Why, doctor, there is the British mail steamer bound south." I looked westward, over the Bay of Sachura; and there, sure enough, apparently a long way off, were two bright orange-coloured lights, each with a conspicuous train, and one just ahead of the other, resembling the flames or light from two smoke-stacks. But, as I regarded them intently, I was struck with the rapidity and inequality of their motion, which seemed to increase and waver from moment to moment. They appeared, indeed, to be chasing each other. They were moving horizontally at almost the same level, only a few feet from the surface of the land or water, and with greater rapidity than it was possible for any steamer to move. At first, supposing them far off at sea, I was surprised at the quickness of their motion and transient variations of relative distance; and soon became convinced that they issued from no steamship's funnels, but were luminous objects of some sort, one following or chasing the other, not many thousand yards, perhaps feet, away. I called the attention of my companions to these points; and they came to similar conclusions. What were these lights? Our curiosity became intensely excited. They would vanish for a moment as the low dunes toward the bay intervened, and appear again moving swiftly southward, sometimes almost coming together, then separating, and never more than ten or fifteen feet apart, and each showing bright yellow luminous trains two or three feet long; both objects strongly brilliant, but not defined with clear outlines. In a word, they resembled large flaming torches without smoke in hot pursuit one after the other, just above the surface of the earth and sea. They were visible many minutes, and suddenly vanished, while yet in full blast, behind what I supposed to be a range of hillocks on the edge of the bay. Of course I was on the *qui vive* for the same or similar phenomena to reappear. We observed in all directions. Some minutes elapsed, when one of my companions detected and followed for a while a small bright light in the south-west. He traced it; but I failed to descry it, to his great surprise. I patiently watched, while my companions busied themselves in preparing for the journey. The grey of dawn was beginning to steal into the eastern night, and I beginning to despair, when to my great delight another luminous object appeared, approaching from the south-west and sailing toward the north, higher in the air than those before described; and almost immediately there appeared another slowly following, but not violently chasing the first, as in the former phenomenon. These strange objects swept along at different heights, without trains, appearing like large irregularly-shaped bladders of light, sometimes near each other, then far apart, rising and falling as if moved by currents of air, or more like slowly sailing birds, and changing their motions in all directions. In aspect they were at first yellow and bright; afterwards, as daylight advanced, growing paler and blueish white, like the fumes of phosphorus seen in the dark. They were more defined in form than the first, less intensely brilliant, yet apparently shapeless, and varying from six to fifteen inches in their various diameters. They were strange "spectral" lights without definite forms or proportions; at moments almost lost to sight, then reappearing again more brightly, and apparently having some relative connexion with each other, like that of gregarious birds. They were a long time visible, and finally were lost in the daylight. They appeared to float over both the shores and waters of the bay. What were they? I know not. The recollection of them is a marvel to me to this day. I rode all that forenoon to the westward toward Point Aguja and over the coast mountain-ridge. No fresh water nor bogs exist in the entire country. It is, on the contrary, a great saline region. The neighbourhood consists of an immense plain of quartzose sand impregnated with salt, and of a lofty broken ridge of slates and old sandstones, capped and flanked with thin beds of marine forms, the most of which are identical with those living in the adjacent sea. On the northern border of Point Aguja, however, is a petroleum spring; and I have been informed that "sleeks of petroleum" are sometimes seen upon Sachura Bay. I have no speculations to offer upon a fact so remarkable and obscure as these strange flying lights, and which, to my knowledge, has no equal, except in the superstitious traditions of the famous "Lumpkin's lights," which the old fishermen of Cape Cod still declare to have been formerly seen "jumping up and down" and flying with a "hissing sound" (to use their own language) over the waters of Provincetown harbour. The sandy surroundings of the two localities are somewhat similar. An exact description and a permanent record of these phenomena are what I wish to make here, in order as much to enlist future observation upon such subjects as appropriately to add another fact to a numerous category, all going to show conclusively that luminosity may be generated in the gaseous envelopes of cosmic bodies independently, in many instances, of heat, and in all cases as direct results of the invisible action of internal mechanical and chemical forces alone.

A reference to certain organic forms, inhabiting both air and sea, which display even dazzling light, and to luminosities generated in decomposing animal and vegetable matter, only more and more reveals such molecular conditions as prove that *internal and local perturbations of force* are the sole requisites for developing luminous phenomena, rather than the impact of foreign masses of a meteoric character, which would render the development of cosmic light subordinate to the prior development of glowing heat. When, indeed, the prodigious amounts of light emitted by many insignificant insects, the glowworm, for instance, are considered, and the inferences derived from their conditions of temperature, magnitude, chemical constitution, &c. are taken into account, it may well be questioned whether any theory can be correctly based which assumes stellar light to be an impossible phenomenon, unless such concomitant conditions exist as must preclude for ever the possible existence of organic forms upon either of the infinite host of worlds which adorns the heavens.

16.

Now, when all these diversified classes of facts, and more even which are not mentioned, are grasped together, compressed into a single category, and viewed in one sense as convertible or as kindred phenomena (since the same reactionary telluric and central forces are, in one way or another, the sources of each, however local or general their display);—when thus grouped and considered, and when the sequences drawn from their analyses are applied to such a magnified condition of things as exists in the sun, we may begin to comprehend how the same mechanical forces at work in the earth would, if at work proportionally in the central orb, be capable of generating, not only its internal thermic, but also its photospheric phenomena; and how they might act so as to allow the former to be generated, and to exist independently of and wholly disconnected from the latter, and *vice versâ* (as revealed by the spectrum), and at the same time to be intimately correlated the one with the other, as light, and heat, and actinism must be correlated under all circumstances, through the radical forces upon which they depend for their individual and general existence. And this induction, at variance with theories of every sort, magnifies itself into a final truth, when we consider that the stupendous forces playing upon and through the solar mass, and ending in developments of central heat and superficial light, are generated by universal reciprocities of *stellar* action and reaction; and that the solar globe, placed in the midst of the rolling masses of the Milky Way, assumes the condition of a mass of iron in a rolling-mill subjected to the impact of countless trip-hammers every instant of time.

Such is the mechanism employed in generating the phenomena of heat and light, which mankind in all ages have believed to exist in the sun; always failing, nevertheless, to detect their causes and their unfailing fountains of supply. The present age, however, more fruitful in inventions and precise microscopic and telescopic study—in a word, in positive experimental and instrumental research—and satisfied with nothing short of absolute knowledge, has happily been aided by accidents more than by theories in reaching certain profound and universal facts and laws. Indeed, it is to the honour of human thought, combated and circumscribed as naturalists have been in times past, and are even now, in all pursuits and speculations demanding the freest inquiry, that a knowledge of these facts and laws has been finally acquired.

17.

Beside the heat and light generated in the solar mass, in virtue of its chemico-elementary constitution, we might well infer the electric and magnetic elements to coexist there also as cosmic agents, in consequence of our knowledge of the fact that the numerical variations of intensity in these elements, so constant as terrestrial functions, are related to and governed by the ellipticity and eccentricity of the earth's orbit. This, nevertheless, as a necessary result of fallacious opinions on cosmical physics, could, up to this time, only be conjectured, without being positively known; and these great facts, besides being undemonstrable upon, can never be reconciled with, present theories of celestial mechanics. A new era, however, has already begun to dawn. The discoveries of Rumford, Faraday, Mayer, and Joule lead us now to understand that where mechanical force, chemical elements, heat and light exist, there also may exist electricity and magnetism, as correlative and equivalent phenomena and elements of dynamic power. The prismatic experiments of Frauenhofer, Kirchhoff, and Bunsen upon the sublimates of various metals, and upon drops of saline solutions illuminated or decomposed by electro-magnetic currents, while clearly confirming the correlation and equivalency of all these different forms of force, and exposing the fact that they are capable of reduction into final principles of polarity and unity of origin—that is to say, that all forms of abstract energy are reducible into those radicals of repulsion and attraction which affect molecules with chemical and mechanical power;—these experiments, while clearly indicating the correlations of electricity and magnetism with all elementary conditions of matter and force, and demonstrating them to exist in the sun as positively and as universally as in the earth, have led those illustrious physicists, nevertheless, to the hasty and unfortunate speculations, that the sun itself must be a boiling mass of metals, and its photosphere a glowing envelope of overheated sublimates, similar to those upon which they experimented; and that therefore that magnificent star, and consequently all the other stars of heaven, so full of mechanical and useful creative energy, were not only unfit to support life, but, furthermore, could never become fit, until the light which gilds them had been for cycles extinguished, allowing them time to cool, &c. In view of preceding data, relative to the thermic characteristics of our planet's surface, and the ejection of molten minerals thereupon, and to the action of telluric forces in generating atmospheric light independently of heat, we can now clearly and safely infer that in a globe so large as the sun, with an atmosphere so capable of refulgence (a condition of matter similar to that observable in comets), its light must necessarily be intense, and its electrified nebulous envelope become a veritable photosphere, whose property, as a light-giving fog, must depend upon conditions of *cold* and condensation rather than upon conditions of absolute incandescence. All inductions from the premises bring us back to the conclusions of Alexander Wilson, which have been confirmed by the observations and opinions of Arago—an immortal scholar, from whose lucid discourses and illustrations I remember to have received, in 1836, my first clear impressions of the physical aspects, magnitude, and motions of the solar orb.

But at last the experiments of M. Edmund Becquerel have lifted philosophy into ampler regions of induction: and while they confirm the discoveries of Kirchhoff and Huggins relative to the identity in chemical constitution of all cosmic bodies, thereby establishing bounds to knowledge in *material* directions, they also show that the electro-magnetic elements are absolute associates of every molecular action, however infinitesimal or distant, engaged in the generation of light and colour; thus identifying the correlation, convertibility, and conservation of force everywhere throughout nature. More even than this: by his experiments it may be shown that electric currents, passing through quantities of water holding any mineral salt, or any number of them, in solution, will, in the evaporation effected by this action, not only generate light without destructive heat, but also reveal, in the spectrum of their aureolæ, the lines of every element, however minutely subdivided, existing in this solution; thus divulging the exact measure of mechanical force with which each element is saturated, and the fixity of the laws which, governing the vibratory functions of the various elements, stamp prismatic pictures with those steadfast characters that not only indicate the compound nature of distant bodies, but also the separate and separable existence and the eternal duration of their atomic being. The analogies and inferences with which this discovery and the deductions therefrom are fraught, applied to the mechanism by which light is generated in the sun or stars, and considered in view of all now known of equivalencies, correlations, transformations, and conservation of force, point to the probable conclusions, that the sun is enveloped with luminous gases; that it is covered with vast saline oceans similar to our own, possibly more composite; that its solid surface, where exposed, may be habitable; that it is a world of absolute utility to itself; that it is not bombarded with meteors, to the certain destruction of everything having birth in its airs and seas; and that its surface may be genial, productive, and mildly or brilliantly illuminated, without necessarily being in a state of incandescence, or enveloped in fiery vapours.

18.

General Conclusions.

Here, happily, we approach the boundaries in *physical* directions, the alpha but not the omega, of our patient studies and laborious task. We see and learn thus far, not only that the forces of attraction and repulsion are, but also why it is necessary for them to be, inseparable associates of molecules and atomicities, and the parents of electricity and magnetism, of heat and cold, of light and darkness; and, furthermore, identical with the force of gravitation and that other force, the cosmical antagonist and co-ordinate of gravitation, *cosmical repulsion*, a nomenclature for which the general term "radiation" might be substituted. We see and learn that, while these two forces constitute mechanical energy, or, as Mr. Rankine would call it, "potential energy," mechanical energy is in turn a principle universally compounded of, and resolvable into, these same two forces, both in cosmic and molecular senses, and under all conditions and circumstances whatsoever of aggregation or subdivision.

Such, then, is *mechanical force*, which, quickened by Divine Providence with subtle and inscrutable properties of subdivision, translation, and instantaneity, unites itself eternally in fixed quantities with each element of matter; and driving one molecule or one mass toward or away from another, thereby assumes secondary forms of force, developing motion and heat, and transforming itself into that measurable function called in rational mechanics *mechanical power*. Mechanical power, as the term is commonly used and practically considered, is really *matter in motion* incessantly seeking to communicate its momentum, that is to say, its contained or derived energy, to other matter. This, often aforetime failing of its ends in space, thereby generated those eccentric circuits exhibited in the heavenly motions; and thus laid the foundations of that universal instability of equilibrium in nature through which central heat, universal light and life, and all other present conditions of things exist, and by which alone they can be maintained. Mechanical force is the abstract spiritual energy itself stored in the sun, stars, and planets, as attraction and repulsion, the radical working agents in general phenomena, and in creative design: agents convertible by vibratory translations through these masses into light and colour, heat, magnetism, and electricity; and, associated with molecules, into endless actinic combinations and molar phenomena, all of which are, indeed, only different expressions of motion and dynamic energy, made manifest to pure consciousness through nervous and ganglionic arrangements of the same molecules. Thus is mind traceable to an absolute connexion with matter; and *vice versâ*.

19.

Cosmo-mechanical energy is, then, *per se*, and in the most absolute manner, motive power; because it is a principle so compounded and active as never to be at rest. Unequally distributed among the elements in the beginning, it must for ever remain so. Confined in given quantities to each and every world, as self-conserving stores of central force, its components endeavour in vain to restore an equilibrium among elements and masses already long lost, to exhaust an inherent instability (so to speak), and gain rest: an effort which ends in pushing these masses respectively around each other and around their individual axes; in effecting internal perturbations; and in urging molecules into various organic forms, segmentations, and successions, under the guidance of a formative intelligence—the true *nisus formativus*—which, lying yet more profound, and everywhere immanent in, being indeed the very soul of, nature, quickens all molecules with a sensitive and vital animus. Thus gemmules, forms, and life bud, unfold, blossom, germinate, die, and succeed each other in endless generations and dynasties upon the face of the earth. Thus the repulsive forces of planets are employed upon their exterior phenomena, both physical and morphological; while their attractive forces maintain the unity of original design, restoring to the surface, or bringing back to the centre, the material elements previously lifted therefrom. In this manner the so-called "radiant heat" of the globe, instead of being lost in the atmosphere and wasted in space, has been engaged in building up vegetable and animal organisms upon the lands and in the waters; and in order to accomplish the purposes of universal providence, has laid them down again, and thus piled up the *débris* of successive life and decay, the depths and diversified forms of which indicate an antiquity in Nature defying calculation, and setting at naught all the traditions of the ancients. Here springs up the wondrous inductive truth, viz. that in the evolutions of heat—that is to say, of mechanical force—from the interior of the globe, which has been so busily employed for countless cycles in fabricating those vast deposits in which revels the palæontologist, the foundations have been laid upon which rest all geographical and geological problems. While the interior of the planet incessantly contributes its stores of force to the mechanical work of its surface, insensibly radiating and appropriating its motive power through every molecule into plant and animal, the earth thus transforming itself into a vast workshop, laboratory, mint, or productive machine, to the useful end of adorning and populating its surface;—while all this manifold and wondrous work is quietly proceeding day and night upon the exterior, the force thus employed upon it is furnished at the transient expense of the interior. The temporary transfer of repulsive force, constituting the important radical of central heat, tends in turn to the shrinking of the molten nucleus, and to the formation of vast voids below the planet's crust. Cavities thus produced and filled with mineral and metallic sublimates, atmospheric gases, or, through fissures and leakage, with water and steam, afford conditions not only for the frequent expression of those phenomena called "subterranean thunders" and for many quakings of the earth in consequence of dynamic perturbations of the nucleus (Humboldt's "reaction of the interior upon the exterior"), but also for the generation from cycle to cycle of such terrific convulsions as end in subsidences or submersions of vast areas of the planet's crust, effecting polar transpositions, universal changes of the sea, and general sudden destruction of existing organisms. The certainty of geological data sets at naught the theories of those astronomers and mathematicians who persist in teaching that the poles of planets are flattened by rotation, and that the surface of our globe has never changed its relations to the plane of the ecliptic. Through the causes just stated, which are even mathematically demonstrable, the geographical aspects of the planet have from time to time been profoundly altered, not by upheavals of continents, but by their slow or sudden submergence, as the different states of the interior have determined. Taught by leading authorities to believe the theory of upheavals, I long ago doubted its correctness in consequence of personal observations over immense expanses of the globe. While the *débris* and wide distribution of identical or analogous organisms in the earliest palæozoic ages indicate the general shallowness of universal seas, and a surprising littoral extension of the planet's surface, this shallowness and these littoral extensions could not have existed without the coexistence of larger diameters and a larger circumference of the planet itself. One fact follows the other as certainly as day follows night; and the successive organic, geological, and geographical changes as clearly follow the consumption of telluric heat on the earth's surface, as the productions of a cotton-mill follow the consumption of water-power or coal. The magnitude of these quantities it is for future science to determine. That oceans, seas, lakes, and rivers, plateaus, mountain-ridges, watersheds, and great river-basins, are formed and outlined by the submersion of continents and the simultaneous breaking up of vast areas of surface—in other words, that general geographical conditions, as they have variously existed in different geological ages, and as they now exist, have been formed by subsidences and not by upheavals—is as certain as the existence of the world itself. Mechanical energy, springing in currents, offshoots, and impulses from the voluminous *central forces of the planet*, and endowed with capacities of specialization for definite purposes in order instantly to meet inductive demands, is the basis of all events and the motive agent in all things. Not even an insect can lift its wing, nor a hammer rise and fall, nor a single molecule stir in the air, nor a cell vibrate, without the influx and efflux of the requisite motive forces which are connected with, and directly derived from, the centre of the globe. And while repulsion lifts plants and separates animals from the surface of the earth in such a manner as even to endow matter with locomotion; while it is the occult agency in segregation, segmentation, growth, and general creation; and while it may expand and lift the atmosphere, the ocean, and even vast areas of the earth's crust, shaking mountain-chains themselves in its periodic throes; still gravitation levels and brings them all down again, and fills the voids which the former has been instrumental in elaborating.

20.

We have not reached these final and positive truths by any *à priori* and theoretical method of treatment, assuming unknown or doubtful data, and then working our way through processes of complex and problematical reasoning to uncertain and incomplete results. We have not taken for granted any of the thousand hypotheses which divert students at every step from direct observations of nature, and which, on account of the commanding positions of their authors or advocates, lead often into paths of hallucination rather than into those of solid philosophy, wasting mental power and years of toil, and disappointing the highest of cultivated aspirations. We commenced these investigations, tedious and uninteresting as they have frequently appeared to be, by planting our steps on certain fundamental facts which were wholly within our grasp, and positively known as far as observation, experiment, and logic can make anything satisfactory and conclusive to the human mind. Ascending step by step from MOLECULES and the *molecular* FORCES *of* ATTRACTION *and* REPULSION—points determined by every method of analysis to be ENTITIES embracing definite attributes and quantities, both physical and algebraical, and governed by specific laws; avoiding symbols, and developing the intricacies and connexions of the various related fields in terms as simple and graphical as the exactness of science would permit; we have at last reached the culmination of our aim. We have seen with no less satisfaction than surprise the gradual unfolding of a great physical fact, viz. the subjugation and obedient behaviour of all matter to two co-ordinate and opposing forces—a subjugation and obedience as sublime and infinite in their ascending increments as they are infinitesimal, and yet never vanishing in their descending series. Every molecule, every element, every mass, wherever located or however surrounded, is endowed with, and acted upon by, *specific* BUT UNEQUAL quantities of attraction and repulsion; and thus all simultaneously and necessarily become so many mechanical agents in nature, whose internal phenomena, external interdependence, and general reactionary reciprocities are determined and diversified by this very *inequality* in *distribution* of both matter and force. Herein is the hidden spring of mechanical motion and mechanical power. Thus the cosmos was instituted in the beginning; and thus it must continue, time without end. In the beginning, whenever that was, light and heat must have appeared with the origin of matter and force. In the end, if it ever come, all things will surely melt with fervent heat. Between these two uncertain, unknown, and unknowable states—the past and the future—the present certain, known, and knowable state exists. It is the existing condition of things which science must boldly explore and faithfully unfold by the freest forms of inquiry: not so much by speculating upon the possibilities of its origin; not at all by setting up dogmas of faith or doubt about its end; not by ignoring and warping truths in physics and psychology in order to patronise or to reconcile academies and consistories; but by honestly tracing minutest, daily, simple facts through observation, experiment, and induction to final laws and the One Great First Cause. The wondrous creations of God may well allure the human mind into their fathomless and cunning depths, and give rise to innumerable conjectures. But this mind of ours, while free to act, and as illimitable in its immortal capacities as the Infinite Being from which it springs, into which it flows, of which it never ceases to be a part, and from which, indeed, it receives its daily light and strength, is, nevertheless, *limited by truth*, and can reach reliable information of the past or future only through precise knowledge of the present. Existing in the midst of time, space, and a universe of correlated objects; constituted geometrically, mechanically, and chemically, both in form and substance, out of the very principles and elements of which the earth and heavens are made—of these and no others; it must be clearly seen that the physical and intellectual forces which come and go and abide within our being are equally a part of Deity, as much as Deity is the living Spring and a part of His universal works. The mission of the sciences is, therefore, not only important because simply practical and adapted to the refinements of social and material civilization, but also sublime and transcendent over all metaphysical dogmas, traditions, and dialectic pretensions, because natural philosophy leads us to the only true and complete knowledge of the adorable Creator of the world. Where the Divine spark may come, there let altars rise, and the Vestals of eternal truth be charged to guard their sacred fires.

The plan of this work has embraced a series of discussions to be developed in two other chapters. The completion of these has been postponed, however, on account of the ill-health of the author.

The twelfth chapter (when published) will treat of endosmosis, exosmosis, dialysis, and of organic force as a vital principle, under the head of "Colloidal Force," in which the principles heretofore unfolded will be applied, it is hoped, in a manner to lead to a better comprehension of the manner in which the planet has been adorned and populated.

The thirteenth chapter is designed to embrace a treatment of the Supreme Energy, in which the union of the physical and metaphysical principles apparent or traceable in the world will be presented from purely natural and scientific points of view.

Although partly written, the health of the writer forbids the immediate completion of the treatises upon these subjects.

APPENDIX.

Note to Line 33, Page 224.

An examination of my catalogues having exposed the record of a violent earthquake in Armenia on "the last of August or first of September, 1859," the *exact* date not appearing, I addressed a letter to Professor Perrey of Dijon, to ascertain if his records could afford me the facts, which I wished to present in connexion with the preceding phenomena. His response is as follows:—

"Here is what I find in my various catalogues for the 31st "August and the 1st and 2d September, 1859.

"*Aug.* 31.—5 A.M. a light shock at Gallipolis, Turkey. The "same day, at 6.30 P.M., two strong shocks at Sophia, Turkey. "A stone edifice at Tchoadjikhan was thrown down. The "direction of the shocks was from S.W. to N.E., following "the direction of the thermal waters of Sophia.

"At 12.30 A.M. of the 31st August, two very light shocks "from E. to S.W. were felt at Saint Pierre, Martinique, at "short intervals and of short duration.

"*Sept.* 1. (Hour not indicated.)—A light shock at Laibach, "Carniola.

"*Sept.* 2.—Light shocks at the Island of Ternaté. The "mountain has smoked only at intervals this month.

"During this month of September, earthquake shocks have "occurred daily at Minahassa, Celebes,—light at their com-"mencement. I have the hours for the strongest only: 19th, "12.30 P.M.; 21st, 10 A.M.; 27th, 10.30 A.M."

My own record contains the following printed notice of the earthquake action recorded for the "last of August or first of September:"—"Letters from Trebizond state that the town "of Chirvan, in the government of Tiflis, had been buried "under a mountain thrown on it by a recent earthquake."

Besides the above facts indicative of the violent "reaction of the interior upon the exterior of the planet" at different points, almost at the moment when Messrs. Carrington and Hodgson were making their observations on the sun spots, I find by my records, that the volcanic activity of the globe, as a general fact, was remarkably great from the middle of August onwards for several weeks.

Note to Section 13, Chapter IX. Page 184.

Since the last pages of this work have been in type I have received from Professor Perrey his pamphlet, entitled "Propositions on Earthquakes and Volcanoes," at the end of which he has appended a memoir of great value to cosmical science: "On the frequency of earthquakes relatively to the moon's "age, during the last half of the eighteenth century; and on "the frequency of the phenomenon relatively to the meridional "positions of the moon."

I have already alluded to the magnitude and importance of Professor Perrey's seismic studies; and have discussed his *theory* with the candour which the present necessities of exact science demand. However we may differ in the interpretation of phenomena, his facts and his numerical results must always remain a precious legacy to physical knowledge, and an enduring monument of his prodigious labours. While no one perhaps may be more indebted to him than myself for his facts and numerical determinations, I will take this opportunity not only to acknowledge his earlier and his recent kindnesses, but also to present the reader with the latest results of his computations, with which he has just favoured me. It will be seen how distinctly his conclusions bear upon our preceding discussions, and fortify the positions reached in our chapter on " Cosmical Repulsion."

In the memoir alluded to, Professor Perrey, after stating his methods of reducing his tables, says: " Finally, after " condensing again the results thus obtained by the new " additions for the two syzygies and the two quadratures " reunited, we find in the present memoir, from A.D. 1751 " to 1800:—

" For the Syzygies	1901·18	Earthquake days.
" ,, Quadratures	1753·82	
" Difference in favour of the Syzygies	147·36.	

" This result is in perfect accordance with those we have " reached in our preceding memoirs. We found, indeed, " in 1847:—

" For the Syzygies	1420·94	Earthquake days.
" ,, Quadratures	1314·06	
" Difference	106·88.	

" We found in 1853 that the number of 5,388 earthquake " days, calculated by the first method mentioned in this " memoir, resulted as follows:—

" For the Syzygies	2761·48	Earthquake days.
" ,, Quadratures	2626·52	
" Difference in favour of the Syzygies	134·96.	

" And by the second method of computation, which gave the " number 6,596, we had:—

" For the Syzygies	3434·64	Earthquake days.
" ,, Quadratures	3161·36	
" Difference in favour of the Syzygies	273·28.	

" We are then compelled to admit that earthquakes are " more frequent at the syzygies than at the quadratures. " The conclusions that we have deduced from our researches "in 1853, and which we have tabulated for the half century, "apply now to the whole century."

Professor Perrey, calculating still more closely, divides his periods successively from 100 years and 50 years into terms of 25 years and 10 years, and invariably reaches the same results. Carrying them still further, and calculating for periods of 5 years, he finds 8 out of 10 periods giving a preponderance of earthquake days to the syzygies.

"Nevertheless," he continues, "these results, the agreement "of which is striking, and which indicate an influence linked "with the moon's orbit, are not the only ones which we can "present." Carrying his computations more closely to the days immediately connected with the periods of the perigee and apogee, and dividing the earthquake days from 1751 to 1800 into two groups, he finds the following results :—

"For the Perigee	526	Earthquake days.
" „ Apogee	465½	
"Difference in favour of Perigee .	60½.	
"For the Perigee	313½	Earthquake days.
" „ Apogee	278½	
"Difference in favour of Perigee .	35."	

"It is the same," he says, "which I found in 1847." Having obtained at great trouble and expense several journals of Italian earthquakes kept for long periods, he gives us also their results, as follows :—

"At Monteleone	475 shocks when the Moon was near the Meridian, and within 45° of it.
"and	453 only when the Moon was more than 45° from Meridian.
"Difference .	22 more when the Moon was nearest Meridian.
"At Messina .	84 shocks when the Moon was nearest Meridian.
"and	60 when most remote from it.
"Difference	24 more when nearest Meridian.

" At Catanzaro . 102 shocks nearest Meridian.
" and 81 when most remote from Meridian.

"21 shocks in favour of the meridional passage of the Moon.

" At Scilla . . 140 shocks within 45° of Meridian.
" and 120 when more remote from Meridian.

"20 shocks, always in favour of the passage of the Moon at Meridian.

" Thus, notwithstanding the perturbations incident to these " phenomena in their general aspects, we still find in the four " positions indicated, that during the 3 h. 6 m. which precede " the passages of the moon at the superior and inferior " meridians, and during the 3 h. 6 m. which follow these " passages, the shocks are more frequent than in the two " other intermediate quarters of the lunar day."

From another catalogue of earthquakes kept at Reggio Calabria, M. Perrey deduced the following results :—

" Relative to the age of the Moon { At the Syzygies . 437 shocks.
„ Quadratures 349 „

"Difference in favour of the Syzygies . . . 88.

"Relative to the passages of the Moon at the superior and inferior Meridians . . . . { 413 shocks when the Moon was 45° from the Meridian ; and
347 only when at a greater angular distance.

" Difference in favour of the passages of the Moon at Meridian . } 66."

Thus it will be seen that while all those data hold definite and remarkable relations to the position of the moon in her orbit, they connect themselves directly with the position of the sun also. While my eminent friend M. Alexis Perrey endeavours to explain their occurrence and periodical intensities upon the prevailing " Theory of the Tides," I regard them in the light of reactionary phenomena, and due to the influences of repulsive, rather than of attractive, force.

Even if we consider for a moment the fearful convulsions which have recently occurred in Peru and Equador, and which extended their pulsating influences to California, Hawaii, and even to Japan and Cochin China (13th to 20th August, 1868), we shall perceive how legitimately they fall within the laws of cosmo-dynamic action, which I have endeavoured to unfold in the preceding work; and how they agree at the same time with the observations both of Aristotle and Alexis Perrey.

The facts are well known that as the earth recedes from perihelion, the moon's orbit contracts, so that at aphelion the two bodies are much nearer together than at any other point in either of their orbits; and, that when the earth descends from aphelion to perihelion, a gradual change in their relative positions begins and increases, so that when the earth is nearest to the sun, the moon is most distant from the earth.

This remarkable perturbation in the lunar motions has always been explained heretofore on the Newtonian hypothesis of solar gravitation. And as a result of these motions, or in some way connected with them, it has been observed by Laplace, Adams, and others (as I have before stated), that the rotary movements of the primary have been possibly retarded from the $\frac{1}{60}$th to the $\frac{1}{64}$th part of a second of time, in the course of the last 2,000 years, while the motions of the earth and moon together around the sun have nevertheless not been in the slightest degree retarded or disturbed. As long ago as 1853 I endeavoured to explain this fluctuating phenomenon in my first treatise on cosmical repulsion, as a result of the reactionary processes taking place within the planet and between the earth and moon during the successive movements of the former from aphelion to perihelion, and from perihelion to aphelion.

The recent terrific convulsions in South America, accompanied as they were by a sudden outburst of Vesuvius, and by earthquakes in Mexico, Hawaii, and elsewhere, indicate an unusual state of reactionary activity in the interior forces of the planet while in the aphelic arc of its orbit. Observation shows such perturbations to be not unfrequent.

Now when it is remembered that the relative position of the earth to the moon and sun was such at the time mentioned as to bring the equatorial points of these bodies to bear upon each other with vertical directness ending in a solar eclipse, and that the proximity of the satellite to the primary was such as to excite that reaction in the latter, which, as I have before shown, is expressed from the centre upon its surface with eruptive results both in relation to the sun and moon, it will be less surprising that the telluric convulsions referred to were of so violent a character.

The cause of the slight retardation in the rotary motions of our planet detected by Laplace, may perhaps lie in the occasional spasmodic and transient backward impulses of the sea produced by seismic reaction, rather than in the daily ebb and flow of the tides, as of late conjectured by Mayer, Delaunay, and others.

INDEX.

INDEX.

T.

U.

V.

W.

Z.

R. CLAY, SONS, AND TAYLOR, PRINTERS, BREAD STREET HILL.

www.ingramcontent.com/pod-product-compliance
Lightning Source LLC
LaVergne TN
LVHW020912110826
845150LV00004B/656

* 9 7 8 1 4 2 5 5 5 6 3 0 3 *